◈ Chemistry in the Laboratory

A Study of Chemical and Physical Changes,
Second Edition

J. A. BERAN
TEXAS A&M UNIVERSITY–KINGSVILLE
KINGSVILLE, TEXAS

John Wiley & Sons, Inc.

New York   Chichester   Brisbane   Toronto   Singapore

PHOTO CREDITS

Common Laboratory Equipment and Special Laboratory Equipment: Yoav Levy/Phototake.
Techniques: Figure T.1: Ken Karp.
Figure T.2x: Courtesy Fisher Scientific.
Figure T.3 (left): Courtesy Corning.
Figure T.3 (right): Courtesy Fisher Scientific.
Figure T.5: Courtesy Jo A. Beran.
Figure T.6: Courtesy Jo A. Beran.
Figure T.7: Ken Karp.
Figures T.8a and T.9: Courtesy Jo A. Beran.
Figure T.10a: Ken Karp.
Figure T.10b-e: Courtesy Jo A. Beran.
Figure T.11: Ken Karp.
Figure T.12: Courtesy Jo A. Beran.
Figure T.13a,b: Courtesy Fisher Scientific.
Figure T.13c: Courtesy Sartorius Co.
Figure T.14a,b,f: Ken Karp.
Figures T.14d,e, T.15 and T.16a,b,c: Courtesy Jo A. Beran.
Figure T.16d: Courtesy Fisher Scientific.
Figure T.17b,c,d: Ken Karp.
Figure T.17e: Courtesy Jo A. Beran.
Experiment 1: Opener: Courtesy Fisher Scientific.
Figure 1.1: Courtesy Jo A. Beran.
Figures 1.x1, 1.x2 and 1.x3: Courtesy Fisher Scientific.
Figure DS.1h: Courtey General Motors Powertrain Division.
Figure DS.1j: Herb Snitzer/Stock, Boston.
Experiment 2: Opener: Bob Daemmrich/Stock, Boston.
Figure 2.1: Courtesy Jo A. Beran.
Figures 2.2, 2.3 and 2.4: Courtesy Fisher Scientific.
Experiment 3: Opener: Courtesy U.S. Forest Service.
Figure 3.1: Courtesy Jo A. Beran.
Figure DS.3a: Dean Abramson/Stock, Boston.
Experiment 4: Opener: Courtesy USDA.
Figure 4.1: Courtesy Jo A. Beran.
Figure 4.x1: Courtesy Fisher Scientific.
Experiment 5: Opener: Kathy Bendo.
Figures 5.x1 and 5.x2: Peter Lerman.
Experiment 6: Opener: Kathy Bendo.
Figure 6.1: Peter Lerman.
Experiment 7: Opener: Courtesy American Museum of Natural History.
Figure 7.1: Yoav Levy/Phototake.
Experiment 8: Opener: Peter Lerman.
Experiment 9: Opener: Michael Watson.
Experiment 10: Opener: Courtesy Icelandic National Tourist Office.
Figure 10.2: Courtesy Jo A. Beran.
Figure 10.x1: Courtesy Fisher Scientific.
Experiment 11: Opener: Christopher S. Johnson/Stock, Boston.
Figure 11.x1: Courtesy CRC Press, Inc.
Experiment 12: Opener: Blair Seitz/Photo Researchers.
Experiment 13: Opener: Bob Daemmrich/Stock, Boston.
Figure 13.x1: Courtesy Edgar Fahs Smith Collection, University of Pennsylvania.
Experiment 14: Opener: Michael Watson.
Figure 14.1: Courtesy VRW Scientific.
Experiment 15: Opener: Courtesy Sung-Hou Kim, University of California at Berkeley.
Figure 15.x1: Library of Congress.
Experiment 16: Opener: Chicago Tribune.
Figure 16.2: Courtesy Jo A. Beran.

Experiment 17: Opener: David R. Frazier/Photo Researchers.
Figures 17.1 and 17.2: Ken Karp.
Figure 17.x1: Courtesy Fisher Scientific.
Experiment 18: Opener: Dorothy Kerper Monnelly/Stock, Boston.
Figure 18.1: Michael Watson.
Figure 18.3: Ken Karp.
Experiment 19: Opener: Courtesy The Port Authority of NY & NJ.
Figure 19.1: Courtesy Edgar Fahs Smith Collection, University of Pennsylvania.
Figure 19.2: Jim Brady and Kathy Bendo.
Experiment 20: Opener: Bob Kalman/The Image Works.
Figures 20.1 and 20.2: Courtesy Fisher Scientific.
Figure 20.3: Ken Karp.
Experiment 21A: Opener: Ken Karp.
Figure 21A.x2: Courtesy Fisher Scientific.
Experiment 21B: Opener: Norm Thomas/Photo Researchers.
Figure 21B.x1: Kathy Bendo.
Figure 21B.x2: Ken Karp.
Experiment 21C: Opener: Ken Karp.
Figure 21.C1: Kathy Bendo.
Experiment 22: Opener: Bob Daemmrich/Stock, Boston.
Figure 22.3: Courtesy Fisher Scientific.
Experiment 23: Opener: Tim Davis/Photo Researchers.
Figure 23.x1: Bortner/National Audobon Society/Photo Researchers.
Experiment 24: Opener: Courtesy The Permutit Co.
Experiment 25: Opener: Bob Krueger/Photo Researchers.
Experiment 26A: Opener: Kathy Bendo.
Figure 26A.1: Courtesy Jo A. Beran.
Experiment 26B: Opener: Ken Karp.
Figure 26.B1: Michael Siluk/The Image Works.
Experiment 26C: Opener: Courtesy Jo A. Beran.
Experiment 27: Opener: Courtesy Royal Crown Cola Company.
Figure 27.2: Ken Karp.
Experiment 28: Opener: Courtesy Union Carbide Co.
Experiment 29: Opener: Courtesy Superior Fireplace Co.
Figure 29.2: Courtesy Jo A. Beran.
Experiment 30: Opener: Lionel Delevingne/Stock, Boston.
Preface: Courtesy Hach Company.
Experiment 31: Opener: Peter Menzel/Stock, Boston.
Figure 31.1: Peter Lerman.
Figures 31.2 and 31.3: Courtesy Jo A. Beran.
Experiment 32: Opener: Ken Karp.
Figure 32.1: OPC, Inc.
Figure 32.2: Courtesy Jo A. Beran.
Experiment 33: Opener: CNRI/Photo Researchers.
Figure 33.1: Ken Karp.
Experiment 34: Opener: Courtesy Cheesebrough Ponds, Inc.
Figures 34.4: and 34.5 Courtesy Jo A. Beran.
Experiment 35: Opener: Courtesy Florida Department of Commerce, Division of Tourism.
Figure 35.1: Richard Megna/Fundamental Photographs.
Figure 35.2: Courtesy Jo A. Beran.
Appendix A: Opener: Courtesy Westinghouse Electric Corp.
Figures A.1, A.2, A.3, A.4 and A.5: Courtesy Jo A. Beran.
Appendix B: Opener: Courtesy Jo A. Beran.
Appendix D: Opener: Courtesy Jo A. Beran.
Color Insert: Front Page: Courtesy Jo A. Beran.
Back Page: (top) Hugh Lieck; (center top) Michael Watson (center bottom and bottom left) OPC, Inc.; (bottom right) Yoav Levy/Phototake.

ISBN 0-471-10952-5

Printed in the United States of America

10 9 8 7 6 5 4 3 2 1

Preface

The chemistry laboratory will be one of the most interesting and rewarding academic experiences you will encounter. It will be one of those experiences that will enable you to interpret the principles and apply the calculations presented in the textbook and lecture to the collected data in the laboratory. The data, quantitative or qualitative, can be observed, discussed, and interpreted in terms of these learned basic chemical principles. Predictions will be made as a result of the assimilation of the data. Solving for "unknowns" will be gratifying when you realize that your accumulated chemical knowledge is correctly used in an analysis of observations.

The chemistry laboratory is also a social experience. Working with others while attempting to gain an interpretation of your data allows you to perceive a conceptual understanding of basic chemical principles from various viewpoints. More simply, sharing the rewards and frustrations of a laboratory assignment promotes lasting friendships. Recollections of the "lab" at class reunions or college social events vary from "the acid holes in the new pair of jeans" to "the percent antacid in Rolaids or the percent vitamin C in a vitamin tablet"—memorable events seem to happen in the "chem lab." Lab is fun, challenging, rewarding, and memorable—enjoy it!

The author truly hopes you find the general chemistry laboratory an enjoyable and memorable experience. Take the time to critically analyze your observations and to clearly and conscientiously complete your calculations.

You can expect to spend an average of three hours for the completion of each experiment. Advance preparation for each experiment, minimally the completion of the Lab Preview, may decrease this time—the lab experience is much more enjoyable when you have a basic, initial understanding of the experiment. On occasion, the analysis of the data will require time outside of the laboratory period. Realize that references to the textbook or discussions with your laboratory instructor may be necessary to effectively complete and understand the experiment.

The experiments in this manual were chosen for you to gain an appreciation and an understanding of chemistry at a level that is necessary for your chosen major field of study, provided it is science related. You need *not* be a chemistry major to understand the "chemistry" of this manual.

The manual has three major sections:

- Proper laboratory technique is necessary for the collection of reliable and reproducible data. The **Laboratory Techniques** section (pps. 1–20) details 18 major techniques that illustrate the proper way to perform various laboratory operations.

- Thirty-nine **Experiments**, utilizing basic laboratory skills and equipment, are presented. A familiarity of basic chemical principles is required.

- Four **Appendices** provide reference information that is often used throughout the manual.

STUDENTS

Expectations and Anticipations

Manual Overview

Organization

An introductory paragraph attempts to place the importance of the experiment in the realm of the real world.

There are five major divisions in each experiment:

- **Objectives**. The objectives focus on the purpose of the experiment and what you are to learn and accomplish from the experiment.

- **Principles**. Several paragraphs discuss the chemical principles that are used in the analysis and interpretation of the data. Appropriate equations, tests, colors, etc., are presented and/or illustrated. A **running glossary** helps to clarify chemical terms and symbols.

- **Procedure.** A detailed procedure for the collection of data includes figures showing the construction of the apparatus, icons as reminders for using the proper laboratory techniques, safeguards, Cautions, and disposal information.

- **Lab Preview**. Several exercises are presented to assist in the understanding of the Principles, Procedure, and Techniques sections of the experiment *before* the laboratory. A careful completion of the Lab Preview reduces the "waste of time and boredom" that the poorly prepared student experiences.

- **Data Sheet**. A structurally designed Data Sheet provides for a careful organization of your data and completion of the necessary calculations. Questions reviewing the experiment and probing your analysis and interpretations of the data appear at the end of the Data Sheet.

Features of the Manual

Several features of the manual help make the lab experience more meaningful:

- **Technique Icons.** A total of 31 icons for the 18 techniques are used as reminders in the Procedure of each experiment. Additionally flow diagram icons appear in Experiments 31, 32, 33 to assist in guiding you through the steps of Cation Identification.

- **Running Glossary.** A running glossary in the margin of each experiment clarifies terms, symbols, and interpretations used in the explanation or description of the Principles and Procedure.

- **Figures.** Line drawings and photographs are used extensively to illustrate the construction of the proper laboratory apparatus and the plotting of data.

- **Safety and Disposal.** Careful attention to safety is an essential component in the design of each experiment. While all potentially hazardous chemicals cannot be eliminated from the laboratory, their numbers are minimal in this manual. Safety is a major concern—a safety icon in the margin cites potential danger spots in the Procedure and a **Caution** citation in the text depicts the actual danger of the chemical. In addition, **Disposal** information for each experiment suggests the method of discarding test solutions and reagents.

LABORATORY INSTRUCTOR

Chemistry in the Laboratory, A Study of Chemical and Physical Changes, Second Edition, is actually the fifth edition of a manual that began as *Laboratory Manual for Fundamentals of Chemistry*. Because of extensive suggestions and insights from users at numerous colleges and universities since that first edition, this manual has developed a reputation as one that enables a student to develop a keen interest and appreciation of chemistry from a background that is relatively void of exposure to chemical phenomena.

This manual covers one year of general chemistry for students who have chosen careers in the sciences or engineering. Although the manual parallels the material in Brady and Holum's, *Chemistry: The Study of Matter and Its Changes, Second Edition,* the experiments are chosen and written so they may easily be adapted to any general chemistry text.

Revisions and Improvements

- **Improved Techniques.** The development of good laboratory techniques continues to be one of the outstanding features of the manual. The **Laboratory Techniques** section, pages 1–20, covering 18 basic laboratory techniques with 47 photographs and line drawings has been revised and refined. Icons indicating the use of appropriate techniques have been designed and are strategically placed in the margin of the **Procedure**. Wherever appropriate, a line drawing or a photograph illustrates the proper apparatus required for the experiment.

- **Safer Experiments.** All experiments have been reviewed and revised for safety. Emphasis on *laboratory safety* is evident throughout the manual: a **Caution** icon sites a warning when a potentially harmful chemical or procedure is used. Within the context of the Procedure, a caution statement clearly identifies the nature of the caution or the consequence of improperly handling the chemical. Disposal information is given for the proper disposal of laboratory chemicals and reagents at the conclusion of the Procedure.

- **New, Revised, Expanded, and Omitted Experiments.**

  New Experiment 11. Chemical Literature. This new experiment introduces the student to the very basic chemical literature that is of most value to the first and second year chemistry student.

  New Experiment 1. Safety and SI. A dry lab segment focusing on SI has returned to this edition; the SI in conjunction with additional safety guidelines now comprise the first experiment. Much of the laboratory measurements (formerly 1) has been moved to Experiment 2, along with additional emphasis on the importance of good laboratory techniques.

  New/Revised Experiment 8. Formula of a Hydrate (formerly 9). Use of the very expensive crucibles for MgO synthesis has been eliminated—the thermal decomposition of copper bromide, using a test tube, has been substituted.

  Revised Experiment 6 (formerly 5, 6). Features of reaction types and the determination of an unknown salt, an analysis of chemical observations/reactions, have been combined into a single experiment.

  Revised Experiment 12. The color slides from previous editions have been eliminated. Students now use the enclosed color plate, showing the same spectra that formerly appeared as the color slides, as a set of data for analyzing spectra.

  Revised Experiments 21A, 21B, 21C (formerly 28, 29, 30) have been sequenced to recognize their dependence on one another and the versatility of a standard NaOH solution. In addition these experiments have moved forward in the manual to reflect upon the importance of an acid-base titrimetric analysis.

Revised Experiments 26A, 26B, 26C (formerly 20, 21) have also been sequenced to recognize the importance of the standardization of a reducing agent and the redox titrimetric analysis. Additionally they appear later in the manual, combined with other redox reaction systems.

Expanded Experiment 7 (formerly 8). More than just the analysis of a hydrate, other aspects of a hydrate (hydration, dehydration, efflorescence and deliquescence) have been introduced.

Omitted Experiments 3 (Kitchen Chemistry) and 40 (Carbohydrates and Proteins)

- **Basic Apparatus.** Lists and photographs of the **Common** and **Special Laboratory Equipment** used in this manual are at the front of the manual. Balances with ±0.001 g sensitivity and visible spectrophotometers are needed on occasion.

- **Instructor's Manual.** This supplement is available to adopters upon request from the publisher. An effective teaching schedule is outlined for each experiment, including: a suggested lecture outline, cautions, representative or expected data, common student questions about the experiment, answers to the Lab Preview and Data Sheet Questions, the chemicals and special equipment that are required, a detailed preparation of all reagent solutions, and the safety rules.

Reviewers

The valuable suggestions and insights provided by the following reviewers were most helpful and greatly appreciated in developing the second edition:

Joan Reeder, Eastern Kentucky University
Robert Kowerski, College of San Mateo
Karen Harding, Pierce College
Rudy Luck, American University

Jane Johnson, University of Colorado, Colorado Springs
Cheryl Baldwin Frech, University of Central Oklahoma
Dale L. Robinson, Palo Alto College

Acknowledgments

The author thanks John R. Amend, Montana State University, for the gathered data (line spectra) presented in Experiment 12 and James E. Brady, St. John's University, N.Y., for his suggestions that appear throughout the manual.

The staff at Wiley has been outstanding: Hilary Newman, Associate Photo Editor, for searching through thousands of photos for use as experiment openers and marginal photos; Rosa Bryant, from the Illustration Department, for being certain that the line drawings and technique icons were of their obviously high quality. *Special recognition* is for Joan Kalkut, Associate Editor with Wiley, for her continued guidance, patience, valuable suggestions, and personal perceptions of what a laboratory manual should be about. My general chemistry students, laboratory instructors, and my colleagues at Texas A&M University, Kingsville campus have been most verbose in providing suggestions for improvement and revision—this I appreciate beyond words.

The author has appreciated the letters and phone calls from users over the years and continues to invite corrections and suggestions for improvements to this manual from colleagues and students.

J. A. Beran
Box 161, Department of Chemistry
Texas A&M University–Kingsville
Kingsville, Texas 78363

Table of Contents

LABORATORY TECHNIQUES

Glassware and Chemicals

Liquid Reagents and Solutions

Solid Reagents

Gases

Handling Data

EXPERIMENTS

Skills and Observations

Chemicals and Stoichiometry

Atoms and Molecules

Gases

Acids, Bases, and Solubility

Oxidation–Reduction

Kinetics

A Summary of Inorganic Chemical Principles

Organic Chemistry

APPENDICES

Common Laboratory Desk Equipment

No.	Quantity	Size	Item	First Term In	First Term Out	Second Term In	Second Term Out	Third Term In	Third Term Out
1	1	10 mL	graduated cylinder						
2	1	50 mL	graduated cylinder						
3	5	—	beakers						
4	2	—	stirring rods						
5	1	500 mL	wash bottle						
6	1	75 mm, 60°	funnel						
7	1	125 mL	Erlenmeyer flask						
8	1	250 mL	Erlenmeyer flask						
9	2	25 x 200 mm	test tubes						
10	6	18 x 150 mm	test tubes						
11	8	10 x 75 mm	test tubes						
12	1	large	test tube rack						
13	1	small	test tube rack						
14	1	—	glass plate						
15	1	—	wire gauze						
16	1	—	crucible tongs						
17	1	—	spatula						
18	2	—	litmus, red and blue						
19	2	90 mm	watch glasses						
20	1	75 mm	evaporating dish						
21	4	—	medicine droppers						
22	1	—	test tube holder						
23	1	large	test tube brush						
24	1	small	test tube brush						

◇ Special Laboratory Equipment

Number	Item	Number	Item
1	reagent bottles	16	crucible and cover
2	condenser	17	mortar and pestle
3	500 mL Erlenmeyer flask	18	glass bottle
4	1000 mL beaker	19	pipets
5	Petri dish	20	ring stands
6	Büchner funnel	21	buret clamp
7	Büchner flask	22	double buret clamp
8	volumetric flasks	23	Bunsen burner
9	500 mL Florence flask	24	buret brush
10	110°C thermometer	25	clay triangle
11	100 mL graduated cylinder	26	rubber stoppers
12	50 mL buret	27	wire loop for flame test
13	glass tubing	28	pneumatic trough
14	U-tube	29	rubber pipet bulb
15	porous cup	30	iron ring

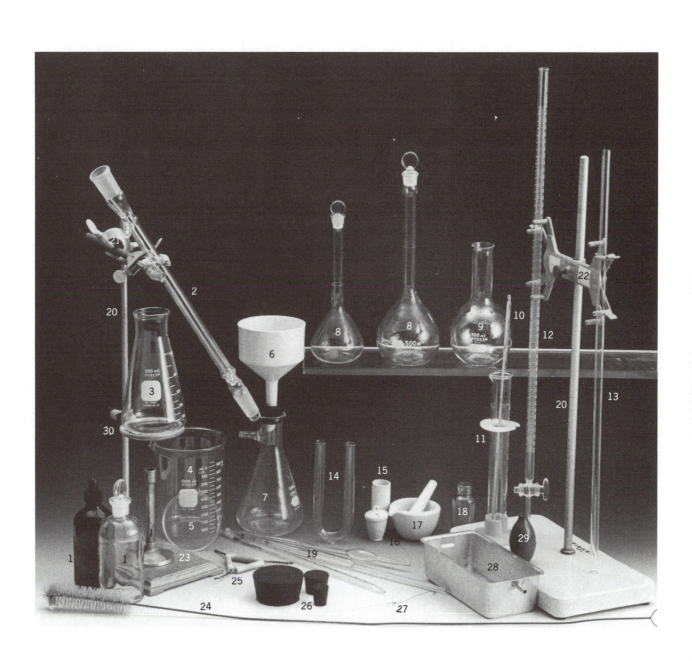

A Beginning
◇ Laboratory Techniques

Glassware is important to any and all chemical apparatus. The proper preparation and handling of glassware for the construction of the apparatus enables the chemist to have more confidence in the collected data.

Chemicals are generally pure. Safe and proper handling of chemicals reduce the chances for their contamination, often producing more reliable and reproducible observations and data. Several very important techniques for handling glassware and chemicals follow.

The procedure for inserting glass tubing through a rubber stopper, when performed incorrectly, causes more serious injuries than any other single operation in the general chemistry laboratory. Please familiarize yourself with this technique before working with glass tubing.

Make sure that the diameter of the glass tubing approximates that of the hole in the stopper. Moisten the previously **fire polished** glass tubing *and* the hole in the stopper with water or glycerol (glycerol works best). Place your hand on the tubing no more than 2–3 cm (≈1 inch) from the stopper. Protect your hands with a towel (Figure T.1) or heavy gloves.

Simultaneously *twist* and *push, slowly* and *carefully*, the tubing through the hole. There *never* should be more than 3 cm of glass tubing between your hand and the stopper. Wash off any excess glycerol.

GLASSWARE AND CHEMICALS

1. Inserting Glass Tubing Through a Rubber Stopper

Fire polish: to round the edges of a sharp cut section of glass with heat. See Appendix A.

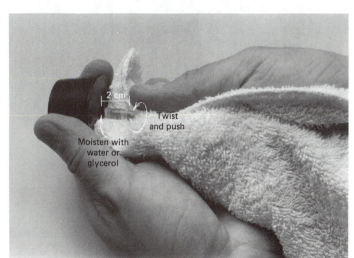

Figure T.1
Insert glass tubing into a rubber stopper with the aid of glycerol

2. Cleaning Glassware and the Lab Bench

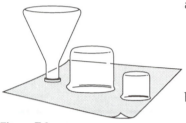

Figure T.2
Invert glassware on a paper towel

A good detergent for washing glassware is essential.

Cleanliness is very important to a chemist. Data is often misinterpreted if an analysis is conducted in contaminated glassware. Clean glassware *immediately* after its use (analogous to the best time to wash dirty dishes in the kitchen). Inspect the glassware for chips, cracks, stars, or other physical defects. Consult with your instructor when defects are observed.

Dispose of broken glassware in marked containers.

a. **Clean glassware.** Clean all glassware with soap and warm tap water. Use brushes of appropriate stiffness and size. Rinse the glassware first with *tap* water and then once or twice with *small* amounts of deionized water. Never use deionized water for washing glassware; it is too expensive.

b. **Dry glassware.** Invert clean glassware such as beakers and flasks on a paper towel or rubber mat to dry (Figure T.2); do not wipe or air-blow dry because of possible contamination. Do not dry calibrated or heavy glassware (graduated cylinders, volumetric flasks, or bottles) in an oven or over a direct flame.

If the glassware is cleaned as described, but is needed for immediate use, rinse the glassware with small amounts of the solvent or solution being used in the experiment and discard.

c. **Clean lab area.** At the end of the laboratory period, wipe clean the surface of the lab bench near your work station clean with a damp towel. Also attend to the area where the laboratory chemicals are placed as well as the hood, balance, and sink areas where you worked.

3. Microscale Analyses

Microscale: a term used to identify relative (very small) amounts of chemicals used for an experiment.

Many of the Procedures in these experiments require a **microscale** technique for analysis. In this technique, small volumes of solutions, generally less than 2 mL are needed. A special apparatus called a **well plate** having either 24, 48, or 96 wells (most often 24 in this manual) has been designed for these experiments. Each well has a volume of about 3.5 mL for the 24 well plate, but only $1/2$ mL for the 96 well plate. Additionally small quantities of test reagent are added to the wells with a small plastic (throw-away) pipet, called a Beral pipet. The Beral pipet is designed to add small diameter drops of test reagent to the well (Figure T.3).

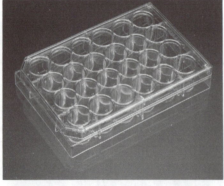

Figure T.3
24 well tray and Beral pipet used for the microscale analyses

4. Handling Chemicals

Chemicals are very safe if handled and disposed of properly. When handling an unfamiliar chemical, ask your instructor for the precautions in handling the chemical and how you are to properly dispose of it. If you wish to know more about the properties and hazards of a chemical, consult the bound collection of Material Safety Data Sheets (MSDS) generally available in the chemical stockroom.

a. **Contact with chemicals**. Avoid direct contact with all laboratory chemicals. Avoid breathing chemical vapors. Never taste or smell a chemical unless specifically directed to do so. Any interaction with a chemical may cause a skin, eye, or mucous irritation. In other words, **play it safe!** The international caution sign is used throughout this manual to cite where extra care should be taken in handling a chemical.

b. **Transfer a chemical**. See Techniques 5 (solution transfer) and 12 (solid transfer). Read the label twice and transfer only the amount needed. The use of the wrong chemical (or wrong concentration of chemical) can lead to an "unexplainable" accident or result (and an argument with your laboratory instructor). *Never* use more reagent than the procedure calls for; *do not return the excess* to the reagent bottle—share it with a friend.

c. **Mix chemicals for preparing solutions**. Add a reagent slowly; never dump it in. While stirring, slowly transfer the solid or pour the *more* concentrated solution into water or into a less concentrated solution, never the reverse! This is especially true when diluting concentrated (conc) sulfuric acid, H_2SO_4, with water.

d. **Dispose of chemicals**. Dispose of waste chemicals as soon as the data collection is complete. Nothing except soap and water and a few identified non-toxic solutions should be discarded in the sink. Each experiment will give you guidelines for the disposal of test chemicals— **Disposal Information**—having the format as follows:

> **Dispose of the waste chemical(s) in the...(for example)**
> - **sink: soap, nonflammable, nontoxic, water-soluble liquids followed by large amounts of water** *as advised by the laboratory instructor*
> - **waste container (properly labeled): water-insoluble liquids, solids, and toxic wastes** *as advised by the laboratory instructor*
> - **waste container (properly labeled): paper products (such as litmus paper, filter paper, and matches) and broken glassware** *as advised by the laboratory instructor*
> - **covered containers (properly labeled): volatile liquids or very reactive chemicals** *as advised by the laboratory instructor*

LIQUID REAGENTS AND SOLUTIONS

Laboratory chemists repeatedly need and use chemicals in liquid and solution form. Because solutions provide a homogeneous distibution of a dissolved chemical, the proper techniques for transferring, heating, evaporating, dispensing, and testing liquids are important and vital for the collection of "good" data.

5. Transferring Liquids

In transferring a liquid reagent from a reagent bottle, remove the glass stopper and hold it between the fingers of the hand that is used to grasp the reagent bottle (Figure T.5a). *Never* lay the stopper from a reagent bottle on a laboratory bench; impurities may be picked up and thus contaminate the solution.

To transfer a liquid from one vessel to another, hold a stirring rod against the lip of the vessel containing the liquid and pour the liquid down the rod which is touching the inner wall of the receiving vessel (Figures T.5a and b). This avoids any splashing of the liquid in the vessel and any loss of reagent down the side of the reagent bottle. *Never* transfer more liquid than is required for the experiment. *Never* return unused chemicals to the reagent bottle.

Figure T.5a
Grasp the stopper with the hand used to hold the reagent bottle and pour the liquid with the aid of a stirring rod

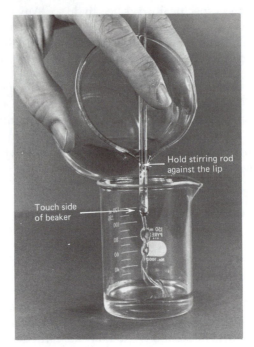

Figure T.5b
Transfer a liquid with the aid of a stirring rod

6. Heating Liquids

Boiling chip: a piece of ceramic containing small, entrapped air pockets that are released with the application of heat.

Bumping: the sudden formation of superheated vapor at the bottom of a test tube, flask, or beaker near the point at which the flame is applied.

a. **Test tube**. (**Caution:** *heating liquids in a test tube must be done carefully.*) The test tube should be no more than one-third full. Place a gentle flame at the same level as the top of the liquid, *not* at the base (Figure T.6a). Move the test tube in and out of the flame while swirling the contents. Never point the test tube toward anyone; the contents may be ejected violently if the test tube is not properly heated.

b. **Erlenmeyer flask**. Small volumes of liquid may be heated with caution in an Erlenmeyer flask using a direct flame. Hold the flask with a piece of tightly folded paper or tongs and gently swirl (Figure T.6b). Do *not* place the hot flask on the laboratory bench; allow to cool by setting the flask on a wire gauze.

c. **Beaker (or flask)**. Support the beaker (or flask) on a wire gauze. Place a support ring around the upper part of the beaker. Place a glass stirring rod or **boiling chip** in the beaker; this minimizes the problem of **bumping**. Position the center of gravity of the assembled apparatus over the base of the ring stand. Place the flame directly beneath the tip of the stirring rod (Figure T.6c).

d. **A hot water bath**. A small volume of solution in a test tube may be heated to a constant temperature in a hot water bath (Figure T.6d). If the solution is in a beaker or Erlenmeyer flask instead of a test tube, place it in a beaker (next available size), one-fourth filled with water, and heat to the desired temperature.

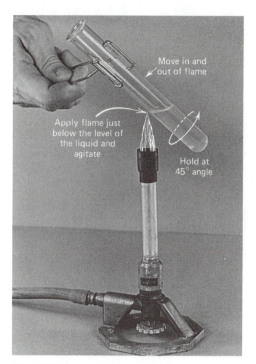

Move in and out of flame

Apply flame just below the level of the liquid and agitate

Hold at 45° angle

Figure T.6a

Carefully heat a liquid in a test tube with a direct flame—apply the flame at the *top* of the liquid

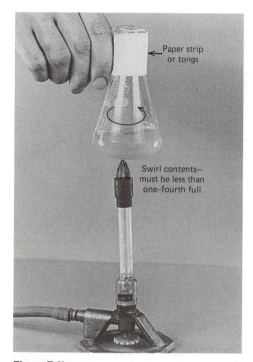

Paper strip or tongs

Swirl contents— must be less than one-fourth full

Figure T.6b

Heat a liquid in a flask directly over the flame while *swirling*

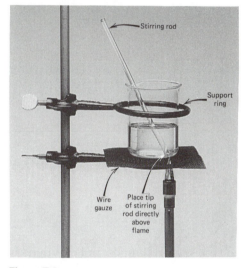

Stirring rod

Support ring

Wire gauze

Place tip of stirring rod directly above flame

Figure T.6c

Place the flame directly beneath the *tip* of the stirring rod in the beaker

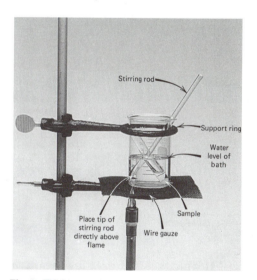

Stirring rod

Support ring

Water level of bath

Place tip of stirring rod directly above flame

Wire gauze

Sample

Figure T.6d

A hot water bath

7. Evaporating Liquids

a. **Nonflammable liquids** may be evaporated from an evaporating dish with a *gentle* direct flame (Figure T.7a) or over a steam bath (Figure T.7b). Gentle boiling is more efficient than rapid boiling.

b. **Flammable liquids** may be similarly evaporated from an evaporating dish using a heating mantle (Figure T.7c). The use of a fume hood (Technique 17c) is recommended for the removal of flammable vapors; consult with your laboratory instructor.

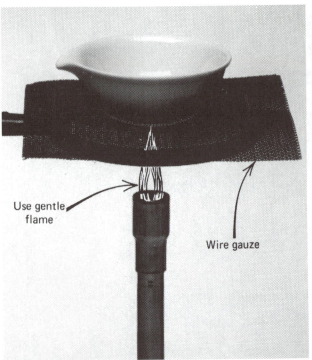

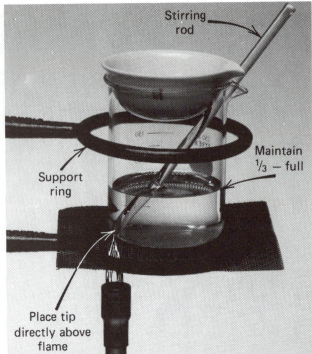

Figure T.7a

Evaporation of a *non*flammable liquid over a low, direct flame

Figure T.7b

Evaporation of a *non*flammable liquid over a steam bath

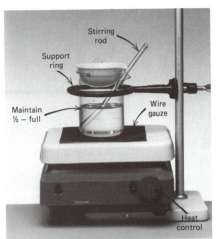

Figure T.7c

Evaporation of a *flammable* liquid over a steam bath

8. Reading a Meniscus

For exacting measurements of clear or transparent liquids and solutions in graduated cylinders, pipets, burets, and volumetric flasks, their volumes are read at the *bottom* of the meniscus. Steady the eye horizontal to the bottom of the meniscus (Figure T.8a); position (horizontally) the top edge of a black mark (made on a white card) just below the meniscus. The reflection of the black background off the bottom of the meniscus better defines its position (Figure T.8b). Substituting a finger for the black mark on the white card is not as effective, but it does help.

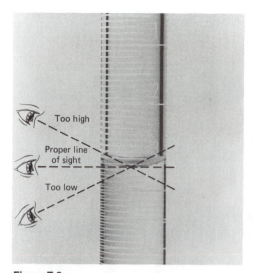

Figure T.8a

Read the meniscus with the eye *horizontal* to the *bottom* of the meniscus

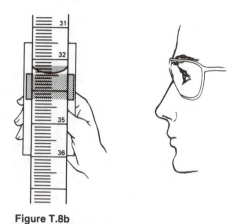

Figure T.8b

Read the meniscus using a black mark on a white card

a. **Prepare the pipet**. Clean the pipet with a soap solution; rinse with several portions of tap water and then with deionized water. *No water droplets should adhere to the inner wall of the pipet.* Transfer the liquid or solution that you intend to pipet from the reagent (see Technique 5) bottle to a beaker. Do *not* insert the pipet directly into the reagent bottle. Dry the pipet tip with a clean, dust-free towel or tissue (e.g., a Kimwipe). Using suction from a collapsed rubber (pipet) bulb (**Caution:** *Never use your mouth*), draw several 2–3 mL volumes into the pipet as rinse. Roll each rinse around in the pipet so that the solution washes the entire surface of the inner wall. Deliver each rinse through the pipet tip into a waste beaker.

b. **Fill and operate the pipet**. To fill the pipet, place the tip well below the surface of the solution in the beaker. Then using the collapsed rubber (pipet) bulb, draw the solution into the pipet until its level is 2–3 cm above the "mark" (Figure T.9a). Remove the bulb and *quickly* cover the top of the pipet with your *forefinger*, *not* your thumb. If you are right handed, operate the rubber bulb with your left hand and use the forefinger of your right hand to cover the pipet.

c. **Deliver the solution.** Remove the tip from the solution, dry the tip with a dust-free towel, and holding in a vertical position over a waste beaker, control the delivery of the excess solution from the pipet with the forefinger until the bottom of the meniscus is at the mark. Practice!! (Figure T.9b). See Technique 8 for reading the meniscus. Remove any drops suspended from the pipet tip by touching it to the wall of the waste beaker. Again wipe off the tip with a clean, dust-free towel or tissue, and deliver the solution to the receiving vessel (Figure T.9c); keep the tip above the level of the liquid and against the wall of the receiving vessel. Do *not* blow or shake out the last bit of solution that remains in the tip; this liquid has been included in the calibration of the pipet. See Experiment 2 for a discussion and procedure for the calibration of pipets.

d. **Clean the pipet.** Once the use of the pipet is complete, rinse the pipet several times with deionized water. Roll each rinse to flush the wall of the pipet and drain through the tip.

9. Pipetting a Liquid or Solution

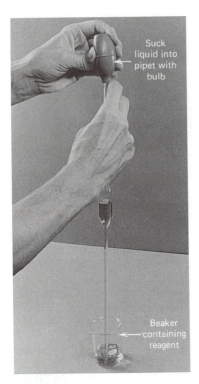

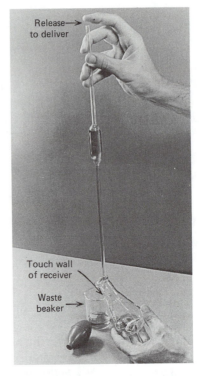

Figure T.9a
Draw the solution into the pipet with a rubber bulb

Figure T.9b
Control the delivery of the solution with the *forefinger*

Figure T.9c
Deliver the solution with the pipet touching the side of the receiving flask

10. Titrating a Solution

Titrant: the solution contained in the buret and added to the analyte.

a. **Prepare the buret**. Clean the buret with a soap solution. If a buret brush is used, prevent the wire handle from scratching the wall. Rinse the soap solution from the buret several times *through the stopcock*, first with tap water and then with deionized water. *No drops should adhere to the inner wall of a clean buret.* Close the stopcock. Rinse the buret with several 3–5 mL portions of **titrant**. Tilt and roll so that the rinse comes into contact with the entire inner surface. Drain each rinse through the buret tip into a waste beaker.

b. **Fill the buret**. Support the buret with a buret clamp (Figure T.10a). Close the stopcock and, with the aid of a funnel, fill the buret with titrant to just above the zero mark. Open the stopcock briefly to remove any air bubbles in the tip. Allow 30 seconds for the titrant to drain from the wall; record the volume (±0.02 mL). Note that the graduations increase in value from top to bottom on the buret.

c. **Prepare for the titration**. Place a piece of white paper beneath the receiving vessel, generally an Erlenmeyer flask, so that the appearance of the endpoint (the point at which the indicator turns color) is more visual. If the endpoint results in the appearance of a light or white color, a black background is preferred.

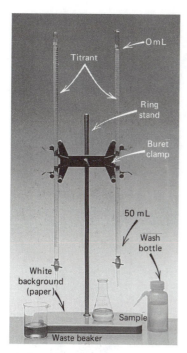

Figure T.10a
A titration apparatus

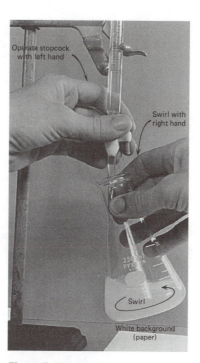

Figure T.10b
Titration technique for *right*-handers

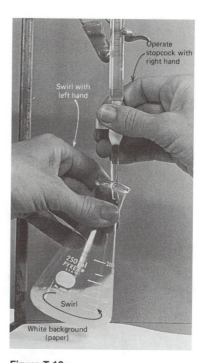

Figure T.10c
Titration technique for *left*-handers

Analyte: the substance being analyzed which is usually in the receiving flask in the titration procedure.

d. **Titrate the analyte.** During the titration operate the stopcock with your left hand if right-handed, Figure T.10b (but with the right hand if left-handed, Figure T.10c) to prevent the stopcock from sliding out of its barrel. Constantly swirl the flask with the right hand. The buret tip should extend 2–3 cm inside the mouth of the receiving vessel. Periodically stop adding the titrant and wash the wall of the flask (Figure T.10d). Near the endpoint (slower color fade, Figure T.10e), slow the rate of titrant addition until a single drop makes the color change of the indicator persist for 30 seconds. **STOP**, allow 30 seconds for the titrant to drain from the wall, read and record the volume (±0.02 mL).

To add less than one drop of titrant, suspend the drop on the buret tip and wash it into the flask with deionized water.

Fill the buret after each titration.

e. **Clean the buret**. After completing your titrations, drain the titrant from the buret, rinse with several portions of deionized water, and drain through the tip. Store the buret as directed by your instructor.

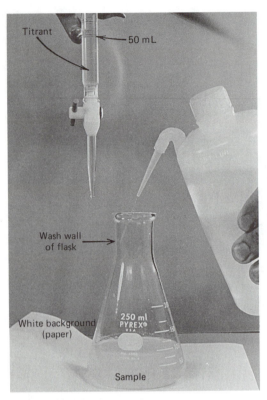

Figure T.10d
Wash the wall of the receiving flask *frequently* **during the titration**

Figure T.10e
Note the slow color fade of the indicator near the stoichiometric point of the titration

11. Testing with Litmus

To test the acidity/basicity of a solution with litmus paper, insert a stirring rod into the solution, withdraw it, and touch it to red or blue litmus (Figure T.11). Acidic solutions turn blue litmus red and basic solutions turn red litmus blue. *Never* place the litmus paper directly into the solution.

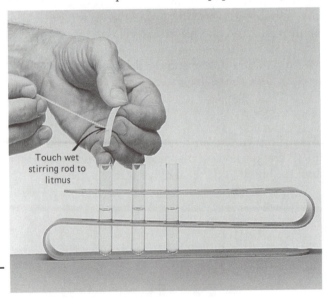

Figure T.11
Litmus test for pH. Touch wet stirring rod to litmus paper

SOLID REAGENTS

Many reagent chemicals, such as salts, are solids. Some solids may readily absorb water, some are highly crystalline, and some have a high degree of purity. Transferring and measuring the mass of solids and separating and purifying solids from mixtures are techniques aimed to minimize contamination.

First, read the label *twice* on the bottle to ensure the use of the correct reagent. Place the lid (glass stopper or screw cap) of the reagent bottle top-side-down. Hold the bottle with the label against your hand, tilt, and roll back and forth (Figure T.12) until the desired amount has been dispensed; try not to dispense more reagent than is needed. If too much reagent is removed, do *not* return the excess, but rather, share it with a friend. Do *not* insert a spatula or other object into the bottle to remove or to return the reagent unless you are specifically told to do so. When finished return the lid to the reagent bottle.

12. Transferring Solids

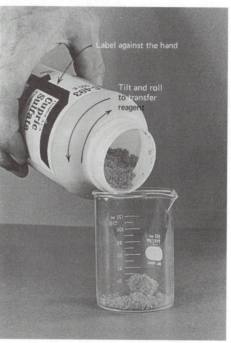

Label against the hand

Tilt and roll to transfer reagent

Figure T.12

Tilt and toll the reagent bottle to dispense a solid reagent

A balance is used to measure the mass of a chemical or a small piece of laboratory equipment. Some various types and models of laboratory balances are shown. Select the balance that provides the sensitivity listed in the Procedure of the experiment.

13. Using the Laboratory Balance

Turn dial to balance, ± 0.01 g

Do not place chemical directly on pan

Figure T.13a

Triple-beam balance, sensitivity, ±0.01 g

Balance/Sensitivity
- triple-beam balance (Figure T.13a)/±0.01 g
- top-loader balance (Figures T.13b)/±0.01 g and/or ±0.001 g
- analytical balance (Figure T.13c)/±0.0001 g

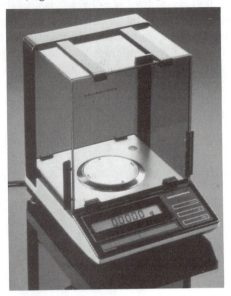

Figure T.13b
Top-loader balance, sensitivity,
±0.01 g and/or ±0.001 g

Figure T.13c
Analytical balance, sensitivity,
±0.0001 g

In using and caring for a balance, follow these guidelines:
- Handle with care; balances are expensive.
- Do not attempt to be a handyman. If the balance is not operating properly, see your instructor.
- Check the level of the balance; see the instructor for assistance.
- Use a beaker, weighing paper, watch glass, or some other container to measure the mass of laboratory chemicals. Do *not* place laboratory chemicals directly on the pan.
- Do not drop anything on the pan.
- After the measurement, return the balance to the zero reading.
- Clean up any spillage of chemicals on the balance or in the balance area.

14. Separating a Liquid from a Solid

14a

Supernatant liquid: the liquid that rests above the solid in a solid-liquid mixture.

14b

a. **Decant a liquid from a solid**. Allow the solid to settle in the beaker (Figure T.14a), flask, or test tube. Transfer the liquid, called the **supernatant**, with the aid of a stirring rod (Figure T.14b) as described in Technique 5. Do this slowly so as not to disturb the solid that has settled.

b. **Gravity filter**. Fold the filter paper in half (Figure T.14c); refold to within about 10° of a 90° fold, tear off the corner unequally, and open. Place the folded filter paper snugly into the funnel. Steady the funnel with a support clamp. Moisten the folded filter paper with the solvent being filtered and press the filter paper against the funnel's top wall to form a seal.

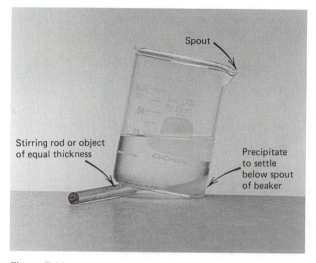

Figure T.14a
Allow the precipitate to settle

Spout

Stirring rod or object
of equal thickness

Precipitate
to settle
below spout
of beaker

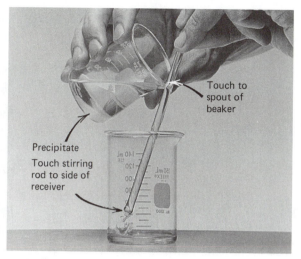

Figure T.14b
**Transfer the supernatant liquid
from the precipitate**

Touch to
spout of
beaker

Precipitate

Touch stirring
rod to side of
receiver

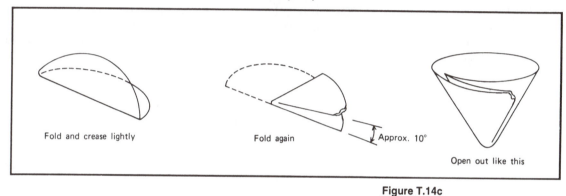

Fold and crease lightly

Fold again

Approx. 10°

Open out like this

Figure T.14c
**Sequence of folding filter paper for
a gravity—filtering funnel**

*Filtrate: the liquid or solution that
passes through a filter.*

Filter as shown in Figure T.14d. The funnel's tip should touch the beaker wall to reduce splashing. *Never* fill the funnel more than two-thirds full. Keep the funnel's stem filled with **filtrate**; the weight of the filtrate creates a slight suction at the base of the barrel of the funnel which hastens the filtration process.

c. **Vacuum filter**. A Büchner funnel substitutes for a common laboratory funnel in vacuum filtration (Figure 14e). A disc of filter paper fits over the flat, perforated bottom of this funnel. Applying a light suction to the filter paper, moistened with the solvent, creates a seal. The solid-liquid mixture is transferred to the filter in the same manner as that for gravity filtration (Technique 5). When applying a vacuum with a water aspirator, open fully the faucet so that a maximum suction can be applied.

Figure T.14d
Gravity filtering apparatus

Decant: to transfer the liquid without disturbing the solid in a solid-liquid mixture by carefully pouring the liquid out the top of the vessel.

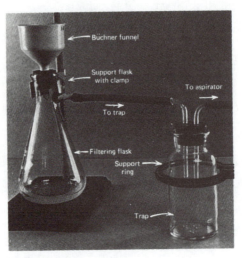

Figure T.14e
Vacuum filtering apparatus

d. **Centrifuge.** Precipitates that form in solution can be made more compact at the bottom of a test tube (or centrifuge tube) by using a centrifuge (Figure T.14f). The supernatant is then easily **decanted** without any loss of the precipitate (Figure T.14g). This quick and efficient separation requires 20–40 seconds.

Observe these precautions while operating a centrifuge:

- Never fill the centrifuge tubes to a height more than 1 cm from the top.
- Label the centrifuge tubes to avoid confusion.
- Always operate with an even number of centrifuge tubes, containing equal volumes of liquid, placed opposite one another; this *balances* the centrifuge and eliminates excessive vibration and wear. If only one tube needs to be centrifuged, balance it with a tube containing the same volume of solvent (Figure T.14h).
- Always close the lid of the centrifuge during its operation.
- STOP the centrifuge immediately if excessive vibration occurs.

Figure T.14f
Centrifuge

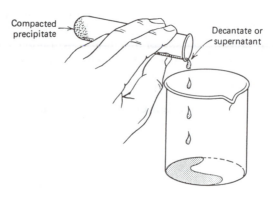

Figure T.14g
Decant the supernatant liquid from a compacted precipitate in the bottom of a centrifuge tube

Flush the precipitate with a wash bottle while holding the beaker or test tube over the receiving container (Figure T.15).

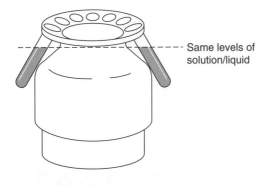

Same levels of solution/liquid

Figure T.14h
Balance the centrifuge. Place tubes with equal volumes of solution opposite each other inside the rotor's metal sleeves

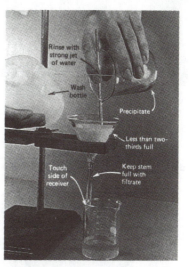

Figure T.15
Flush a precipitate from a beaker with a wash bottle

a. **Dry and/or fire the crucible**. Support the crucible on a **clay triangle** (Figure T.16a) and heat the crucible in a hot flame until it glows red. Rotate the crucible with tongs to ensure complete ignition. Allow the crucible and lid to cool to room temperature. Use crucible tongs for any transfer of the crucible and lid. If the crucible still remains dirty, add 2 mL of 6 M HNO$_3$ (**Caution**: *avoid skin contact*) and evaporate to dryness.

b. **Ignite the contents in the absence of air**. Set the crucible upright in the clay triangle; adjust the lid slightly off the lip of the crucible (Figure T.16b). Use the tongs for adjustment.

c. **Ignite the contents for complete combustion**. Tilt the crucible and adjust the lid so that only about half of the crucible is covered (Figure T.16c).

d. **Cooling in a desiccator/desicooler.** Moisture commonly condenses on the surface of a crucible and lid that are set to cool in the laboratory, adding to their mass. If a mass measurement is critical, the cooling process is completed in a desiccator/desicooler (Figure T.16d). A desiccant, typically **anhydrous** CaCl$_2$, absorbs the moisture contained within the desiccator.

16. Igniting a Crucible

Clay triangle: a porcelain covered wire triangle used to heat porcelain labware to high temperatures over a direct flame. The porcelain diffuses the heat to minimize hot spots.

Desiccator (or desicooler): a laboratory apparatus that provides a dry atmosphere.

Anhydrous: means "without water"

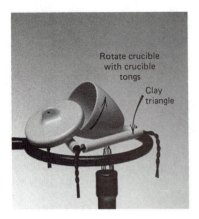

Figure T.16a
Dry and/or fire a crucible and cover with a hot flame

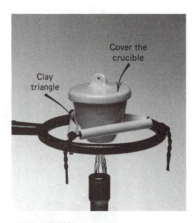

Figure T.16b
Ignition of a crucible's content _without_ air

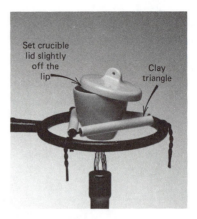

Figure T.16c
Ignition of a crucible's content for complete combustion

Figure T.16d
A laboratory desicooler (left) and desiccator (right)

GASES

Gases are the most elusive state of matter. Collecting and transferring quantitiative amounts of a gas that is transparent and colorless can be difficult for an inexperienced chemist. A special knowledge of the techniques for handling gases is valuable.

17. Handling Gases

a. **Test for odor**. An educated nose is an important and a very useful asset in the chemistry laboratory. Use it with caution, however, because some vapors are toxic. Fan the vapor toward your nose (Figure T.17a); never hold your nose directly over the vessel.

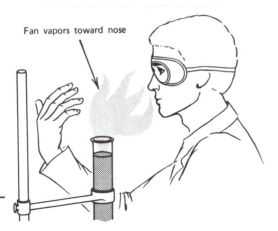

Fan vapors toward nose

Figure T.17a
To test the odor of a gas, fan the vapor gently toward the nose

b. **Collect gases.**

i. *By air displacement*. Gases more dense than air may be collected by using the experimental setup shown in Figure T.17b. Gases less dense than air are collected using the apparatus in Figure T.17c.

ii. *By water displacement.* Gases that are relatively insoluble in water are collected using the apparatus shown in Figure T.17d.

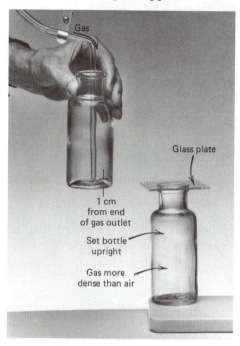

Figure T.17b
Collection of gases *more* dense than air

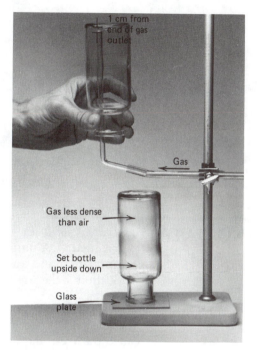

Figure T.17c
Collection of gases *less* dense than air

c. **Remove irritating or toxic gases.**

i. *Laboratory fume hood*. When fume hood space is available, use it; do not substitute the improvised hood.

ii. *Improvised hood.* If space in the fume hood is inadequate to remove *small* quantities of the toxic or nauseating vapors, use the improvised hood (Figures T.17e). Use this method *only* with the advice of the instructor.

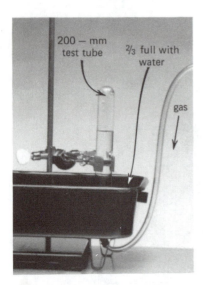

Figure T.17d
Collection of a *water-insoluble gas* by displacement

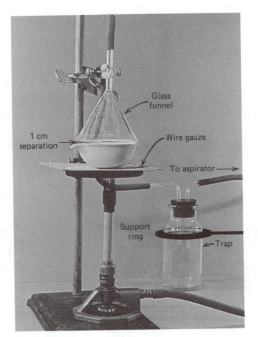

Figure T.17e
The use of an improvised hood over an evaporating dish

HANDLING DATA

Data analysis is the final, but very important, step in completing an experiment. A knowledge of organizing and portraying data for interpretation and analysis can conclude a most meaningful and intense work effort in the laboratory.

18. Graphing the Data

The most common graph used in the chemical laboratory is the line graph. The line graph has two lines drawn perpendicular to each other; these are called the axes: the x-axis (called the **abscissa**) is horizontal and the y-axis (the **ordinate**) is vertical. Each axis is scaled according to the units and range of the measurements; the scale on each axis need not be the same units or the same divisions—they only need to be consistent with the measurements. Each division or subdivision on the graph must have a constant integral value along each axis.

To illustrate the proper construction of a line graph, let's plot the following data for helium gas:

Volume, mL	20	24	30	35	37	41	46
Temperature, K	100	120	150	175	185	205	230

a. **Draw and label the axes.** Generally the dependent variable (that which you measure) is plotted along the y-axis and the independent variable (that which you control in the experiment) is plotted along the x-axis. For example, if we control the temperature of the system and observe the new volume, we study volume (V) as a function of temperature (T), or V *vs.* T, and plot volume along the y-axis and temperature along the x-axis. The label for each axis should identify the parameter that is being plotted and its units (Figure T.18a).

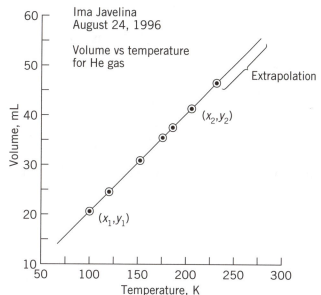

Figure T.18a
A line graph of volume (V) vs. temperature (T) with labeled and scaled axes; a "best" smooth curve is drawn through (or near) the data points; the data is also extrapolated for further interpretation

b. **Select scales so that the range of data points fill, or nearly fill, the entire space allotted for the graph**. The subdivisions of the scale should be easy to interpret. The intersection of the two axes does not have to be the zero point for each. In our example, the volume units measure from 20 to 46 mL; let's select an ordinate scale from 10 to 60 mL with 5 mL subdivisions. The temperature units range from 100 to 230 K; let's select an abscissa scale from 50 to 250 K with 25 K subdivisions. This selection of ranges allows us to not only fill the space allotted for the graph with data but also allows us to easily interpret the value of each subdivision along each axis. Also our scale selection allows for some **extrapolation** of units beyond the experimental range of our data. (See Figure T.18a)

Extrapolation: to extend the experimental relationship beyond the collected data points of the graph.

c. **Place each data point at the appropriate place on the graph**. Plot a point for the observed y-measurement at the selected x-value from the data. Draw a circle around each point. Ideally, the size of the circle should approximate the error in the measurement.

d. **Draw the best smooth line through the data points**. The line does not have to pass through all of the points or, for that matter, any of the points—it must merely represent the best averaging of all the data points. Notice that the line passes "undrawn" through the circled data points. The extrapolation of data (the dashed lines) extends the data for additional interpretations.

e. **Place a title on the graph well away from the plot**. If possible, the title, along with your name and date, should be placed in the upper portion of the graph.

f. **For straight-line graphs**, the drawn line is represented algebraically by the equation

$$y = mx + b$$

m is the slope and **b** is the intersection of the line along the y-axis at x = 0. The slope is the ratio of the change in the ordinate (y-axis) values to that of the abscissa (x-axis) values, $\Delta y / \Delta x$ (Figure T.18b).

$$\text{slope, } m = \frac{(y_2 - y_1)}{(x_2 - x_1)} = \frac{\Delta y}{\Delta x}$$

In our example, the slope can be calculated from

$$x_2 = 200 \text{ K}, \ y_2 = 40 \text{ mL}; \ x_1 = 100 \text{ K}, \ y_1 = 20 \text{ mL}$$

$$m = \frac{(40 - 20)\text{mL}}{(200 - 100)\text{K}} = 0.20 \ \text{mL/K}$$

We can conclude that the volume changes 0.20 mL for each 1 K.

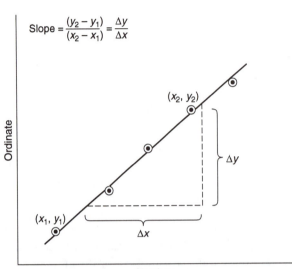

Figure T.18b
Determination of the slope of a straight line

NOTES ON LABORATORY TECHNIQUES

Experiment 1

◇ Safety and SI

Laboratory experimentation and data analyses are the core of all science. It is in the laboratory where new discoveries are sought; progress toward cures of diseases such as cancer, sickle cell anemia, and AIDS is slow and meticulous, but only through careful and reproducible experimentation or *research* can we ever expect to eliminate such devastating ailments in our society. The vaccine for polio, a very debilitating ailment prior to the mid-1950s, was discovered in the laboratory. Other plagues to modern society, such as anthrax and small pox, have been virtually eliminated through research.

High-purity silicon for semiconductors, synthetic fibers (such as nylon, acrilan, and polyester), and prescription drugs for heart disease, the common cold, and birth control were developed in the laboratory environment. These examples are only a small sampling of the effect that chemistry has had on modern society. You are about to begin a very exciting discipline where chemistry (and science in general) has produced materials that have changed the course of history more than once. New, yet to be discovered breakthroughs, will make it happen again. Will you be ready?

OBJECTIVES

- To check-in to the laboratory
- To learn various safety rules and procedures
- To identify and locate common laboratory and safety equipment
- To develop manipulative skills in the SI

PRINCIPLES

The chemical principles discussed in lecture are interpretations and conclusions of experimental data. Years of observation, experimentation, interpretation, and prediction are needed before a "principle" finds its way into a textbook.

The techniques, procedures, and equipment used in this laboratory assist in providing a clearer understanding and interpretation of those principles. No textbook or lecturer can substitute for the observations that are made, the data that are collected and analyzed, and the interpretations that evolve. An understanding of chemical principles can help us interpret the real world of science and the science of everyday living.

You will use laboratory equipment in designing experiments aimed at gathering reliable data. As you work, record the data and use your knowledge of chemical principles to explain what is seen. A good scientist is a thinking scientist. Cultivate self-reliance and confidence in your work, even it doesn't look right; this is how scientific breakthroughs occur.

In the first few laboratory periods, you are introduced to basic rules, tools, and techniques of chemistry and situations requiring their use. The Techniques section (pages 1–20) illustrates many laboratory techniques that can make your laboratory experience more meaningful. The Procedure for each experiment indicates the techniques (with icons) that are immediately appropriate. Other techniques are introduced as the course advances.

Safety

The chemistry laboratory is a *very safe* place *if* known tried and tested procedures for handling chemicals and conducting experiments are followed. If the suggested laboratory procedures and techniques for completing an experiment and the cited safety precautions are followed, then your experiences as a chemist will enable you to collect and analyze data without fear of injury or exposure. Where a danger in handling a chemical is possible, a boldface **Caution** statement is noted and a caution icon appears in the margin. Refer to Technique 4.

Laboratory safety instructions for self-protection and general laboratory safety rules are stated in this experiment and should be referred to often as you proceed through the manual. A safety presentation/lecture will outline those laboratory safety rules that are most important to your particular laboratory environment.

Le Système Internationale d'Unités (SI)

Chemists, and scientists in general, throughout the world use the **SI** (French: **Le Système Internationale d'Unités)** as the standard units for weights and measurements of scientific data. This system, adopted by the International Union of Pure and Applied Chemistry (IUPAC), provides a logical and interconnected framework for all scientific measurements. In the SI, base units are defined from which all other expressions of measurements are derived; derived units (from the base units) are established for the reporting of specific data.

The *base* units of measurement in SI (for our purposes in this course) represent mass, length, time, temperature, and amount of substance. The **kilogram** is the base unit of mass, the **meter** is the base unit of length, the **second** is the base unit of time, the **kelvin** is the base unit of temperature, and the **mole** is the base unit for the amount of substance. For volume, a common measurement in chemistry, the *derived* unit is the **cubic meter**. Because the cubic meter is a very large volume in the chemistry laboratory, the cubic decimeter, dm^3, commonly known as the **liter,** is the more common unit. The **gram** is the common derived mass unit used by chemists.

Subdivisions and multiples of base and derived units are recorded in powers of ten, designated by scientific notation. Prefixes for the common powers of ten are listed in Table 1.1.

Units for the expression of scientific measurements can be changed from SI to English (or vice versa) or within SI. A knowledge of SI prefixes and an understanding of the application of conversion factors in a calculation makes nearly any change of units possible. In Table 1.1, the prefix means the power of ten. For example, 8.3 *centi*meters means 8.3 $x\ 10^{-2}$ meter; *centi* represents $x\ 10^{-2}$.

Expressed as a conversion factor, $\dfrac{centi}{10^{-2}} = \dfrac{10^{-2}}{centi} = 1$; also, $\dfrac{cg}{10^{-2}\,g} = \dfrac{10^{-2}\,g}{cg} = 1$.

For example, to convert 6.6 kilograms to nanograms, first convert the unit of measurement (kilograms, kg) to a unit without the prefix (the gram, g) by using the conversion factor. For this example, kilogram is converted to gram by the conversion factor, $[10^3\ g/kg]$:

$$6.6 \text{ kg} \times \left[\frac{10^3 \text{g}}{\text{kg}} \right] = 6.6 \times 10^3 \text{ g}$$

Next convert from the unit without the prefix to the desired unit of expression. For this example, gram is converted to nanogram by the conversion factor, $[\text{ng}/10^{-9} \text{ g}]$:

$$6.6 \times 10^3 \text{ g} \times \left[\frac{\text{ng}}{10^{-9} \text{ g}} \right] = 6.6 \times 10^{12} \text{ ng}$$

These two steps can be combined in consecutive operations:

$$6.6 \text{ kg} \times \left[\frac{10^3 \text{ g}}{\text{kg}} \right] \times \left[\frac{\text{ng}}{10^{-9} \text{ g}} \right] = 6.6 \times 10^{12} \text{ ng}$$

kg cancel   g cancel

Note that the conversion factors, $[10^3 \text{ g}/\text{kg}]$ and $[\text{ng}/10^{-9} \text{ g}]$, do not affect the magnitude of the measurement since 10^3 means *kilo(k)* and 10^{-9} means *nano(n)*--only the unit for expressing the measurement is changed.

Table 1.1. The Meaning of Prefixes in SI

Prefix	Abbreviation	Meaning (Power of Ten)	Example using "grams"
femto-	f	10^{-15}	$10^{-15} \text{ g} = \text{fg}$
pico-	p	10^{-12}	$10^{-12} \text{ g} = \text{pg}$
nano-	n	10^{-9}	$10^{-9} \text{ g} = \text{ng}$
micro-	μ	10^{-6}	$10^{-6} \text{ g} = \mu\text{g}$
milli-	m	10^{-3}	$10^{-3} \text{ g} = \text{mg}$
centi-	c	10^{-2}	$10^{-2} \text{ g} = \text{cg}$
deci-	d	10^{-1}	$10^{-1} \text{ g} = \text{dg}$
kilo-	k	10^3	$10^3 \text{ g} = \text{kg}$
mega-	M	10^6	$10^6 \text{ g} = \text{Mg}$
giga-	G	10^9	$10^9 \text{ g} = \text{Gg}$

The English system of measurement is still in common use in the United States and a few other countries. Some relationships are listed in Table 1.2 and Appendix B.

Being able to covert measurements between SI and the English system are quite beneficial, especially since more and more international trade is based on SI measurements. See Figure 1.1.

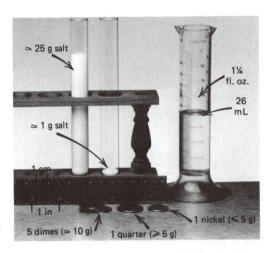

Figure 1.1
Comparisons of SI and English Quantities

Table 1.2 Comparison of the SI and English Systems of Measurement

Measurement	SI Base Unit	SI Derived Unit	Conversion Factors
Length	meter (m)		1 km = 0.6214 mi 1 m = 39.37 in 1 in = 0.0254 m = 2.540 cm (exactly)
Volume		cubic meter (m^3)	$1 L = 10^{-3} m^3 = 1 dm^3 = 10^3 mL$ 1 L = 1.057 qt 1 oz (fluid) = 29.57 mL
Mass	kilogram (kg)	gram (g)	1 lb = 453.6 g 1 kg = 2.205 lb
Pressure		Pascal (Pa) $(kg/m{\cdot}s^2)$	$1 Pa = 1 N/m^2$ 1 atm = 101.325 kPa = 760 Torr $1 atm = 14.70 lb/in^2$ (psi)
Temperature	kelvin (K)	degrees Celsius (°C)	K = 273 + °C
Energy		joule (J) $(kg{\cdot}m^2/s^2)$	1 cal = 4.184 J 1 Btu = 1055 J

Carefully study how each of the following change of units is made using Tables 1.1 and 1.2 and the appropriate conversion factors.

Example 1. Express 3.84 millimeters (mm) in centimeters (cm).

Step 1. Use Table 1.1 to write conversion factors for mm and cm.
$mm = 10^{-3} m$; $cm = 10^{-2} m$

These equalities gives rise to the following conversion factors:
$$\frac{mm}{10^{-3} m} \text{ and } \frac{10^{-3} m}{mm}; \qquad \frac{cm}{10^{-2} m} \text{ and } \frac{10^{-2} m}{cm}$$

Step 2. Set up the sequence of steps starting with the "desired unit of expression (cm)" set equal to the "unit of measurement (mm)" —use the above conversion factors arranged so that the unit of measurement cancels, followed by the cancellation of the unit without the prefix:

$$cm = 3.84 \text{ mm} \times \left[\frac{10^{-3} m}{mm}\right] \times \left[\frac{cm}{10^{-2} m}\right] = 3.84 \times 10^{-1} cm = 0.384 \text{ cm}$$

desired unit of expression   unit of measurement   mm cancel   m cancel

Example 2. What is the volume of 250 mL expressed in pints?

Step 1. Write the meanings of all units that are needed to make the conversion.
mL = 10^{-3} L; 1 qt = 2 pt. We also need an SI to English volume conversion; from Table 1.2, 1 L = 1.057 qt.
The corresponding conversion factors are:

$$\frac{mL}{10^{-3}\,L} \text{ and } \frac{10^{-3}\,L}{mL}; \qquad \frac{1\,qt}{2\,pt} \text{ and } \frac{2\,pt}{1\,qt}; \qquad \frac{1\,L}{1.057\,qt} \text{ and } \frac{1.057\,qt}{1\,L}$$

Step 2. Set up the sequence of steps starting with the "desired unit of expression (pt)" set equal to the "unit of measurement (mL)"—use the above conversion factors arranged so that the unit of measurement cancels, followed by the appropriate conversion factors:

$$pt = 250\,mL \times \left[\frac{10^{-3}\,L}{mL}\right] \times \left[\frac{1.057\,qt}{1\,L}\right] \times \left[\frac{2\,pt}{1\,qt}\right] = 0.529\,pt$$

desired unit   unit of   mL cancel   L cancel   qt cancel
of expression   measurement

Notice again that the conversion factors do not change the magnitude of the measurement, only its form of expression. Refer to the Data Sheet and complete the SI and SI-English conversions in the same manner as outlined in the previous two examples.

PROCEDURE

A. Laboratory Check-In

For this experiment you are assigned to a work station/drawer containing common laboratory equipment. Place the glassware and hardware on the laboratory bench and, with the check-in list (page x); check off the presence and determine the condition of the equipment. If you are unsure of the names of the items, refer to the chemical "kitchenware" on page xi. See your instructor about any item that may be defective or not present, especially glassware that is chipped, cracked, starred, or has any other observed defects.

A good scientist is always neat and well-organized; keep your glassware and hardware clean and neatly arranged so that it is ready for instant use in an experiment. Always have a dishwashing detergent available for cleaning glassware and paper towels on which to set cleaned items for drying. Have your instructor approve your completed check-in procedure.

B. Laboratory Safety

You and your laboratory partners must *always* practice laboratory safety. It is your responsibility, not the instructor's, to *play it safe*. On the inside front cover of this book, there are spaces to list the location of important safety equipment, telephone numbers, and reference information. Fill this out now.

Chemicals can enter the body via four routes:
- through the respiratory system
- through the digestive tract
- through the pores of the skin
- through cuts and scratches of the skin

Each of these avenues of invasion must be protected while you are working in the laboratory.

Study the following safety rules and laboratory guidelines and answer the questions on the Lab Preview.

1. Self-Protection

a. *Safety glasses, goggles, or eye shields must be worn* at all times to guard against the laboratory accidents of others as well as your own. Contact lenses should not be worn, especially soft contact lenses since they

readily absorb and retain chemical vapors; if contact lenses are worn for therapeutic reasons, fitted eye protection is absolutely necessary.

b. Laboratory aprons or coats (with snap fasteners only) should be worn to protect clothing. Above all, wear old clothing. Slacks (or jeans) and skirts that extend below the knee is permitted; however, shorts, cutoffs, and short skirts are not permitted. Loose clothing, especially long loose-fitting sleeves, should either not be worn or be constrained.

c. Shoes must shed liquids. Sandals, high-heeled, open-toed or canvas shoes are *not* permitted.

d. Jewelry, e.g., rings, bracelets, watches, should be removed. A chemical that seeps and becomes entrapped under jewelry can cause severe skin damage before the jewelry can be removed or before the danger is even realized.

e. Confine long hair. Hair will ignite near an open flame!

f. Wash your hands and arms periodically with soap and water during the laboratory session and especially before leaving the laboratory. Toxic chemicals may be inadvertently transferred to the mouth or cause a delayed reaction on the skin.

g. Whenever your skin (hands, arms, face,...) comes into contact with laboratory chemicals, wash it quickly and thoroughly with soap and warm water. Use the eyewash fountain to flush chemicals from the eyes or face. Move the eyelids away from the eye during the cleansing process. Do *not* rub the affected area, especially the face or eyes, with your hands before washing.

h. If a chemical is spilled over a large part of the body, use the safety shower and flood the affected area for at least 5 minutes. Do *not* attempt to wipe the chemicals from the clothing. Instead remove all contaminated clothing, shoes, and jewelry. No time is to be wasted because of modesty! Get medical attention as soon as possible.

i. Report any accident or injury to your instructor, even if it is perceived to be minor.

j. Identify the location and learn how to use the fire extinguisher and other safety devices.

2. **Laboratory Rules**

a. Do *not* work alone in the laboratory. At least the instructor must be present.

b. Unauthorized experiments, including variations of those in the laboratory manual, are forbidden without prior consent of your instructor.

c. *Prepare* for the lab. Complete the Lab Preview and study the Objectives, Principles, Procedure, and Techniques for the experiment *before* lab. Always try to understand what you are doing and to *think* (whistle if you like) while working. Note beforehand the safety precautions and the need for any extra equipment that is to be checked out from the stockroom.

d. No smoking, drinking, eating, or chewing is permitted. Also do not apply makeup during the laboratory session. Chemicals may enter the body through the mouth, lungs, or skin.

e. The chemistry laboratory is a serious place for learning and working. Horseplay or other careless acts are prohibited. Do not entertain guests. Do *not* waste time.

f. Maintain an orderly, clean lab bench, drawer, and work area, including the balance area. Keep drawers closed while working and the aisles free of any obstructions, such as book bags and athletic equipment. When glassware is to be no longer used for the experiment, immediately clean and return it to the lab drawer.

g. Clean up all chemical and water spills with a damp towel (and discard); discard paper scraps in marked containers. Clean the sinks of all debris. Only soap and water is to be discarded in the sink without the prior consent of the laboratory instructor.

h. Be aware of your neighbors' activities; you may be the victim of their mistakes. Advise them of improper techniques or unsafe practices. If necessary, report your concern to the instructor.

i. Record the experimental data on the Data Sheet as you complete the each part of the Procedure for each experiment. Believe in your data. A scientist's most priceless possession is integrity. If the results "do not look right", repeat the experiment; don't alter the data to make it look better—be a scientist!

Record data *in ink* as you perform the experiment. Data on scraps of paper will be confiscated. Where calculations are required, be orderly for the first set of data analysis. Do not clutter the Data Sheet with arithmetic details.

j. *Think* while you perform the experiment. Questions at the end of the Data Sheet are designed to test your understanding of the experiment, the analysis of the data, and the principles involved. Your text, various reference books (such as the Chemical Rubber Company's *Handbook of Chemistry and Physics*), or discussions with other scientists (your lecturer, laboratory instructor, or other students) are often reliable sources of information.

Have your instructor approve your review and completion of Part B.

1. Familiarity with SI is gained from a practice of converting units of measurement to desired units of expression. For example, the mass of dirt in a wheelbarrow is measured *not* in milligrams, but rather kilograms; the distance between atoms is measured in nanometers rather than kilometers. Refer to the Data Sheet and complete the suggested exercises to gain some "feeling" of various units of measurement.

C. Le Système International d'Unités (SI)

2. The conversion of a unit of measurement in the laboratory into a desired expression of the unit is a common laboratory technique. For example, if the laboratory balance reads 0.022 g, it is often more practical or convenient to express this measurement as 2.2 cg or 22 mg; to transfer two one-millionth of 1.0 L of an acid solution is *not* expressed as 0.000 002 0 L by the chemist, but rather as 2.0 μL. Refer to the Data Sheet to complete the unit of the measurement to a desired expression of the unit. Refer to your textbook for manipulations of scientific notation (powers of ten).

3. Acquire a meter stick.

   a. What prefix of the meter is used to mark the major subdivisions of the meter stick? What prefix of the meter indicates the length of each marked division?

   b. Use the meter stick to make the required measurements indicated on the Data Sheet. Use appropriate prefixes to represent the magnitude of the measurement.

NOTES, OBSERVATIONS, AND CALCULATIONS

◇ Safety and SI

Date _____ Name _____ Lab Sec. _____ Desk No. _____

1. **True or False**. Review the Techniques section and the Principles of Experiment 1 to answer the following questions.

    _____ a.   Paper towels are to be used for wiping dry your clean glassware.

    _____ b.   Sandals are *not* to be worn in the laboratory.

    _____ c.   Eye protection is worn only while you are conducting an experiment.

    _____ d.   Slacks, jeans, or skirts that extend below the knee are considered appropriate attire in the laboratory.

    _____ e.   You are encouraged to conduct experiments beyond those presented in the laboratory manual.

    _____ f.   Record data in pencil so that any errors in measurement can be corrected without presenting a messy Data Sheet.

    _____ g.   Do *not* discuss data, calculations, or the interpretations of the data with others.

    _____ h.   Eating is permitted in the laboratory *after* lunch.

    _____ i.   To determine the safety of a chemical, read the label.

    _____ j.   Do *not* invite friends into the laboratory to show them your work.

    _____ k.   A glass surface is considered clean when, after a rinse with deionized water, no water droplets cling to the surface.

    _____ l.   Always wear flammable clothing to the laboratory.

    _____ m.   Bracelets and rings are *not* to be worn in the laboratory.

    _____ n.   Immediately wipe up any chemical spills.

2.   Describe the proper clothing (head to foot) for a chemist that is working in the laboratory.

3. Why can't all chemicals be discarded down the drain of the laboratory sink?

4. Name the SI base unit for

    a. length:

    b. temperature:

    c. mass:

Safety and SI

Date _____Name _____ Lab Sec. _____Desk No. _____

A. Laboratory Check-In

Instructor's approval _____

B. Laboratory Safety

Complete the information on the inside front cover of the manual.

Instructor's approval_____

C. Le Système International d'Unités (SI)

1. Predict a suitable SI unit, with prefix, for reporting the following measurements.

    Distance across the state of Ohio   _____

    Diameter of a penny   _____

    Volume of a coffee cup   _____

    Mass of a pea   _____

    Mass of a particle of dust   _____

2. Convert each of the following using the definitions in Tables 1.1 and 1.2. Show the cancellation of units.

    a. $16.2 \text{ cm } \times \left[\dfrac{m}{cm} \right] \times \left[\dfrac{\mu m}{m} \right] =$ _____ μm

    b. $4670 \text{ mL } \times \left[\rule{1cm}{0pt} \right] \times \left[\rule{1cm}{0pt} \right] =$ _____ kL

    c. $42 \text{ mg } \times \left[\rule{1cm}{0pt} \right] \times \left[\rule{1cm}{0pt} \right] =$ _____ cg

    d. $6.64 \times 10^{-5} \text{ g} \times \left[\rule{1cm}{0pt} \right] =$ _____ μg

    e.   Express the size of a $^{9}/_{16}$-inch wrench in millimeters.

    $^{9}/_{16}$-inch $\times \left[\rule{1cm}{0pt} \right] \times \left[\rule{1cm}{0pt} \right] \times \left[\rule{1cm}{0pt} \right] =$ _____ mm

f. The 1500 meter race in international track competition approximates that of a mile. Express 1500 meters in miles.

g. A soft drink can holds 12.0 fluid ounces of your favorite beverage. Express 12.0 fluid ounces in milliliters.

h. A popular sports car is powered by a 5.7 L engine. Assuming the sports car to be an 8-cylinder engine, calculate the displacement volume, in mL, for *each* cylinder.

i. The amount of heat required to raise the temperature of 12 fluid ounces of water from room temperature to boiling is approximately 26.6 kilocalories. Express this quantity in joules, kilojoules, and British thermal units.

j. A basketball is inflated to 9.0 pounds per square inch (psi) above the atmospheric pressure of 14.7 psi (a total pressure of 23.7 psi relative to vacuum). The diameter of a basketball is 10 inches.
    i. What is the pressure of air in the basketball, expressed in atmospheres?

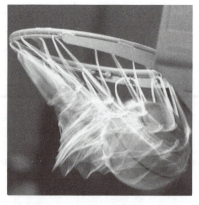

    ii. What is the volume (in liters) of the basketball?
    The volume of a sphere = $\frac{4}{3}\pi r^3$.

    iii. Assuming air occupies the volume of a basketball at 20°C and 23.7 psi of pressure, determine the mass of air in the basketball. The density of air is 1.94 g/L at 20°C at 23.7 psi pressure.

3. a. Prefix for major subdivisions on the meter stick: _____

    Prefix for each marked division of the meter stick: _____

    b. Use a measuring (meter, ruler, or yard) stick to measure your lab partner's height. Express the height in nanometers.

    c. Use a measuring (meter, ruler, or yard) stick to determine the volume, expressed in cm^3, of this laboratory manual. Calculate the volume of each page of the manual.

    d. Measure the diameter, expressed in units of the measurement, of your lab partner's head. Assume the head to be a sphere, what is the volume of your lab partner's head, expressed in dm^3?

    e. Assuming the density of water to be 1.0 g/mL and that your lab partner's head is filled with water(!), what mass of water, expressed in kilograms, is in your lab partner's head? See Part 3.d.

QUESTIONS

1. Cite at least two advantages that SI has over the traditional English system that is used in the U.S.

2. How might the following English expressions be stated in SI using *whole number* SI expressions, not exact conversions.

   a.   . . . can't be touched with a 10 foot pole"

   b.   . . . doesn't have an ounce of sense"

   c.   . . . a footlong hotdog"

   d.   . . . give him an inch and he will take a mile"

   e.   . . . as big as a 10 gallon hat"

   f.   . . . Dallas Cowboys' ball "1st and 10"

Experiment 2

◇ Techniques and
Measurements

Accurate measurements are necessary for fair purchasing practices. For example, the costs per gallon (or liter) of gasoline, pound (or kilogram) of potatoes (see photo), or yard (or meter) of cloth depend upon an amount of consumer product, measured with a "most practical" but reliable instrument.

A recent Environmental Protection Agency (EPA) report states that the concentration of lead in a water sample taken from a river flowing through a rural area is 0.03 ppm. What does that level of lead mean to the scientist and the general public? The report generates more questions than answers. For example, is the concentration too high? Is *any* level of lead too high and should its concentration be reduced to zero? If so, what does zero concentration mean—does it mean *absolutely* zero, or does it mean too low to measure with the best available scientific techniques and equipment? Practically speaking, is there any telltale difference?

Quantitative, reliable, and reproducible measurements of the amounts and concentrations of potentially hazardous chemicals are the key to a public understanding of environmental issues. Scientists must not only be trained to obtain accurate, reliable data, but also be able to interpret the data with regard to trends and cycles.

OBJECTIVES

- To properly select and operate a balance
- To calibrate laboratory glassware
- To determine the density of a pure liquid or solid

PRINCIPLES

The best available equipment and experimental techniques to collect the data for analysis require a knowledge and understanding of the scope and limits of the equipment that is being used. For example, a bathroom scale might be "good enough" for collecting data on the average weight of a student in a class of students, but would be useless in determining the average weight of an ant in a colony of ants.

The proper selection of the laboratory apparatus for a scientific investigation can only be made only if the scientist knows the extents and limits of the equipment.

Laboratory Techniques

Once the appropriate chemical apparatus has been selected for the experiment, the next step is to use it properly so that the best data can be obtained. Introductory to this manual is a large Laboratory Techniques section that describes the proper techniques for constructing apparatus, for handling gases, liquids (and solutions), and solids, for conducting various tests, for analyzing data, and for using calibrated equipment. Please review these techniques—an icon for the appropriate technique appears in the margin of the Procedure in each experiment.

Laboratory Balance

The laboratory balance is perhaps the most common and most often used piece of laboratory apparatus. Balances, *not* scales, are of different sizes, makes, and sensitivities. The selection of the appropriate balance for a mass measurement depends upon the degree of accuracy and precision required from the analysis. Read Technique 13 carefully.

Laboratory Glassware

A large assortment of laboratory glassware is available for collecting the appropriate data from an experiment. Glass tubing is required for the transfer of gases and liquids from one vessel to another, glass beakers and Erlenmeyer flasks are used to transfer and/or collect liquids, and calibrated glassware is necessary for measuring, transferring, and collecting accurate volumes of liquids and gases.

Calibrated glassware has been calibrated at the factory to deliver (or contain) an accurate volume of liquid or gas. Graduated cylinders, pipets, burets, and volumetric flasks are available for use in the experiments described in this laboratory manual.

An assortment of calibrated glassware is used and studied in this experiment, in conjunction with the continued use of SI, introduced in Experiment 1.

Solute: the substance dissolved in the solvent.

The amount of chemical that is needed for a procedure in the laboratory is often measured and transferred as the **solute** of a solution. The volume of a solution is measured with glassware that has been previously calibrated by the supplier. Since the glassware is "mass produced" at the factory, occasionally the graduations on the glassware are not as exact as what the supplier has quoted. A scientist who strives (or is required) to do very quantitative work in the laboratory always calibrates the glassware before it is used. In this experiment, you will calibrate test tubes and a pipet, but the procedure can be used to calibrate any glassware.

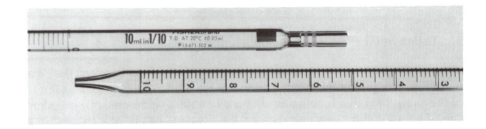

Figure 2.1
Pipet calibrated to deliver "TD"

Volumetric (calibrated) glassware is labeled as TD 20°C (Figure 2.1) or TC 20°C. A pipet labeled TD 20°C (**to d**eliver at 20°C) means that the volume of the pipet is calibrated according to the volume it delivers from gravity flow only, at 20°C. The liquid that is retained in the pipet tip is considered in the calibration of the pipet and therefore is *not* to be blown out. Glassware labeled TC 20°C (*to* **c**ontain, or rinse out, at 20°C) means that *all* of the liquid in the vessel is considered in the calibration. Therefore, in a transfer, all of the solution must be removed, generally by the addition of a rinse with the solvent. Most graduated cylinders and volumetric flasks are calibrated as TC 20°C glassware.

Density

Intensive property: a property of a substance that is independent of sample size.

Density is an **intensive property** of all substances. Its measurement is expressed as a ratio of mass to volume; substances having a large density have a large mass occupying a small volume. Water has a density of 1.0 g/cm^3, or 62.4 lb/ft^3; osmium metal (atomic number 76) has the greatest mass of any of the elements, 22.6 g/cm^3. In this experiment the density of a liquid and/or solid is determined.

The density of a liquid is determined by measuring the mass of a known volume of the liquid; the density of a water-insoluble solid is determined by measuring its mass and then the volume of water it displaces.

$$Density = \frac{mass}{volume}$$

PROCEDURE

1. The Laboratory Techniques section outlines the proper procedure for performing various tasks in the laboratory. Read Technique 13 carefully—this technique is perhaps the most often used technique in this manual.

2. Read Techniques 8 and 9 before completing the Lab Preview.

3. After studying Techniques 8, 9, and 13, obtain your instructor's approval before continuing the Procedure.

A. Laboratory Techniques

1. Several types of balances may be found in the chemistry laboratory. The triple-beam balance and the top-loading balance are the most common. Use the top-loading and the analytical balances only *after* the instructor explains their operation. As suggested on the Data Sheet, determine the mass of several objects.

B. The Laboratory Balance

1. **Graduated cylinders**. Graduated cylinders (Figure 2.2) are used to measure small volumes of liquids and solutions for experiments.

   a. Check your glassware: what volumes do your graduated cylinders measure? What is the smallest volume increment marked on the graduated cylinder?

   b. Half fill a graduated cylinder with deionized water. Notice the shape of the meniscus of the water in the cylinder. Sketch a vertical cross section of the meniscus on the Data Sheet. Describe how to read the volume of the water in the cylinder, i.e., how is the eye to be positioned relative to the surface of the water?

   c. How precisely can you estimate the volume of water in each graduated cylinder?

C. Laboratory Glassware

Figure 2.2
A graduated cylinder is often used for measuring volumes of solution

2. **Pipets.** Pipets (Figure 2.3) are long narrow pieces of glassware that measure more precisely the volumes of liquids or solutions. Pipets may be calibrated with a single, etched mark on the upper stem of the pipet or graduated, but all dispense a calibrated volume. The technique for filling and dispensing a liquid requires practice. Technique 9 describes the correct procedure for the use of pipets in transferring liquids.

   a. Practice filling and dispensing deionized water from a pipet. Practice pipetting an exact volume of water with both a calibrated pipet and a graduated pipet. When you have mastered the technique of filling and dispensing a water from a pipet, ask your instructor for approval.

   b. What are the smallest volume increments marked on the graduated pipet? How precisely can you estimate the volume of a solution in a graduated pipet?

   c. Fill a (5 mL, 10 mL, or 25 mL) pipet with deionized water and transfer it to a graduated cylinder (10 mL, 25 mL, or 50 mL, respectively). Use a ruler (marked in inches or a meter stick) to measure the height of

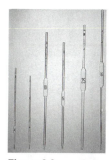

Figure 2.3
Pipets are used for measuring small volumes accurately

the liquid in the graduated cylinder. *Calculate* the inside radius of the graduated cylinder.

3. **Burets.** A buret (Figure 2.4) is a piece of glassware used to transfer a partial, but precise, volume of solution. The 25 mL and 50 mL burets are most common. Read Technique 10 for details of using a buret for an analysis.

a. Examine a buret—notice that the major divisions are marked in ascending order as you move down the buret. What is the smallest volume increment marked on the buret?

b. Partially fill the buret with deionized water. Read and record the volume of water in the buret—ask your colleague and your laboratory instructor to confirm the precision of your measurement.

How precisely can you estimate the volume of water in a buret?

Dispense, through the stopcock of the buret, about 15 mL of water. Again, read and record the volume of water. How much water was dispensed from the buret?

c. Collect the appropriate data that are required to calculate the radius of the buret.

Figure 2.4
Burets are used for many analyses of solutes in solutions

D. Calibration of Glassware

1. **Commercial calibration of graduated cylinders.** Look closely at the 10 mL and 50 mL graduated cylinders. Are the graduated cylinders calibrated as TC 20°C or TD 20°C? How accurately can each be read? Record this data on the Data Sheet as, for example, TC 20°C or TD 20°C and ±0.5 mL, ±0.1 mL, ±0.05 mL, or ±0.01 mL.

2. **Volumes of common test tubes.**
   a. Fill a clean, dry 150 mm test tube with water. Transfer the water to a 50 mL graduated cylinder, read and record the volume of water as accurately as the calibration allows.

   b. Repeat with a clean, dry 75 mm test tube. Determine its volume in the same manner.

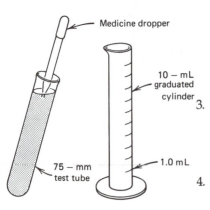
Medicine dropper
10 – mL graduated cylinder
75 – mm test tube
1.0 mL

3. **Drops in 1.0 mL of water.** With a clean, dry dropping pipet (or Beral pipet) transfer 1.0 mL of water to a clean, dry 10 mL graduated cylinder (Figure 2.5). Count and record the number of drops. What is the average volume per drop?

4. **Calibration of a pipet.**
   a. Obtain a pipet with any volume between 5 and 25 mL. Look at its calibration. How accurately can its volume be recorded? If there is no indication of accuracy, it is called a Class B pipet having an accuracy range from ±0.012 to ±0.06 mL. If the pipet is labeled "A", for Class A pipet, the accuracy range is ±0.006 to ±0.03 mL.

Figure 2.5
Determination of the number of drops in 1.0 mL of water

b. Determine the mass (±0.001 g) of your smallest-volume, clean, dry beaker. Fill the pipet to the etched mark with deionized water. Wipe the tip dry with a lint-free tissue, and transfer the water (the tip should be touching the beaker wall) to the beaker. Again determine the mass of the beaker and water on the same balance. Assuming the density of water to be 0.998203 g/mL at 20°C, calculate the TD volume of the pipet. Calculate the percent error in the calibration of the pipet.

Ask your instructor which balance is assigned to you for Part E. Write the balance number on the Data Sheet.

E. Density

1. **Density of a liquid.**

   a.   Determine the mass (±0.001 g) of your smallest beaker (with a volume ≥ 25 mL) in your drawer. Use the pipet that was calibrated in Part C.4. Rinse the pipet several times with your unknown sample and discard each rinse.

   b.   Fill the pipet to the mark. Deliver the liquid to the beaker. Measure the combined mass of the liquid and beaker on the same balance. Calculate the density of the liquid. Note: use the corrected volume of the pipet (Part C.4) in your calculation.

2. **Density of a solid.**

   a.   Determine the mass of a sheet of weighing paper. Place the solid on the weighing paper and measure the combined masses.

   b.   Half-fill a 50 mL graduated cylinder with water and record its volume, ± 0.02 mL. See Figure 2.6.

   c.   Carefully transfer the solid to the graduated cylinder. Roll the solid around in the water to remove any trapped air bubbles that adhere to the solid. Record the new volume of water in the graduated cylinder. Calculate the density of the solid.

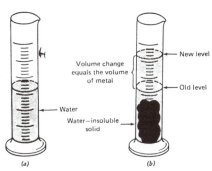

Figure 2.6
Apparatus for measuring the density of a solid

Check with your instructor for the proper disposal of the liquid and solid samples.

NOTES, OBSERVATIONS, AND CALCULATIONS

Techniques and Measurements

Date _____ Name _____ Lab Sec. _____ Desk No. _____

1. Diagram the cross section of a graduated cylinder, illustrating how a meniscus is read (Technique 8).

2. a. Explain how to fill a pipet to the mark (Technique 9).

    b. Which finger is used to control the delivery rate of the liquid from a pipet?

3. A micropipet delivers 153 drops of alcohol for each milliliter. Calculate the volume, in mL, of each drop.

4. A 4 g sample of zinc is measured on a triple-beam balance. Which mass should be recorded (Technique 13)?
    a. 4.0 g
    b. 3.99 g
    c. 3.9941 g
    d. 3.994 g

5. If 50.0 mL of an alcohol has a mass of 40.3 g, what is its density?

6. The density of diamond is 3.51 g/cm^3 and the density of lead is 11.3 g/cm^3. If, according to the Procedure, Part E.2, equal masses of diamond and lead were transferred to equal volumes of water in separate graduated cylinders, which graduated cylinder would have the highest volume reading? Explain.

◇ Techniques and Measurements

Date _____ Name _____ Lab Sec. _____ Desk No. _____

A. Laboratory Techniques

Instructor's approval of review of Techniques 8, 9, and 13 _____

B. The Laboratory Balance

1. Measure the mass of the following objects on the triple-beam balance. Express your answer to ± 0.01 g.

    a. 250 mL beaker (*g*)   _____

    b. 150 mm test tube (*g*)   _____

    c. a piece of glass tubing (*g*)   _____

    d. a quarter (*g*)   _____

    e. a key (*g*)   _____

2. Measure the mass of the following objects on the top-loading balance. Express your answer according to the sensitivity of the balance.

    a. a dime (*g*)   _____

    b. a dollar bill (*g*)   _____

    c. an empty 10 mL graduated cylinder(*g*) _____

    d. a 10 mL graduated cylinder
        containing 6 mL of water (*g*)   _____

    e. 150 mm test tube (*g*)   _____

C. Laboratory Glassware

1.

Volume of selected graduated cylinders	Smallest volume increment	Accuracy of a volume measurement
a.		
b.		

Sketch *at right* the vertical cross section of a meniscus.

Brief description of the technique for determining the volume of a liquid in a cylinder.

2. a. Laboratory instructor's approval for mastering the technique of filling and dispensing a liquid from a pipet.

_____

b.

Volume of selected graduated pipets	Smallest volume increment	Accuracy of a volume measurement

c. A _____ mL pipet was used to transfer water to a _____ mL graduated cylinder. What is the calculated inside radius, expressed in millimeters, of the graduated cylinder. The volume of a cylinder is $V = \pi r^2 l$.

3.

a.

Volume of buret	Smallest volume increment	Accuracy of a volume measurement

b. Volume of water, final (mL)   _____

   Volume of water, initial (mL)   _____

   Volume of water dispensed (mL)   _____

c. What is the radius, expressed in millimeters, of the buret? Record the appropriate data and show the calculation.

D. Calibration of Glassware

1. **Commercial calibration of graduated cylinders**

	TD or TC?	Accuracy of Measurement
a. Calibration of 10 mL graduated cylinder (*mL*)		
b. Calibration of 50 mL graduated cylinder (*mL*)		

2. **Volumes of common test tubes**

   a. Volume of 150 mm test tube (*mL*)   _____

   b. Volume of 75 mm test tube (*mL*)   _____

3. **Drops in 1.0 mL of water**

    a. Drops in 1.0 mL water   _____

    b. Volume of each drop (*mL*)   _____

4. **Calibration of pipet**

    a. Volume of pipet (*mL*)   _____

    b. Accuracy of measurement (*mL*)   _____

    c. Mass of beaker (*g*)   _____

    d. Mass of beaker + water delivered from pipet (*g*)   _____

    e. Mass of water delivered (*g*)   _____

    f. Calculated volume of pipet (*mL*)   _____

    g. Percent error in calibration (%)*   _____

$$\text{*\% error in calibration} = \left[\frac{|\,\text{calculated volume} - \text{labeled volume}\,|}{\text{calculated volume}} \right] \times 100$$

E. Density

Number of assigned balance   _____

1. **Density of a liquid**

    a. Corrected volume of pipet, Part D.4 (*mL*)   _____

    b. Mass of beaker (*g*)   _____

    c. Mass of beaker + liquid delivered from pipet (*g*)   _____

    d. Mass of liquid delivered (*g*)   _____

    e. Density of liquid (*g/mL*)   _____

2. **Density of a solid**

    a. Mass of weighing paper (*g*)   _____

    b. Mass of weighing paper + solid (*g*)   _____

    c. Mass of solid (*g*)   _____

    d. Volume of water in graduated cylinder (*mL*)   _____

    e. Volume of water after solid is transferred (*mL*)   _____

    f. Density of solid (*g/mL*)   _____

QUESTIONS

1. Water drops will adhere to the inner wall of a dirty pipet. Does a dirty pipet deliver more or less than its calibrated volume? Explain.

2. Suppose that in calibrating the volume of a pipet (TD 20°C), sufficient time is not allowed for the liquid to drain from the pipet. Does this cause a higher or lower calculated volume for the pipet? Explain.

3. Suppose that in calibrating the volume of the pipet (TD 20°C), the temperature of the water is not 20°C, but instead is 25°C. How does this affect the calibrated volume of the pipet? Hint: the density of water decreases with increasing temperature.

4. If, in the determination of the density of the solid (Part E.2), air bubbles adhered to its surface, will the reported density of the solid be reported as high or low? Explain.

Experiment 3

◇ Combustion

Fire is a fascinating phenomenon. The burning of wood has been known throughout recorded history, and its fascination has continued to intrigue each generation since then. Nearly everyone has experienced the consequences of "playing with fire" even though no disastrous event (see photo) may have occurred as a result. Fire was recognized as being so important to the early development of science that it was designated as one of the four basic "elements," the others being earth, air, and water.

The rapid burning of wood (or any fossil fuel) is called **combustion**, but slower types of combustion also occur, such as the "combustion" of iron, although we commonly call that corrosion or rusting, and the "combustion" of foods, such as the sugars in our bodies for energy, or in its spoilage (or oxidation).

- To determine the various temperature regions of a flame
- To determine the combustion products of a candle and a Bunsen flame

OBJECTIVES

PRINCIPLES

Fuel: the substance that burns.

Oxidizer: the substance that supports combustion.

Luminesce: to give off a glowing light.

This experiment characterizes a flame. A flame requires a **fuel**, often the wood of a match, the wax of a candle, gasoline, hydrogen, or natural gas, and an **oxidizer**, most commonly the oxygen in air although solid chemicals that function as oxidizers are common to the laboratory. We will study the flames of a candle and a laboratory burner.

The candle provides a very inefficient, low temperature flame, with the candle wax serving as the fuel. When there is an insufficient supply of oxygen, small particles of carbon are produced which, when heated to incandescence, produce a yellow, **luminous** flame. The appearance of a flame can be changed by affecting air currents, the shape of the wick, and the candle wax.

Laboratory burners come in many shapes and sizes, but all accomplish one main purpose: to provide a combustible gas-air mixture that produces a high temperature flame. Because Robert Bunsen (1811-1899) was the first to construct this type of burner, his name is commonly given to all burners of this type (Figure 3.1).

The natural gas used in most laboratories has methane, CH_4, as its major component. With a proper mixture of methane and oxygen, a blue, *non*luminous flame yields carbon dioxide and water at temperatures approaching $1500^\circ C$ in certain regions of the flame.

$$CH_4(g) + 2\,O_2(g) \rightarrow CO_2(g) + 2\,H_2O(g)$$

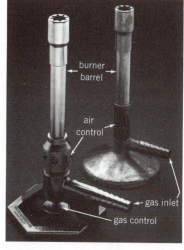

Figure 3.1
Bunsen–type burners

Small concentrations of carbon monoxide are also produced. With an insufficient supply of oxygen, small particles of carbon form from the unburned fuel, which then luminesce resulting in a yellow flame.

Refer to the Data Sheet as you progress through the Procedure to record and/or account for your observations.

PROCEDURE

A. The Candle

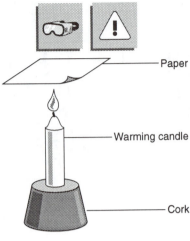

Paper

Warming candle

Cork

Figure 3.2
A candle flame tested for heat zones

1. a. Mount a warming candle on a large cork. Measure the mass of the candle and large cork.

    b. Light the candle. Study the flame closely: Why is the entire flame not the same color? Account for the color at the base of the flame as being different from that at the top of the flame.

    c. Position a wire gauze horizontally at the half-height position of the flame. Explain why the flame appears as it does.

    d. Position a 3" x 5" white index card horizontally at the half-height position of the flame until the card is scorched (see Figure 3.2). **Caution**: *do not let the card catch fire!*

    e. *Quickly* (so as to not catch fire) insert and withdraw the same index card positioned vertically in the flame. Perform this procedure several times to obtain a profile of the flame. Record and account for these observations. Sketch the horizontal and vertical profiles of the flame.

2. Place a slotted piece of aluminum foil around the wick just below the base of the flame, but above the melted wax. Leave the foil there for 30–60 seconds. How does the shape and appearance of the flame change? Why?

3. With a lighted match in hand, blow out the flame of the candle; hold the lighted match 2–3 cm from the wick in the column's "smoke". Account for your observation.

4. a. Place an inverted wide-mouth 500 mL Erlenmeyer flask over the candle flame. What substance collects on the wall of the flask? Determine the time required to extinguish the flame.

    b. Repeat with a wide-mouth 250 mL Erlenmeyer flask. Why is the flame extinguished?

5. a. Exhale into a wide-mouth 500 mL Erlenmeyer flask until the air has been expelled ($\approx$ 30 s); invert the flask over the candle flame. Determine the time to extinguish the flame.

    b. Repeat the procedure with a wide-mouth 250 mL Erlenmeyer flask.

6. a. Set upright the 250 mL Erlenmeyer flask from Part 5b and quickly add 25 mL of limewater; stopper the flask, and shake. Describe the change in appearance of the limewater?

    b. Place 25 mL of limewater in a second, clean 250 mL Erlenmeyer flask. Use a soda straw to exhale into the limewater.

    c. Use the straw to *gently* exhale into the candle flame.

    d. Based upon your data in Parts A.4-6, write a balanced equation for the combustion of candle wax. See the Data Sheet.

7. a. Half-fill a pie plate (or any flat dish) with limewater. Float the lighted candle (on the cork) on the limewater. Cover the lighted candle with an 800 mL beaker.

   b. Note changes of the water level in the beaker and note the *appearance* of the water.

   c. Repeat, substituting tap water for the limewater. Account for the differences in Parts 7b and 7c.

8. a. Place a spatula-full sample of baking soda into a 150 mL beaker. Add 10 mL of vinegar. Smell the gas that is evolved.

   b. "Pour" the gas generated by the reaction over the flame of a candle. What happened to the flame? Why?

9. Extinguish the candle flame and measure the mass of the candle and cork. From the difference in mass measurements and the balanced equation from Part A.6, determine the masses of carbon dioxide and water vapor that were released to the atmosphere during the experimentation.

B. The Bunsen Burner

1. **Lighting the burner**.
   a. Attach the burner's tubing to the gas outlet on the lab bench. Close the gas control at the base of the burner (Figure 3.1) and open the gas valve at the outlet. Close the air holes at the base of the burner and open slightly the gas control valve.

   b. Bring a *lighted* match or striker up the outside of the burner tube until the escaping gas at the top of the burner barrel ignites. Once the flame appears, open slightly the air holes at the base of the burner barrel. Open the gas control *at the base of the burner* until a nonluminous, pale blue flame appears and has two or more distinct cones.

   c. Further opening of the air control valve produces a slight hissing sound characteristic of a burner's hottest flame. The addition of too much air may blow out the flame. When the best adjustment is reached, three distinct cones are visible (Figure 3.3) and the flame should "sing". Obtain an instructor's approval of your well-adjusted flame.

   d. *If the flame goes out*, immediately turn off the gas valve at the outlet and repeat the procedure for lighting a burner.

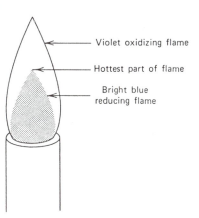

— Violet oxidizing flame

— Hottest part of flame

Bright blue reducing flame

Figure 3.3
Flame of a properly adjusted Bunsen burner

2. **Flame temperatures using a wire gauze.**
   a. Test the temperature in the different zones of a nonluminous flame by holding (**Caution**: *use crucible tongs*) a wire gauze horizontally about 1 cm above the burner (Figure 3.4). Note the color and appearance of the gauze. Now move it up through the flame until it no longer glows.

   b. Position the wire gauze vertically in the flame. This shows a vertical profile of the temperature regions of the flame.

   c. Briefly place an 800 mL beaker over the flame.

   d. Close the holes of the *air* control valve and repeat Part B.2a, b, c with a luminous flame.

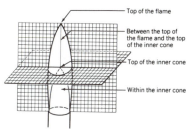

Top of the flame

Between the top of the flame and the top of the inner cone

Top of the inner cone

Within the inner cone

Figure 3.4
Regions of temperature measurement in a Bunsen flame

3. **Flame temperatures using the melting points of metals.**
   The melting points of iron, copper, and aluminum are 1535°C, 1083°C, and 660°C, respectively. Hold the metals, with crucible tongs, in the four zones of a nonluminous flame shown in Figure 3.5. Estimate the temperatures at each zone.

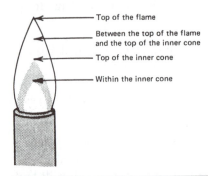

Top of the flame

Between the top of the flame
and the top of the inner cone

Top of the inner cone

Within the inner cone

Figure 3.5
Regions of the flame for
temperature measurements

Return the metal strips to the appropriately marked containers.

NOTES,
OBSERVATIONS,
AND
CALCULATIONS

◇ Combustion

Date _____Name _____ Lab Sec. _____Desk No. _____

1.   What is the dominant color of the nonluminous flame from a Bunsen burner? Explain.

2.   The flame of your Bunsen burner sputters and goes out. What should you do immediately?

3.   a.   What is the fuel for the burning of a candle?

     b.   Why are candle flames yellow?

4.   Arrange the following steps in the proper sequence for lighting a Bunsen burner:

     a.   Open slightly the gas control valve

     b.   Turn on the gas outlet valve

     c.   Close the air control valve

     d.   Close the gas control valve

     e.   Open the air control valve

     f.   Connect the burner's tubing to the gas outlet

     g.   Light the escaping gas

     h.   Open the gas control valve even further

     _____ , _____, _____, _____, _____ , _____ , _____

5. Consider the fire in a fire place. Assume that the wood in the following situations is not a factor in discussing the observation.
   a. Why does the fire burn yellow rather than blue?

   b. The flame of burning embers is nearly colorless, but the embers are yellow. Can you explain why?

6. The candle flame is typically yellow whereas the properly adjusted Bunsen flame is blue. Account for these differences.

*7. A fire in the "pits" at the Indianapolis 500 motor speedway is especially dangerous because the flame is nearly colorless and nonluminous, unlike that of a gasoline fire. How might the fuel used in Indianapolis 500 race cars differ from that of gasoline?

◇ Combustion

Date _____Name _____ Lab Sec. _____Desk No. _____

A. The Candle

1. a.   Mass of candle and cork (g)   _____

    b.   Briefly account for the various color regions of the candle flame.

    c.   Characterize the candle flame when using the wire gauze. What is the "smoke" escaping from the candle?

    d and e. At right, sketch the burn pattern for the horizontal cross section of the flame and the vertical cross section of the flame.

                                                        Horizontal   Vertical

Write a short description of the temperature characteristics of the flame, based on the above tests.

2. Account for the observations resulting from the placement of the aluminum foil.

3. What is the source and nature of the smoke? Explain the effect of the smoke in your observation.

4. a. time to extinguish flame (500 mL Erlenmeyer flask) _____

    b. time to extinguish flame (250 mL Erlenmeyer flask) _____
    Account for the time differential and explain why the flame was extinguished.

    What substance collects on the wall of the beaker? _____

5. a. time to extinguish flame (500 mL Erlenmeyer flask) _____

    b. time to extinguish flame (250 mL Erlenmeyer flask) _____
    Account for the time differences from those in Part 4.

6. a. What does the limewater test indicate about a combustion product of the candle flame?

    b. How does exhaled air relate to the combustion reaction for the candle?

    c. What analogy exists between the air exhaled through the straw and Part A.5?

    d. Assuming the formula of candle wax to be $C_{20}H_{42}$, write the balanced equation for the combustion of candle wax. You may need to ask your instructor for assistance.

7. a. Does the water level rise or lower? Why?

    b. What is the reaction occurring in the water? Where have you seen a similar reaction in this experiment?

c. What is observed in this procedure that is different from the observations in Part 7b? Account for these observations. In your discussion consider the following facts:
- the candle does *not* combine with *all* of the O_2 in the air;
- the formation of CO_2 from O_2 does *not* produce a volume change;
- the change in the water level is *not* as great as in Part 7.

8. What happened when the gas was "poured" onto the candle flame?

Because of the observations in Part 8b, what can you conclude about the nature of the gas? Cite evidence from earlier in the experiment to substantiate your prediction.

9. Mass of candle and cork (g) _____

Based upon your data in Parts A.4-6 and the balanced equation for the combustion reaction in Part A.6, calculate the masses of carbon dioxide and water vapor released to the atmosphere during the experiment. You may need to ask your instructor for assistance.

B. The Bunsen Burner

1. Instructor's approval of a well-adjusted Bunsen flame. _____

2. a and b. Use your observations from the horizontal and vertical wire gauze tests to sketch the profile of a nonluminous Bunsen flame; label the "cool" and "hot" regions.

Flame Regions

c. What appears on the inner surface of the beaker? Is this consistent with the observations in Part A? Explain.

d. What differences in the flame temperatures are observed between the nonluminous and luminous flames. Use supporting data to substantiate your conclusions.

3. At right, sketch Figure 3.5 for a nonluminous flame and indicate the approximate temperature in each of the four zones.

QUESTIONS

1. a. What are the products of combustion of candle wax? Of a Bunsen flame?

   b. Describe the physical properties of the product(s).

   c. Describe the chemical properties of the product(s).

2. Why is it possible to extinguish the flame of a candle or match by *very gently* blowing on it?

3. What substance is "burning" and what substance is supporting the combustion for the candle? for the Bunsen burner?

*4. Referring to the calculations of Parts A.6d and A.9 and assuming the density of carbon dioxide gas to be 1.96 g/L at STP, what volume of carbon dioxide was released to the atmosphere?

Experiment 4

◇ Water Analysis: Solids

The raw water from surface reservoirs (see photo) and underground aquifers, that is used for public drinking water supplies, contains dissolved salts, mostly soluble carbonate and bicarbonate salts. The water is piped into a water treatment facility where impurities are removed and bacteria are killed before the water is placed into the distribution lines. The contents of the surface water must be known and predictable so that the treatment facility can properly and adequately remove these impurities. Tests are used to determine the contents of the surface water.

When you are thirsty and need a drink of water, do you take it for granted that the tap water is safe to drink? Do you think about what might be dissolved or suspended in the water? The tap water of public drinking water supplies must meet certain federal health standards; for example, the water at the tap must have a chlorine concentration of one part per million to eliminate bacteria. The U.S. Public Health Service recommends that drinking water not exceed 500 mg total solids/kg water. However, in some localities, the total solids content may range up to 1000 mg/kg of potable water; that's 1 g/L!! An amount over 500 mg/kg water does not mean the water is unfit for drinking; an excess of 500 mg/kg is merely not recommended.

OBJECTIVE

- To determine the total, dissolved, and suspended solids in a water sample

PRINCIPLES

Watershed: the land area over which the natural drainage of water occurs.

Water in the environment has a large number of impurities with an extensive range of concentrations. **Dissolved solids** are water soluble substances, usually salts. Naturally-occurring dissolved solids generally result from the movement of water over or through mineral deposits, such as limestone. These dissolved solids, characteristic of the **watershed**, generally consist of the sodium, calcium, magnesium, and potassium cations and the chloride, sulfate, bicarbonate, carbonate, bromide, and fluoride anions. The dissolved solids are responsible for the "hard" water that exists in some locales. Anthropogenic (human-related) dissolved solids include nitrates from fertilizer runoff and human wastes, phosphates from detergents and fertilizers, and organic compounds from pesticides and industrial wastes. **Salinity**, a measure of the dissolved solids in a water sample, is defined as the grams of dissolved solids per kilogram of water.

Suspended solids are very finely divided particles that are water *insoluble* but are filterable. These particles are kept in suspension by the turbulent action of the moving water. Examples of suspended solids include decayed organic matter, sand, salt, and clay.

Suspended solids: solids that exhibit colloidal properties or solids that remain in the water because of turbulence.

Total solids are the sum of the dissolved and suspended solids in the water sample. In this experiment the total solids and the dissolved solids are determined directly; the suspended solids are assumed to be the difference, since

total solids = dissolved solids + suspended solids

PROCEDURE

Turbid sample: a cloudy suspension due to stirred sediment.

A. Dissolved Solids

Filtrate: the solution that passes through the filter into a receiving flask.

Cool flame: a Bunsen flame of low intensity—a slow rate of natural gas is flowing through the burner barrel.

B. Total Solids and Suspended Solids

Obtain 100 mL of a water sample from your instructor. Preferably the water sample is high in turbidity. Record the sample number on the Data Sheet. With approval, bring your own "environmental" water sample to the laboratory for analysis. The water sample may be from the ocean, a lake, a stream, or from an underground aquifer.

Check out one additional evaporating dish from the stockroom.

Technique 9 Check: Use tap water to practice the technique for drawing a liquid into (using a pipet bulb) and dispensing a liquid from a pipet (into a beaker). Have your instructor approve your technique.

1. Clean, dry, and determine the mass (±0.001 g) of two evaporating dishes. Be certain that you can identify each. Use the same balance for the remainder of the experiment.

2. Gravity filter about 50 mL of a thoroughly stirred or shaken water sample into a clean, dry 100 mL beaker. While waiting for the filtration to be completed, proceed to Part B.

3. a. Pipet a 25 mL aliquot (portion) of the **filtrate** into one evaporating dish. Determine the combined mass of the evaporating dish and sample. Place the dish/sample on a wire gauze (Figure T.7a) and *slowly* heat—do not boil—the mixture to dryness.

   b. As the mixture nears dryness, cover the evaporating dish with a watch glass and reduce the intensity of the flame.[1] If spattering does occur, allow the dish to cool to room temperature, rinse the adhered solids from the watch glass (Figure 4.1) and return the rinse to the dish.

4. Again heat slowly, being careful to avoid further spattering. After all of the water has evaporated, maintain a **"cool" flame** beneath the dish for 3 minutes. Allow the dish to cool to room temperature and determine its final mass. Cool the evaporating dish and sample in a desiccator, if available.

1. a. Thoroughly stir or agitate 100 mL of the original water sample; pipet a 25 mL aliquot of this sample into the second evaporating dish. Evaporate the sample to dryness as described in Part A.3.

   b. Collect and record the appropriate data needed to calculate the mass of total and suspended solids in the original water sample, using data from Parts A and B.

---

[1]This reduces the spattering of the remaining solid and its subsequent loss in the analysis.

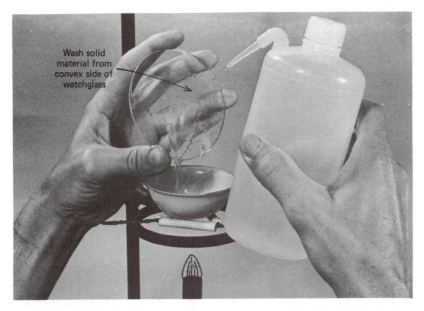

Wash solid
material from
convex side of
watchglass

C. Analysis of Data

1. Compare your data with three other chemists in your laboratory who have analyzed the *same* water sample. Record their results on the Data Sheet.

2. Calculate the standard deviation of the suspended solids data from the four analyses on the water sample. The standard deviation, σ, for n sets of data is calculated using the equation

$$\sigma = \sqrt{\frac{d_1^2 + d_2^2 + d_3^2 + \ldots + d_n^2}{(n-1)}}$$

d_1 is the difference between your suspended solids value (Data set #1) and the average suspended solids value, x, obtained from your data and the three other chemists. The meaning of the standard deviation value is that 68% of all subsequent solid determinations for that particular water sample will fall within the range of $x + \sigma$ to $x - \sigma$.

**NOTES,
OBSERVATIONS,
AND
CALCULATIONS**

Portable water quality test kits can be used to perform many "quick" tests on a water sample. The following portable water quality test kits are commercially available from the Hach Company. The parameters for testing, along with their range of sensitivity, follow.

Tests	Concentration Range	Tests	Concentration Range
aluminum	0-0.7 mg/L	manganese (EPA)	0-10 mg/L
arsenic (EPA)	0-0.2 mg/L	mercury	0-3 µg/L
barium	0-250 mg/L	molybdenum	0-25 mg/L
benzotriazole	0-15 mg/L	nickel (EPA)	0-1.5 mg/L
boron	0-15 mg/L	nitrogen, ammonia (EPA)	0-2 mg/L
cadmium	0-70 µg/L	nitrogen, nitrate	0-30 mg/L
chloride	0-20 mg/L	nitrogen, nitrite (EPA)	0-200 mg/L
chlorine, free (EPA)	0-1.7 mg/L	nitrogen, total	0-200 mg/L
chlorine, total (EPA)	0-1.7 mg/L	oil in water	0-60 ppm
chlorine dioxide	0-0.055 mg/L	oxygen, dissolved	0-10 mg/L
chromium (EPA)	0-0.5 mg/L	oxygen demand, chemical (EPA)	0-800 mg/L
cobalt	0-1.2 mg/L	ozone	0-1.3 mg/L
color, true	0-70 units	pH	pH 4-10
copper (EPA)	0-150 µg/L	phenols (EPA)	0-0.2 mg/L
cyanide (EPA)	0-0.12 mg/L	phosphonates	0-20 mg/L
cyanuric acid	0-50 mg/L	phosphorus, reactive (EPA)	0-2 mg/L
N,N-diethylhydroxylamine	0-300 µg/L	polyacrylic acid	0-20 mg/L
detergents, anionic	0-0.2 mg/L	potassium	0-6 mg/L
erythorbic acid	0-1000 µg/L	residue, suspended solids	0-500 mg/L
fluoride (EPA)	0-2 mg/L	selenium	0-1.0 mg/L
formaldehyde	0-650 µg/L	silica	0-15 mg/L
hardness	0-2.5 mg/L	silver	0-0.5 mg/L
hydrazine	0-0.015 mg/L	sodium chromate	0-1000 mg/L
iodine	0-7 mg/L	sulfate (EPA)	0-100 mg/L
iron, ferrous	0-2 mg/L	sulfide (EPA)	0-0.5 mg/L
iron, total (EPA)	0-2 mg/L	tannin and lignin	0-5 mg/L
lead (EPA)	0-150 µg/L	tolyltriazole	0-15 mg/L
		turbidity	0-400 FTU
		volatile acids	0-2500 mg/L
		zinc (EPA)	0-1 mg/L

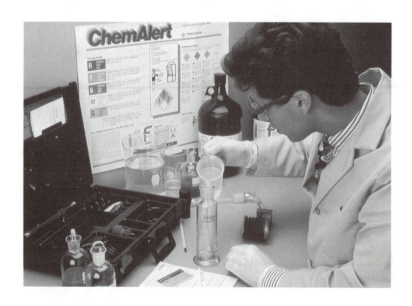

◇ Water Analysis: Solids

Date _____ Name _____ Lab Sec. _____ Desk No. _____

1. a. What are the characteristics of a suspended solid?

    b. Identify three kinds of suspended solids.

    c. How do suspended solids stay suspended?

2. List several cations and anions, by formula, that contribute to the salinity of a water sample.

3. A 25 mL aliquot of a well-shaken sample of river water is pipetted into a 27.211 g evaporating dish. After the mixture is evaporated to dryness, the dish and dried sample has a mass of 43.617 g. Determine the total solids in the sample; express total solids in units of g/kg sample. Assume the density of the river water to be 1.01 g/mL.

4. In evaporating a solution to dryness in an evaporating dish, why must the heating rate be decreased as the mixture nears dryness?

5. The reported concentration of suspended solids in a water sample is 460 g/kg. The standard deviation for six trials is 12 g/kg.

    a. List the range in g/kg of precision for the suspended solid determinations.

    b. What percent of all subsequent determinations will lie within this range?

6.   a. What is an **aliquot** of a sample?

    b. What is the **filtrate** in a filtration procedure?

    c. How full (the maximum level) should a funnel be filled with solution in a gravity filtration procedure?

    d. What finger should be used in controlling the delivery of a solution from a pipet?

    e. How is a suspended drop removed from the tip of a pipet after delivery of the solution (filtrate) to the evaporating dish in Part A.3?

Water Analysis: Solids

Date _____ Name _____ Lab Sec. _____ Desk No. _____

Technique 9 Check: Instructor approval of your pipeting technique: _____

Sample Number: _____ Describe the nature of your water sample, i.e., its color, turbidity, etc.

A. Dissolved Solids

<u>Trial 1</u>

1. Mass of evaporating dish (g) _____

2. Mass of evaporating dish + water sample (g) _____

3. Mass of water sample (g) _____

4. Mass of evaporating dish + *dried* sample (g) _____

5. Mass of dissolved solids in 25 mL aliquot of filtered sample (g) _____

6. Mass of dissolved solids per total mass of sample (g solids/g sample) _____

7. Dissolved solids or salinity (g solids/kg sample) _____

B. Total Solids and Suspended Solids

1. Mass of evaporating dish (g) _____

2. Mass of evaporating dish + water sample (g) _____

3. Mass of water sample (g) _____

4. Mass of evaporating dish + *dried* sample (g) _____

5. Mass of total solids in 25 mL aliquot of unfiltered sample (g) _____

6. Mass of total solids per total mass of sample (g solids/g sample) _____

7. Suspended solids or salinity (g solids/kg sample), ss_1 _____

C. Analysis of Data

	Chemist #1 (you)	Chemist #2	Chemist #3	Chemist #4
Dissolved Solids (g/kg)				
Total Solids (g/kg)				
Suspended Solids (g/kg)	$ss_1 =$	$ss_2 =$	$ss_3 =$	$ss_4 =$

1. Average value of suspended solids from four chemists (x) = _____

	$d_n = x - ss_n$	$d_n^2 = (x - ss_n)^2$
Chemist #1		
Chemist #2		
Chemist #3		
Chemist #4		

2. Standard deviation, σ = _____
   Show work here.

QUESTIONS

1. If the sample in the evaporating dish in Part B had not been heated to total dryness, how would this have affected the reported value for
   a. total solids? Explain.

   b. suspended solids? Explain.

2. Suppose that the water sample has a relatively high percent of volatile solid material. How would this affect the reported mass of
   a. dissolved solids? Explain.

   b. total solids? Explain.

   c. suspended solids? Explain.

3. Suppose that in Part A.1 of the Procedure, the evaporating dish had not been properly cleaned of a volatile material before the mass of the sample was determined. How would this error in technique affect the reported mass of dissolved solids in the water sample? Explain.

Experiment 5

◇ Nomenclature: Elements and Compounds

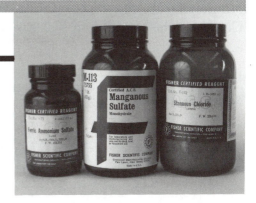

All matter consists of chemicals. Foods and plants consist of naturally occurring chemicals; automobiles and our bodies are built with and powered by chemicals; plastics, pharmaceuticals, and building insulation are manufactured chemicals. Each chemical interacts with its environment in its own unique way; foods spoil, automobiles rust, bodies age, and plastics deteriorate. A passing knowledge of chemicals and their properties is important in forming the basis for an understanding of the chemical world.

Every profession has its own language; physicians easily communicate technical information to other physicians, but may have difficulty communicating similar information to patients; lawyers write legal jargon for insurance policies or deeds which is readily understood by other lawyers, but not necessarily by the person buying the policy. Similarly chemists have a language that is used to explicitly characterize a reaction or identify a compound—other chemists may understand the language, but the nonchemist is often left in the dark.

For chemists to communicate research data and results internationally, a standardization of the technical language (see photo) has been necessary. Historically, common names for many compounds were used and are still universally understood—for examples, water, sugar, and ammonia—but with new compounds being prepared daily, a random or "convenient" system for naming compounds is no longer viable. In this assignment you will learn a few systematic rules, established by the International Union of Pure and Applied Chemistry (IUPAC), for naming and writing the formulas for some inorganic compounds.

OBJECTIVES

- To recognize and to write the symbols for the elements
- To name common inorganic compounds
- To write formulas for common inorganic compounds

PRINCIPLES

A chemist's basic language consists of *symbols* for elements, *formulas* for compounds, and *equations* to describe chemical *reactions*.

Symbols of Elements

Symbol: an abbreviation that represents the name and also one atom of an element.

Elements discovered early in the history of chemistry are generally given Latin names that describe their appearance or chemical properties. The **symbols** for these elements are abbreviations of the Latin names; for example, **Au** represents aurum (gold), **Cu** is the symbol for cuprum (copper), and **Fe** is the symbol for ferrum (iron). However, the more recently discovered elements were given English names with corresponding symbols, such as **Ni** for nickel and **As** for arsenic. The periodic table (see back cover) lists the symbols for the elements, their atomic numbers (the whole number), and their relative atomic masses.

Formulas of Compounds

Formula: combination of the symbols of elements that represents the name and also one molecule (or formula-unit) of a compound.

Subscripts in the **formula** for a compound indicate the number of atoms of each element in the compound. For example, H_2O represents water, a chemical combination of 2 H atoms and 1 O atom; $C_{12}H_{22}O_{11}$ is the formula for sucrose (table sugar), a chemical combination of 12 C atoms, 22 H atoms, and 11 O atoms. A change in subscript represents a change in the chemical composition of the compound. For example, $2\,C_2H_4$ is *not* the same as C_4H_8: $2\,C_2H_4$ represents two molecules of C_2H_4 (ethylene) whereas C_4H_8 represents one molecule of C_4H_8 (butene).

Nomenclature of Compounds

The naming of chemical compounds can be categorized into the nomenclature for inorganic compounds and for organic compounds. This assignment focuses on the inorganic compounds; Experiment 35 introduces organic compounds and its nomenclature.

Inorganic compounds can be classified into several groups:

Salt: an ionic compound existing in the solid state.

Cation: an atom or group of atoms with a positive charge.

Anion: an atom or group of atoms with a negative charge.

Binary Compounds. These compounds consist of only two elements. Three major subclassifications of binary compounds are:
- **Ionic compounds** (called **salts**) consist of a **cation** and an **anion**.
- **Molecular compounds** consist of two nonmetals or a metalloid and a nonmetal.
- **Acids** consist of hydrogen and a more electronegative nonmetal dissolved in water.

Polyatomic ion: a group of covalently bonded atoms that carry a charge.

Compounds with Polyatomic Groups of Atoms. Salts of these compounds consist of a cation and an anion, one or both of which is a **polyatomic ion**. The polyatomic ion may exist in the solid state and/or in aqueous solution. Three major subclassifications of compounds are:
- **Ionic compounds** (called salts) consist of ions, one or both of which may be a polyatomic ion.
- **Acids** consist of hydrogen and a polyatomic group of atoms which assume a negative charge in aqueous solution.
- **Acid salts** are salts in which one or more of the metal cations in the salt has been replaced by hydrogen.

Nomenclature of Binary Compounds

Ionic Compounds (Salts), a Cation and an Anion. A binary ionic compound is a **salt** that consists of a metal cation and a nonmetal anion. For a binary salt, the metal cation (more electropositive) is named first. The root of the nonmetal anion is then named with the suffix *-ide* added to it. NaBr is sodium brom*ide*; Al_2O_3 is aluminum ox*ide*. The names of some binary salts have the *-ide* ending, but are not binary compounds; these compounds contain the polyatomic ions, NH_4^+, OH^-, and CN^-. NH_4Cl is ammonium chlor*ide*; KOH is potassium hydrox*ide,* and NaCN is sodium cyan*ide*.

Metal cations of the elements from Groups 1A and 2A invariably carry charges of 1^+ and 2^+ respectively in ionic compounds. The charge of other metal cations in ionic compounds is not always so predictable.

Alfred Stock (1876-1946), a German chemist, who was severely affected by mercury poisoning, spent much of his life designing experimental techniques for its detection at low concentrations.

Many transition and post-transition metals form more than one cation resulting in the formation of more than one compound with a given anion. Depending upon reaction conditions, copper may form Cu^+ and Cu^{2+} cations and therefore can form two salts with, for example, chlorine: CuCl and $CuCl_2$. Two systems are used to distinguish the names of the cations in the two salts. The **Stock** system uses the English name for the metal cation followed immediately by a Roman numeral in parentheses which indicates the charge of the ion. An older system uses the Latin root name of the metal and applies a suffix that is dependent upon the charge on the ion: an *-ous* suffix designates the lower of the two possible charges of the metal cation and an *-ic* suffix designates the higher of the two charges. The CuCl and $CuCl_2$ salts, as well as examples for iron and tin, are:

- CuCl is copper(I) chloride and cupr*ous* chloride
- $CuCl_2$ is copper(II) chloride and cupr*ic* chloride
- $FeBr_2$ is iron(II) bromide and ferr*ous* bromide
- $FeBr_3$ is iron(III) bromide and ferr*ic* bromide
- SnO is tin(II) oxide and stann*ous* oxide
- SnO_2 is tin(IV) oxide and stann*ic* oxide

When naming salts, the Stock system is preferred.

Molecular Compounds, Two Nonmetals or a Metalloid and a Nonmetal.
Two nonmetals (or metalloid and a nonmetal) may combine to form more than one molecular compound. For example, carbon and oxygen combine to form carbon monoxide, CO, and carbon dioxide, CO_2; sulfur and fluorine combine to form sulfur difluoride, SF_2, sulfur tetrafluoride, SF_4, and sulfur hexafluoride, SF_6. Nitrogen and oxygen combine to form N_2O, NO, NO_2, N_2O_3, N_2O_4, and N_2O_5. To distinguish between the nitrogen oxides, we use the same format as is used for the sulfur fluorides—Greek *prefixes* are used to indicate the number of atoms of each element in the compound. The common prefixes are in Table 5.1.

Table 5.1 Greek Prefixes

mono-*	one	hexa-	six
di-	two	hepta-	seven
tri-	three	octa-	eight
tetra-	four	nona-	nine
penta-	five	deca-	ten

*mono- is seldom used, since "one" is generally implied

Acids, Hydrogen and a More Electronegative Nonmetal in Water. Acids produce acidic (aqueous) solutions. A binary acid can be considered as the substitution of hydrogen for the metal in a binary salt. To name a binary acid, the prefix *hydro-* and the suffix *-ic* are added to the root name of the anion; the word "acid" is added to identify the chemical nature of the compound. Notice that the *-ide* suffix of the anion becomes *hydro-__-ic* as the acid. Remember that acids are not named as "acids" unless they are dissolved in water.

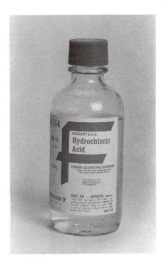

- the chloride ion, Cl^-, becomes *hydro*chloric *acid*, HCl when dissolved in water
- the bromide ion, Br^-, becomes *hydro*brom*ic acid*, HBr when dissolved in water
- the sulfide ion, S^{2-}, becomes *hydro*sulfuric *acid*, H_2S when dissolved in water

Ionic Compounds (Salts), a Polyatomic Cation and/or a Polyatomic Anion. Most often the cation is a metal cation and the anion is the polyatomic ion for these salts. For the polyatomic anion, one element is almost always oxygen and the other is *usually* a nonmetal. The metal cation is named as it was for the binary salts (either using the Stock system or the "old" system); the polyatomic anion is named using the root of the "other" element, *not the* oxygen, and the suffix *-ate* or *-ite*, depending on the number of oxygens— the polyatomic anion with the greater number of oxygen atoms receives the *-ate* suffix. The sulf*ate* ion is SO_4^{2-}, the nit*rate* ion is NO_3^-, and the arsen*ate* ion is AsO_4^{3-}; the sulf*ite* ion is SO_3^{2-}, the nit*rite* ion is NO_2^-, and the arsen*ite* ion is AsO_3^{3-}.

- Na_2SO_4 is sodium sulf*ate*; Na_2SO_3 is sodium sulf*ite*
- KNO_3 is potassium nit*rate*; KNO_2 is potassium nit*rite*

Acids, Hydrogen and a Polyatomic Group (Oxoacids). An **oxoacid** can be considered as the substitution of hydrogen for the metal in a salt that contains a polyatomic anion. To name these acids we do *not* name the

Nomenclature of Compounds with Polyatomic Groups of Atoms

Oxoacid: a hydrogen combined with a polyatomic group and having the formula $H_wX_yO_z$.

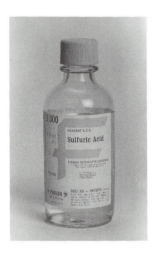

hydrogen, but rather change the name of the polyatomic anion—the -*ate* name of the polyatomic anion is changed to an "*-ic acid*" or an -*ite* name is changed to an "*-ous acid*".

- Na_2SO_4 is sodium sulf*ate*; H_2SO_4 is sulfur*ic acid*
- Na_2SO_3 is sodium sulf*ite*; H_2SO_3 is sulfur*ous acid*
- KNO_3 is potassium nitr*ate*; HNO_3 is nitr*ic acid*
- KNO_2 is potassium nitr*ite*; HNO_2 is nitr*ous acid*

Notice that the -*ate* suffix for the polyatomic anion becomes the -*ic* suffix for the acid, while the -*ite* suffix becomes the -*ous* suffix for an acid.

Some polyatomic anions, most notably those of the halogens, form more than just two polyatomic anions. For these polyatomic anions the prefixes *per-* and *hypo-* are appropriately added. For example, the polyatomic anions of chlorine and their respective chloro oxoacids are

Per-: a prefix meaning completely; a radical or polyatomic ion containing an element with its maximum oxidation number.
Hypo-: a prefix meaning below; a radical or polyatomic ion containing an element with a very low oxidation number.

- perchlorate anion, ClO_4^-, becomes *per*chlor*ic* acid, $HClO_4$
- chlor*ate* anion, ClO_3^-, becomes chlor*ic* acid, $HClO_3$
- chlor*ite* anion, ClO_2^-, becomes chlor*ous* acid, $HClO_2$
- *hypo*chlor*ite* anion, ClO^-, becomes *hypo*chlor*ous* acid, $HClO$

Bromine and iodine form similar oxoacids and salts. Notice again the -*ate*, -*ic* and the -*ite*, -*ous* relationships between the polyatomic anions and the acids; the prefixes remain unchanged.

Acid Salts, Cation and a Polyatomic Group Containing Hydrogen. An acid salt can be considered as a substitution of a metal cation for less than all of the hydrogens of an acid. The remaining hydrogens become a part of a polyatomic anion; the number of hydrogens are indicated by a Greek prefix. The suffix for the anion is the same as that used for salts (binary or those with a polyatomic group).

- $NaHS$ is sodium hydrogensulfide
- NaH_2PO_4 is sodium dihydrogenphosphate
- $CaHPO_4$ is calcium hydrogenphosphate
- $Ca(H_2PO_4)_2$ is calcium dihydrogenphosphate

An older system of naming acid salts substitutes the prefix *bi-* for a single hydrogen of the polyatomic anion; for example $NaHSO_4$ is sodium *bi*sulfate and $NaHCO_3$ is sodium *bi*carbonate.

Writing Formulas

Ionic Compounds. The formulas of ionic compounds are written such that the sum of the charges for the cations and anions equals zero. Table 5.2 lists the common monoatomic and polyatomic cations and Table 5.3 lists the common monoatomic and polyatomic anions. You should learn most, if not all, of those listed—others you will learn with experience. Lets try writing formulas for a few ionic compounds (we'll need to use Tables 5.2 and 5.3):

- Barium fluoride. Barium is Ba^{2+} and fluoride is F^-. For the sum of the charges to be zero, one Ba^{2+} ion must combine with two F^- ions; the formula must be BaF_2.
- Calcium nitride. Calcium is Ca^{2+} and nitride is N^{3-}. Three Ca^{2+} provides a $6+$ charge; this balances two N^{3-} (a 6^- charge). The formula is Ca_3N_2.
- Potassium oxalate. Potassium is K^+ and oxalate is $C_2O_4^{2-}$. Two K^+ balances one $C_2O_4^{2-}$; the formula is $K_2C_2O_4$.

- Ferric chromate. Ferric is Fe^{3+} and chromate is CrO_4^{2-}. Two Fe^{3+} (a 6^+ charge) balances three CrO_4^{2-} (a 6^- charge); the formula is $Fe_2(CrO_4)_3$.

Table 5.2. Name and Charge of Monoatomic and Polyatomic Cations

Charge of 1^+

NH_4^+	ammonium	H^+	hydrogen	K^+	potassium
Cu^+	copper(I), cuprous	Li^+	lithium	Ag^+	silver(I)
Au^+	gold(I), aurous			Na^+	sodium

Charge of 2^+

Ba^{2+}	barium	Fe^{2+}	iron(II), ferrous	Ni^{2+}	nickel(II)
Cd^{2+}	cadmium	Pb^{2+}	lead(II), plumbous	Sr^{2+}	strontium
Ca^{2+}	calcium	Mg^{2+}	magnesium	Sn^{2+}	tin(II), stannous
Cr^{2+}	chromium(II), chromous	Mn^{2+}	manganese(II), manganous	UO_2^{2+}	uranyl
Co^{2+}	cobalt(II), cobaltous	Hg^{2+}	mercury(II), mercuric	VO^{2+}	vanadyl
Cu^{2+}	copper(II), cupric	Hg_2^{2+}	mercury(I), mercurous	Zn^{2+}	zinc(II)

Charge of 3^+

Al^{3+}	aluminum	Co^{3+}	cobalt(III), cobaltic	Fe^{3+}	iron(III), ferric
Cr^{3+}	chromium(III), chromic	Au^{3+}	gold(III), auric	Mn^{3+}	manganese(III), manganic

Charge of 4^+

Pb^{4+}	lead(IV), plumbic	Sn^{4+}	tin(IV), stannic

Charge of 5^+

V^{5+}	vanadium(V)

Table 5.3. Name and Charge of Monoatomic and Polyatomic Anions

Charge of 1^-

$CH_3CO_2^-$	acetate	F^-	fluoride	NO_2^-	nitrite
Br^-	bromide	OH^-	hydroxide	ClO_4^-	perchlorate
ClO_3^-	chlorate	H^-	hydride	IO_4^-	periodate
Cl^-	chloride	ClO^-	hypochlorite	MnO_4^-	permanganate
ClO_2^-	chlorite	I^-	iodide		
CN^-	cyanide	NO_3^-	nitrate		

Charge of 2^-

CO_3^{2-}	carbonate	O^{2-}	oxide	SO_3^{2-}	sulfite
CrO_4^{2-}	chromate	$C_2O_4^{2-}$	oxalate	S^{2-}	sulfide
$Cr_2O_7^{2-}$	dichromate	O_2^{2-}	peroxide	$S_2O_3^{2-}$	thiosulfate
SiO_3^{2-}	silicate	SO_4^{2-}	sulfate		

Charge of 3^-

AsO_4^{3-}	arsenate	N^{3-}	nitride	PO_3^{3-}	phosphite
AsO_3^{3-}	arsenite	PO_4^{3-}	phosphate	P^{3-}	phosphide
BO_3^{3-}	borate				

Molecular Compounds. For molecular compounds, the Greek prefixes used in identifying the compound also identifies the number of atoms of each element in the compound.
- Tetraphosphorous decaoxide. Four phosphorous and ten oxygen atoms gives the formula P_4O_{10} for tetraphosphorous decaoxide.

Acids. Binary acids are **hydro- ___-ic acids** and those containing polyatomic groups are either **-ic acids** or **-ous acids**.

- Hydrosulfuric acid. Since this is a hydro- ___-ic acid, the acid only contains hydrogen and sulfur—two hydrogens for each sulfur and the formula is H_2S.
- Phosphoric acid. Since this is an -ic acid, then the corresponding polyatomic anion must be an -ate anion. The phosphate anion, PO_4^{3-}, becomes phosphoric acid with hydrogens added for neutrality. Therefore, the formula of the acid is H_3PO_4.

PROCEDURE

Answer all questions on a separate sheet of paper.

A. A Survey of the Ions

1. List and name (Stock system and the "old"-ic, -ous system) all of the cations having two common charges.

2. List and name all anions that have a chlorine atom.

3. List and name all anions that have a sulfur atom.

4. a. What is the common charge of the Group 1A metals?
   b. What is the common charge of the Group 2A metals?
   c. What is the common charge of the monoatomic anions of Group 7A

B. Nomenclature

Your instructor will indicate which of the following (parts of) exercises to complete.

1. Name the following as binary compounds.

a. $NaCl$	e. Li_2S	i. SrS	m. Li_3N	q. IF_7
b. $CsOH$	f. Al_2O_3	j. V_2O_5	n. Ag_2O	r. $TiCl_4$
c. $CaBr_2$	g. LiH	k. CaC_2	o. KCN	s. N_2O
d. MgO	h. AgI	l. BaI_2	p. NH_4Cl	t. NI_3

2. Name the following salts with polyatomic anions.

a. Na_2SO_4	g. K_2CrO_4	m. K_3PO_4	s. $VOCl_2$
b. KNO_3	h. KCH_3CO_2	n. $(NH_4)_3PO_4$	t. UO_2Cl_2
c. Li_2CO_3	i. $NaMnO_4$	o. KIO_4	u. $NaClO$
d. $CdCO_3$	j. $Li_2S_2O_3$	p. $KClO_3$	v. $NaNO_2$
e. $Ca_3(PO_4)_2$	k. $Ba(NO_2)_2$	q. $Ca(CH_3CO_2)_2$	w. $(NH_4)_2S_2O_3$
f. $K_2Cr_2O_7$	l. NH_4ClO_3	r. $MgSO_4$	x. $K_2C_2O_4$

3. Name the following salts using the Stock system.

a. CrS	f. FeS	k. PbO_2	p. $FeCl_3$
b. Fe_2O_3	g. Cu_3P	l. $ZnSO_4$	q. SnO_2
c. AuI_3	h. $CuBr_2$	m. SnF_4	r. $AgCH_3CO_2$
d. $Hg(NO_3)_2$	i. $Sn(NO_2)_4$	n. $Ni(NO_3)_2$	s. $Co_3(PO_4)_2$
e. $CuCO_3$	j. $Co_2(CO_3)_3$	o. $CuClO_4$	t. $PbSO_4$

4. Name the following salts using the Stock system *and* the "old"-ic, -ous system.

a. $FeSO_4$	g. CoS	l. $Pb(CH_3CO_2)_4$	q. $MnSO_4$
b. $Au(NO_3)_3$	h. $Fe(OH)_3$	m. Au_2O_3	r. $Hg_2(ClO_2)_2$
c. $CuCN$	i. $Cr(CN)_2$	n. $FePO_4$	s. $Hg(ClO_2)_2$
d. $CrCl_3$	j. $CoBr_3$	o. CoO	t. $PbCl_2$
e. PbO	k. $SnCl_2$	p. $Hg(NO_3)_2$	u. Mn_2O_3
f. Hg_2Cl_2			

5. Name the following acids.
   a. HBr   e. H_3PO_4 i. H_3BO_3 m. H_2SO_4 q. CH_3COOH
   b. HClO   f. H_3AsO_3 j. HCN   n. HNO_3 r. H_3PO_3
   c. H_2S g. $HMnO_4$ k. HNO_2 o. $H_2S_2O_3$ s. $H_2C_2O_4$
   d. H_2SO_3 h. H_2CrO_4 l. HIO_4 p. $HClO_4$ t. H_2CO_3

6. Name the following acid salts. In addition, use the "bi-" method wherever appropriate.
   a. $KHCO_3$ d. $Ca(HCO_3)_2$ g. NaHS   i. $NH_4H_2PO_4$
   b. $NaHSO_4$ e. $KHSO_3$ h. $NaHCO_3$ j. KH_2AsO_4
   c. KHC_2O_4 f. Li_2HPO_4

7. Write formulas for each of the following compounds.
   a. ferrous sulfate   r. barium acetate
   b. potassium permanganate   s. silver(I) thiosulfate
   c. calcium carbonate   t. nitrogen triiodide
   d. iron(III) oxide   u. potassium iodide
   e. cupric hydroxide   v. sodium silicate
   f. aluminum sulfide   w. calcium hypochlorite
   g. mercury(II) chloride   x. potassium chlorate
   h. cadmium sulfide   y. cuprous iodate
   i. copper(I) chloride   z. sodium dihydrogen phosphite
   j. ammonium cyanide   aa. ammonium oxalate
   k. sodium chromate   bb. potassium dichromate
   l. nickel(II) nitrate   cc. copper(I) sulfate
   m. manganese(II) oxide   dd. cobalt(II) phosphate
   n. manganese (IV) oxide   ee. chromium(III) phosphate
   o. lead(II) carbonate   ff. iron(II) oxalate
   p. stannous chloride   gg. copper(I) sulfide
   q. sodium nitrite   hh. potassium periodate

8. Write formulas for each of the following compounds.
   a. hydrogen fluoride   r. gold(III) nitrate
   b. sulfuric acid   s. nitric acid
   c. iodine pentafluoride   t. hydrocyanic acid
   d. hydrobromic acid   u. carbon monoxide
   e. hypobromous acid   v. chlorous acid
   f. hydrogen sulfide   w. uranyl sulfate
   g. sulfur trioxide   x. potassium phosphide
   h. phosphorus pentafluoride   y. oxalic acid
   i. potassium borate   z. thiosulfuric acid
   j. perchloric acid   aa. sodium bicarbonate
   k. nitrous acid   bb. calcium hydride
   l. silicon dioxide   cc. boric acid
   m. acetic acid   dd. carbonic acid
   n. vanadyl acetate   ee. zinc hydroxide
   o. tetraphosphorus decaoxide   ff. dinitrogen pentaoxide
   p. dichlorine heptaoxide   gg. hypochlorous acid
   q. mercurous cyanide   hh. xenon hexafluoride

9. Write formulas for the following "everyday" compounds.
   a. baking soda, sodium bicarbonate
   b. calamine, zinc oxide
   c. calcite, calcium carbonate
   d. calomel, mercurous chloride
   e. Chilean saltpeter, sodium nitrate
   f. lye, sodium hydroxide
   g. milk of magnesia, magnesium hydroxide
   h. potash, potassium carbonate
   i. quartz, silicon dioxide
   j. rouge, ferric oxide

◇ Nomenclature: Elements and Compounds

Date _____ Name _____ Lab Sec. _____ Desk No. _____

1. Write the English and Latin names for elements with the following symbols.

Symbol	Match the Latin Name (at right) with the Symbol (at left)	English Name	
K	_____	_____	natrium
Pb	_____	_____	ferrum
Ag	_____	_____	aurum
Fe	_____	_____	stibium
Hg	_____	_____	kalium
Sb	_____	_____	plumbum
Sn	_____	_____	hydragyrum
Au	_____	_____	argentum
Na	_____	_____	stannum

2. At least 109 elements are known; however, only a few make a significant contribution to the Earth's crust. Complete this table of elements that are of greatest abundance in the Earth's crust.

Name of Element	% of total atoms	Atomic Number	Symbol	Atomic Mass	Metal or Nonmetal
oxygen	46.6				
silicon	27.2				
aluminum	8.13				
iron	5.00				
calcium	3.63				
sodium	2.83				
potassium	2.59				
magnesium	2.09				

3. All of the naturally occurring gases (except the noble gases), bromine, and iodine are diatomic molecules. Write formulas for the elements that occur as diatomic molecules in their natural state.

_____, _____, _____, _____, _____, _____, _____

4. Write the formula for the following compounds.

    a. ammonia   1 N-atom, 3 H-atoms   _____

    b. lye   1 Na-atom, 1 O-atom, 1 H-atom _____

    c. calcium carbonate   1 Ca-atom, 1 C-atom, 3 O-atoms _____

    d. glucose   6 C-atoms, 12 H-atoms, 6 O-atoms _____

    e. hydrogen peroxide   2 H-atoms, 2 O-atoms   _____

    f. alcohol (ethanol)   2 C-atoms, 6 H-atoms, 1 O-atom _____

    g. octane   8 C-atoms, 18 H-atoms   _____

    h. ammonium perchlorate 1 N-atom, 4 H-atoms, 1-Cl atom, 4-O atoms _____

Nomenclature: Elements and Compounds

Date _____ Name _____ Lab Sec. _____ Desk No. _____

A. A Survey of the Ions

1.   List and name (Stock system and the "old" *-ic, -ous*) the cations having two common charges.

2.   List and name all of the anions of chlorine.

3.   List and name all of the anions of sulfur.

4.   a.   Common charge of Group 1A cations: _____

     b.   Common charge of Group 2A cations: _____

     c.   Common charge of the monoatomic anions of Group 7A: _____

B. Nomenclature

Problems assigned by laboratory instructor: _____, _____, _____, _____. Use the following table to answer the problems that you were assigned. Use your own paper as necessary.

Formula	Name	Formula	Name

Instructor's signature. _____

Experiment 6

◆ Reaction Types and Analysis

Isn't salt just those white crystals that come from a salt shaker, you know, the stuff we use to enrich the flavor of foods? Yes, but it is more than that to a chemist! Do we mean to say that salt is not just "salt"?

To a chemist a salt is a combination of cations and anions in a ratio that results in a neutral, solid compound. The salt may be colored, toxic, insoluble in water, and/or chemically reactive. Each physical and chemical property of the salt may be dependent upon the cation, the anion, or both. Therefore, "salt" is not only the salt from a salt shaker; it may also be the rust on our automobiles, a lawn fertilizer, or a form of limestone. In this experiment we will determine the chemical properties of common anions that are present in a large number of salts, anions that have a common existence in nature.

OBJECTIVES

- To write balanced equations for observed chemical reactions
- To observe the chemical properties of SO_4^{2-}, CO_3^{2-}, PO_4^{3-}, Cl^-, and S^{2-}
- To identify the SO_4^{2-}, CO_3^{2-}, PO_4^{3-}, Cl^-, and S^{2-} anions in various salts

PRINCIPLES

Chemical reactions. The product of a chemical reaction has chemical and physical properties unlike those of the reactants. For example, sodium is a very reactive, silvery-white metal and chlorine is a greenish-yellow, very toxic gas; and yet, when they react chemically, sodium chloride, a white, crystalline solid, that is a part of our diet, is produced. Clearly, sodium chloride is unlike the elements from which it forms.

Chemical reaction: the production of a substance having properties different from the substance or substances from which it was formed.

How does a chemist recognize chemical change? Generally, a change of color, the appearance of a change of state (gas, liquid, solid), the detection of a different odor, and/or the absorption or evolution of heat accompany a chemical reaction. Other evidence of chemical change may be more subtle.

A **chemical equation** is used to describe the events of a chemical reaction in the general form,

Chemical equation: a brief description of a chemical reaction in which products form from reactants.

$$\text{reactants} \rightarrow \text{products}$$

For example, the chemical reaction that occurs between the reactants, gaseous HCl and gaseous NH_3, to produce the product, solid NH_4Cl is represented by the equation,

$$HCl(g) + NH_3(g) \rightarrow NH_4Cl(s)$$

More specifically, the chemical equation states that one molecule of gaseous hydrogen chloride reacts with one molecule of gaseous ammonia to produce one formula-unit of ammonium chloride–that's a pretty wordy statement! Notice how the equation simplifies the representation of the chemical reaction.

Balanced equation: a chemical
equation that indicates the same
number atoms of each element on
both sides of the equation.

A chemical equation not only simplifies a statement of a chemical reaction, it also indicates the number of atoms involved in the reaction and how they rearrange to form products. Another interpretation of the above equation is that 4 H atoms, 1 Cl atom, and 1 N atom rearrange to form a product that also contains 4 H atoms, 1 Cl atom, and 1 N atom—the atoms are conserved and the equation is said to be **balanced**; since each atom has its own mass, then mass is also conserved. A correct equation, then, must have the same number of atoms of each element appearing on both sides of the "→".

Simple reactions can be categorized into several groups.

- A single displacement reaction occurs when an element, A, (usually a metal) displaces another element, B, from a compound.

$$A + BY \rightarrow AY + B$$

- A double displacement (also called metathesis) reaction occurs when two elements substitute for each other in their respective compounds.

$$AX + BY \rightarrow AY + BX$$

- A direct combination reaction occurs when two elements combine to form a single compound.

$$X + Y \rightarrow XY$$

- A decomposition reaction occurs when heat or some other influence causes a compound to dissociate or decompose.

$$AQ \rightarrow A + Q$$

In Parts A–D, a number of chemical reactions will be observed and chemical equations will be written to express the reaction.

Salts and Solubility

Cation: an atom or group of atoms
with a positive charge.
Anion: an atom or group of atoms
with a negative charge.
Solubility: a measured amount of salt
that dissolves in a known mass of
water.

A salt consists of a **cation** and an **anion**. The **solubility** of a salt is dependent upon the temperature of the solution and the cation and anion of the salt; for example, sodium salts of the carbonate and phosphate anions are considered soluble, whereas the calcium salts of the carbonate and phosphate anions are considered to be only very slightly soluble.

The acidity of an aqueous solution also affects the solubility of a salt. A sulfate salt, such as $CaSO_4$, generally has a relatively low solubility in water and in an acidic solution; on the other hand, a carbonate salt, such as $CaCO_3$, also has a relatively low solubility in water, but readily dissolves in an acidic solution. The solubility of the carbonate salt increases because of the reaction of the carbonate ion with the hydrogen ion of the acid (a displacement reaction); carbon dioxide gas is evolved as a reaction product (a decomposition reaction, Figure 6.1):

$$CaCO_3(s) + 2 H^+(aq) \rightarrow H_2CO_3(aq) + Ca^{2+}(aq)$$
$$H_2CO_3(aq) \rightarrow H_2O(g) + CO_2(g)$$

Therefore, an acid can be used to quickly distinguish a sulfate salt from a carbonate salt, both having low solubilities as the calcium salt. Many phosphate salts, such as calcium phosphate, $Ca_3(PO_4)_2$, also have a limited solubility and, like the carbonate salts, readily dissolve in an acidic solution.

$$Ca_3(PO_4)_2(aq) + 4 H^+(aq) \rightarrow 3 Ca^{2+}(aq) + 2 H_2PO_4^-(aq)$$

Figure 6.1

Carbon dioxide gas is evolved from the reaction of carbonate ions with hydronium ions

How then could we distinguish between a carbonate salt and a phosphate salt if both were present in a mixed solid? One distinguishing chemical property is that phosphate salts, those having the PO_4^{3-} anion, do not produce a gas in the presence of hydrogen ion (as does the CO_3^{2-} anion), but merely form the dihydrogen phosphate ion, $H_2PO_4^-$.

Occasionally we find that an additional test is necessary to confirm our prediction for the presence of an anion in a salt. Ammonium molybdate, $(NH_4)_2MoO_4$, is often used to identify the PO_4^{3-} anion; its addition to a solution containing the $H_2PO_4^-$ ion causes the formation of a yellow **precipitate**:

Precipitate: an insoluble substance that forms in water.

$$H_2PO_4^-(aq) + 12\ (NH_4)_2MoO_4(aq) + 22\ H^+(aq) \rightarrow$$
$$(NH_4)_3PO_4(MoO_3)_{12}(s) + 21\ NH_4^+(aq) + 12\ H_2O(l)$$

Sulfate salts and carbonate salts do not behave in a similar manner. Therefore, we can distinguish between the SO_4^{2-}, CO_3^{2-}, or PO_4^{3-} anions in salts that are very slightly soluble.

Obviously many other anions form salts that have a low solubility. One of the most common anions is the chloride ion, Cl^-. Most chloride salts are considered soluble, but the silver salt, $AgCl$, is an exception.

$$Ag^+(aq) + Cl^-(aq) \rightarrow AgCl(s)$$

The sulfide anion, S^{2-}, is also common to many salts. A quick test for the S^{2-} anion is to acidify a concentrated solution of its salt and note the odor of hydrogen sulfide gas; its smell is characteristic—you'll know!! If the acid is nitric acid, then the characteristic yellow color of sulfur appears.

In Part E, the SO_4^{2-}, CO_3^{2-}, PO_4^{3-}, Cl^-, and S^{2-} anions present in an unknown salt and salt mixture are determined by conducting the tests that have been described.

Set up the following experiments in a 24 well plate.

PROCEDURE

A. Single Displacement Reactions

24 well plate: chemical apparatus used to contain about 3.5 mL of solution per well.

1. Well A1: *Polish* a 2 cm Zn strip with steel wool and insert it into 6 *M* HCl (**Caution**: *avoid skin contact; flush with water*). Record your observations on the Data Sheet.

2. Well A2: Dissolve several crystals of $CuSO_4 \cdot 5H_2O$ in 1 mL of water and insert a polished Zn strip. Allow the solution to set through the completion of Part B. Wipe off and examine the reddish-black deposit and the zinc surface for pits or pock marks. Record.

3. Well A3: Dissolve several crystals of $ZnSO_4$ in 1 mL of water and insert a polished Cu strip. Allow the solution to set through the completion of Part B. Do you see evidence of a chemical change?

B. Double Displacement Reactions

"**M**": *represents the concentration of the respective solute in an aqueous solution, in concentration units of moles solute per liter solution.*

Beral pipet: a small plastic apparatus used for extracting and dispensing small volumes of solutions.

Pinch: a volume about the size of a grain of rice.

1. Set up wells C1-C5 and D1–D4 in the 24 well plate and transfer the following quantities of solutions.

   Well C1: 1 mL of 0.1 M NH_4Cl
   Well C2: 1 mL of 0.1 M Na_2CO_3
   Well C3: 1 mL of 0.1 M $BaCl_2$
   Well C4: 1 mL of 0.1 M $CuSO_4$
   Well C5: 1 mL of 0.1 M $FeSO_4$
   Well D1: 1 mL of 0.1 M Na_3PO_4
   Well D2: 1 mL of 6 M H_2SO_4 (**Caution**: *handle with care and avoid skin contact*)
   Well D3: 1 mL of 6 M NaOH (**Caution**: *handle with care and avoid skin contact*)
   Well D4: 1 mL of 6 M NH_3 (**Caution**: *handle with care and avoid skin contact*)

2. Using clean **Beral pipets** (one per solution) combine the contents of the wells as directed. Record your observations on the Data Sheet.

   a. Add $^1/_2$ of D3 to C1. Smell *cautiously*. Place your finger on the bottom of C1 to warm the solution. Test the *vapor* with moistened red litmus paper.[1]
   b. Add drops of C5 to D1.
   c. In another well add $^1/_2$ of C2 to $^1/_2$ of D2. Smell *cautiously*. Test the *vapor* with blue litmus paper.
   d. Add C4 to D4.
   e. Add C3 to the remainder of C2.
   f. Insert a thermometer into the remainder of D2 and add the remainder of D3. This, an acid-base reaction, is called a **neutralization** reaction.[2]

3. **Caution:** *complete this part on the experiment in a fume hood.* Place a **pinch** of FeS in a 75 mm test tube and add $^1/_2$ mL of 3 M HCl. Cautiously and with proper technique test the odor, the odor of rotten eggs. (**Caution**: *the odor is due to* H_2S, *a very poisonous gas and should be tested only long enough to detect the odor.*) Test the $H_2S(g)$ with moistened blue litmus paper.

> **Dispose of the test solutions in the "Waste Salts" container. Return the solid pieces of zinc and copper (Part A) to the "Waste Solids" container.**

C. Combination Reactions

1. Grip a 2 cm piece of Mg ribbon with a pair of crucible tongs. Heat it directly in a Bunsen flame until it ignites. (**Caution**: *do not look directly at the burning Mg ribbon!*)

2. **Caution:** *a fume hood is suggested for this observation.* Place a pinch (no more!) of sulfur on the end of a spatula and heat directly with a Bunsen flame. Test the gas (SO_2) with moistened blue litmus paper.

---

[1]To test the vapors with litmus paper, moisten it with water and hold it across the diameter of the well. If the blue litmus turns red, then the vapors are acidic; if the red litmus turns blue, the vapors are basic.

[2]A more quantitative investigation of the heat evolved in a neutralization reaction is performed in Experiment 10.

1. a.   Place several crystals of $NaHSO_3$ in a 200 mm test tube and heat cautiously with a Bunsen flame. Test and describe the odor (**Caution:** *smell cautiously*—use the proper technique, Technique 17a, for testing!). Test the vapor with blue litmus paper. Where, during this experiment, have you previously detected this odor?

    b.   Now heat the contents more strongly until a red-brown color appears. Allow the contents to cool to room temperature. Add deionized water (the test tube should be $^1/_3$ to $^1/_4$ full) and agitate the test tube until the contents dissolve. Divide the solution into 2 equal volumes.

    *   To the first aliquot, add several drops of 0.1 M $BaCl_2$.
    *   To the second aliquot, add several drops of 3 M HCl. Test for odor.

    c.   Dissolve several crystals of $NaHSO_3$ in water and repeat the test with 0.1 M $BaCl_2$. and 3 M HCl. How do the tests differ? Record.

Dispose of the test solutions in the "Waste Salts" container.

2. **Demonstration only**. Place about 3 g of sugar, $C_{12}H_{22}O_{11}$, into a porcelain evaporating dish. Place the dish in a fume hood. Fill a Beral pipet with conc H_2SO_4 (**Caution:** *conc H_2SO_4 causes severe skin burns and clothing to disappear! Immediately flush the skin with water*) and add it to the sugar. Conc H_2SO_4 is a strong dehydrating agent, extracting H_2O molecules from the $C_{12}H_{22}O_{11}$ molecule. Record your observations and write a balanced equation.

D. Decomposition Reactions

Your instructor will issue two unknown samples. Sample A consists of a single salt and Sample B consists of two salts. In each sample you are to identify the anion(s) present.

E. Identification of a Salt

Analyzing your unknown. The test salts and the unknown samples are *pure* samples. You must realize that the observations in these tests would not be as obvious if impurities were present in the salt. Sample A contains one salt, a salt that has only one of the five anions. Sample B, having two or more anions will be more difficult. For example, if a solid in Part E.2 dissolves in Part E.3, you may not know immediately whether the PO_4^{3-} anion or CO_3^{2-} anion is present and you will need to complete Parts E.4–5 to determine the presence of the PO_4^{3-} anion. Also the $CO_2(g)$ formation may be difficult to observe—you must watch closely. It would be advisable for you to check Sample B two or three times to ensure confidence in your data.

1. To wells A1–A5 of a 24 well plate, transfer a small amount (20–40 crystals) of the salts Na_2SO_4, Na_2CO_3, Na_3PO_4, NaCl , and Na_2S. Place your two unknowns in two distant wells, such as C1 and C2.

2. Half-fill each of the seven wells with deionized water and agitate and swirl the solutions for about 30 s. Note how quickly the salts dissolve. Record on the Data Sheet.

3. Add 1 drop of 6 M NH_3 (**Caution:** *do not inhale*) to each solution; then add several drops of a 0.2 M $Ba(NO_3)_2$ and stir or swirl. Which anions form a precipitate with the Ba^{2+} cation? Record.

4. Add 5 drops of 6 M HNO$_3$ (**Caution**: *do not allow* HNO$_3$ *to contact the skin. If it does, wash immediately with plenty of water*.) to each solution. Look at each solution; is there any evidence of a chemical reaction? *Carefully* test for any odor. Record all of your observations.

5. Use Beral pipets to transfer about one-half of each test solution to an adjacent well (such as B1–B5, D1 and D2); avoid transferring any precipitate.
   a. To wells A1–A5, C1 and C2, add 5 drops of 0.1 M AgNO$_3$.
   b. To wells B1–B5, D1 and D2, add 5 drops of 0.1 M (NH$_4$)$_2$MoO$_4$, and agitate. Record evidence of any precipitate (the precipitate may be slow in forming). Record all of your observations on the Data Sheet.

Dispose of all test solutions in the "Waste Salts" container.

NOTES AND OBSERVATIONS

◇ Reaction Types and Analysis

Date _____Name _____ Lab Sec. _____Desk No. _____

1. a. Describe the technique for testing the odor of a chemical.

    b. Describe the technique for testing a vapor with litmus paper.

2. What chemical test(s) can you use to distinguish between calcium chloride, $CaCl_2$, and calcium carbonate, $CaCO_3$, both of which are white solids?

3. Write the formula for the *product* (or resultant) compound that confirms the presence of each of the five anions tested in this experiment.

    a. $SO_4{}^{2-}$ _____

    b. $CO_3{}^{2-}$ _____

    c. $PO_4{}^{3-}$ _____

    d. Cl^- _____

    e. S^{2-} _____

◇ Reaction Types and Analysis

Date _____ Name _____ Lab Sec. _____ Desk No. _____

A. Single Displacement Reactions

	Chemical Reactants	Evidence of Reaction	Chemical Products	Balanced Equation
1.			$ZnCl_2 + H_2$	
2.			$ZnSO_4 + Cu$	
3.			no reaction	

B. Double Displacement Reactions

	Chemical Reactants	Evidence of Reaction	Chemical Products	Balanced Equation
2a.	$NaOH +$ NH_4Cl		$NH_3 + H_2O +$ $NaCl$	

Result of litmus test

2b.			$Fe_3(PO_4)_2 +$ Na_2SO_4	
2c.			$Na_2SO_4 +$ $CO_2 + H_2O$	

Result of litmus test

2d.			$[Cu(NH_3)_4]^{2+}$ $+ SO_4^{2-}$	
2e.			$BaCO_3 +$ $NaCl$	
2f.			$Na_2SO_4 +$ H_2O	
3.			$FeCl_2 + H_2S$	

Result of litmus test

C. Combination Reactions

	Chemical Reactants	Evidence of Reaction	Chemical Products	Balanced Equation
1.			MgO	
2.			SO_2	

D. Decomposition Reactions

1. a. Odor _____. Conclusion from litmus test_____.

   Formula of gas _____.

   b. Observation from $BaCl_2$ test _____. $BaSO_4$ is an insoluble salt; therefore what is one of the decomposition products of $NaHSO_3$?_____

   Observation from 3 M HCl test _____. From a detection of the odor, what is a second decomposition product of $NaHSO_3$?_____

   Write a balanced equation for the thermal decomposition of $NaHSO_3$.

   c. Observation from $BaCl_2$ test _____. How does the appearance of the system differ from that in Part D.1b?

   Observation from 3 M HCl test _____. How does the odor differ from that in Part D.1b?

   Write a balanced equation for the reaction of HCl(aq) with $NaHSO_3$(s).

2. What evidence of a chemical reaction is indicated?

   The products of the reaction are carbon and water. Write a balanced equation for the decomposition of sugar.

E. Identification of a Salt

Sample A: Identification Number_____

Sample B: Identification Number _____

	Na_2SO_4	Na_2CO_3	Na_3PO_4	NaCl	Na_2S	Sample A	Sample B
Soluble in Water?							
Precipitate with $Ba(NO_3)_2$?							
Reaction with HNO_3?							
Odor Test							
Precipitate with $AgNO_3$?							
Precipitate with $(NH_4)_2MoO_4$?							

Sample A: Anion present _____

Sample B: Anions present _____

Write (in an organized format) balanced equations for all reactions that were observed.

QUESTIONS

1. In Parts A.2 and A.3, the relative chemical reactivities of Zn and Cu are determined. Which metal is more reactive? _____ Explain.

2. Predict the color of a cotton shirt or jeans after having spilled conc H_2SO_4 on them. (Note: cotton is chemically similar to sugar.)

3. What single test reagent can distinguish a soluble Cl^- salt from a soluble SO_4^{2-} salt? Explain.

4. What single test reagent can distinguish $BaSO_4$ from $BaCO_3$? Explain.

*5. Describe a procedure for identifying the presence of *both* Na_2CO_3 and $BaCO_3$ in a salt mixture?

*6. Describe a procedure for identifying the presence of *both* Na_3PO_4 and $Ba_3(PO_4)_2$ in a salt mixture.

◇ Formula of a Hydrate

Water is a part of nearly everything that we see or touch! Water is the component of our atmosphere that brings rainfall and the medium by which nutrients and salts are transported both within our bodies and in the world's waterways. Water vaporizes and condenses, melts and freezes, sublimes and crystallizes. Water **adsorbs** to solid surfaces and binds to the internal network of a salt's structure.

It is this latter property that we study in this experiment. Many salts that crystallize in nature do so with the accompaniment of water molecules. In fact, many salt crystals, for example those that are "grown" in physical science laboratories, have the shiny appearance and the many brilliant facets because of the (internal) presence of water molecules (see photo). When these *hydrated* salt crystals are heated, the water molecules escape from the solid structure, the salt crumbles and its brilliance is lost.

- To determine the percent by mass of water in a hydrated salt
- To establish the mole ratio of salt to water in a hydrated salt

OBJECTIVES

PRINCIPLES

Many naturally occurring salts and salts purchased from chemical suppliers are **hydrated**; that is, water molecules (called waters of hydration) are bound to the ions in the salt structure. Hydrated salts are generally highly crystalline, showing many reflective facets. Salts that are not hydrated are called **anhydrous** and by comparison appear to be **amorphous**, even though under a microscope they do have a crystalline structure. Salts bind the waters of hydration to varying degrees. Salts that spontaneously lose water molecules to the atmosphere are **efflorescent,** whereas salts that readily absorb water from the atmosphere are **deliquescent**.

Adsorb: to physically bind to the surface of a substance.
Hydrate: water molecules are chemically bound to the salt as part of the internal structure of the compound.
Amorphous: having no regular structure, noncrystalline.

The number of moles of water per mole of salt is usually constant for a hydrated salt. For example, iron(III) chloride is normally purchased as $FeCl_3 \cdot 6H_2O$, not as $FeCl_3$; copper(II) sulfate as $CuSO_4 \cdot 5H_2O$ or $CuSO_4 \cdot H_2O$, not as $CuSO_4$. Epsom salt (Figure 7.1) has the formula $MgSO_4 \cdot 7H_2O$. Some salts, such as washing soda, $Na_2CO_3 \cdot 10H_2O$, readily **dehydrate** with gentle heat:

$$Na_2CO_3 \cdot 10H_2O(s) \; -\overset{\Delta}{\longrightarrow} \; Na_2CO_3(s) \; + \; 10\,H_2O(g)$$

"$\overset{\Delta}{\longrightarrow}$": a symbol indicating that heat is required for the reaction.

In sodium carbonate decahydrate, $Na_2CO_3 \cdot 10H_2O$, a ratio of 10 moles of H_2O to 1 mole of Na_2CO_3 or 180 g of H_2O to 106 g of Na_2CO_3 exists within its crystalline structure. The percent H_2O by mass in the hydrated salt is

$$\frac{180 \text{ g } H_2O}{(180 \text{ g } H_2O + 106 \text{ g } Na_2CO_3)} \; \text{X} \; 100 \; = \; 62.9 \; \% \; H_2O$$

In other salts the water molecules cannot be easily removed, no matter how intense the heat, e.g., $FeCl_3 \cdot 6H_2O$.

Figure 7.1

Epsom salt , internally, is a purgative and externally, in concentrated solutions, is used to reduce inflammation

PROCEDURE

A. Hydration and Dehydration

1. Place about 0.1 g (an amount needed to fill the rounded bottom portion of the test tube) of the following compounds (or those designated by your instructor) in separate 75 mm test tubes: $KAl(SO_4)_2 \cdot 12H_2O$, $CuSO_4 \cdot 5H_2O$, $CoCl_2 \cdot 6H_2O$, $NaCl$, $C_{12}H_{22}O_{11}$

2. Heat each sample directly over a *gentle* flame. Observe the upper, cooler portion of the test tube for condensed moisture. Record your observations.

3. After the sample has cooled, add several drops of deionized water. What can be concluded?

B. Efflorescence and Deliquescence

1. Place several crystals of $Na_2CO_3 \cdot 10H_2O$ and $CaCl_2$ in separate watchglasses. Observe their initial appearance and then again at the end of the laboratory period.[1]

C. Analysis of a Hydrate

Anhydrous: without water.

Complete 3 trials for Part C. Obtain an unknown hydrated salt sample from your instructor.

1. Support a clean crucible and lid on a clay triangle and heat with an intense flame for 5 minutes. Cool. If the crucible remains dirty, add 1–2 mL of 6 M HNO$_3$ (**Caution:** *avoid skin contact, flush immediately with water*) and evaporate to dryness. Handle the crucible and lid with the crucible tongs for the rest of the experiment; do not use your fingers. Determine the mass (±0.001 g) of the cooled crucible and lid.[2]

2. Add *at most* 3 g of your unknown hydrated salt to the crucible and again determine the mass (±0.001 g) of the crucible, lid, and sample.

3. Return the crucible with the sample to the clay triangle and set the lid off the crucible's edge to allow evolved gases to escape.

4. At first, heat the sample slowly and then gradually intensify the heat. Do not allow the crucible to become red hot. This can cause the **anhydrous** salt to decompose. Heat the sample for 15 minutes. Cover the crucible with the lid, cool to room temperature (in the desicooler, if available), and determine the mass of the crucible, lid, and sample.

5. Reheat the sample for 5 more minutes. Again measure the combined mass. If the second mass measurement disagrees (>1%) with the first, repeat the heating until a mass of <1% is achieved.

Discard the anhydrous salt in the "Waste Salts" container.

---

[1]Anhydrous calcium chloride is used as a desiccant (a substance that absorbs water) in desicoolers and desiccators. See Technique 16d.

[2]Place the crucible and lid in a desicooler or desiccator (if available) for cooling. Cool the crucible and lid in the same manner for the remainder of the experiment.

Formula of a Hydrate

Date _____ Name _____ Lab Sec. _____ Desk No. _____

1. Distinguish the meanings of adsorbed water and hydrated water.

2. Naturally occurring gypsum is a hydrate of $CaSO_4$. A 1.803 g sample of gypsum is heated in a crucible until a constant mass is obtained. The mass of the anhydrous (without water) salt, $CaSO_4$, is 1.426 g.

   a. Calculate the percent by mass of water in the gypsum sample.

   b. Calculate the moles of H_2O removed and moles of $CaSO_4$ remaining in the crucible.

   c. What is the formula of gypsum?

3. In today's experiment, what error in the data is likely to occur if the hydrated salt is heated too strongly? (Hint: read the Procedure.)

4. After the mass of the crucible is determined for the first time in Part C.1 of the Procedure, you are advised to handle the crucible and lid with the crucible tongs only, *not* your fingers. Explain why this is good technique.

5. The following data were collected for determining the mole ratio of $MgCl_2$ to H_2O in magnesium chloride hexahydrate:

Mass of crucible and lid   18.733 g
Mass of crucible, lid and sample   21.171 g
Mass of crucible, lid, and anhydrous sample   20.122 g

Determine the mole ratio of $MgCl_2$ to H_2O as determined from the data. Suggest a reason as to why the data is in error and experimentally how the error can be corrected.

◇ Formula of a Hydrate

Date _____ Name _____ Lab Sec. _____ Desk No. _____

A. Hydration and Dehydration

Observations or Appearance	Solid Before Heating	Solid After Heating	Did Moisture Condense?	Solid After Addition of Water
$KAl(SO_4)_2 \cdot 12H_2O$				
$CuSO_4 \cdot 5H_2O$				
$CoCl_2 \cdot 6H_2O$				
NaCl				
$C_{12}H_{22}O_{11}$				

B. Efflorescence and Deliquescence

1. Describe the changes in appearances of $Na_2CO_3 \cdot 10H_2O$ and $CaCl_2$ over the duration of the laboratory period.

C. Analysis of a Hydrate

Name of salt_____	Trial 1	Trial 2	Trial 3
1. Mass of crucible and lid (g)			
2. Mass of crucible, lid, and hydrated salt (g)			
3. Mass of hydrated salt (g)			
4. Mass of crucible, lid, and anhydrous salt 1st mass measurement (g)			
2nd mass measurement (g)			
3rd mass measurement (g)			
5. Mass of anhydrous salt (g)			
6. Mass of water lost (g)			

Calculations

1. Percent by mass of H_2O lost from hydrated salt (%)

2. Average % H_2O in hydrated salt (%)

3. Moles of anhydrous salt (*mol*)

4. Moles of water lost (*mol*)

5. Mole ratio of anhydrous salt to water (whole numbers)

     *

6. Formula of hydrate

*Calculation for Trial 1. Show your work.

QUESTIONS

1. If some volatile impurities are not burned off the crucible in Part C.1, but are removed in Part C.4, is the recorded mass of anhydrous salt too high or too low? Explain.

2. If the hydrated salt spatters out of the crucible during the heating process in Part C.4, will the reported mole ratio of the salt to water in the hydrated salt be too high or too low? Explain.

3. Anhydrous $CaCl_2$ is used as a desiccant in desiccators—it removes water from the atmosphere within the desiccator.
   a. Explain the mechanism by which $CaCl_2$ removes the water vapor.

   b. A 4.232 g $CaCl_2$ sample has a mass of 5.604 g after being left in a desiccator for several weeks. What is the formula of the hydrated $CaCl_2$ salt?

Experiment 8

◇ Formula of a Compound

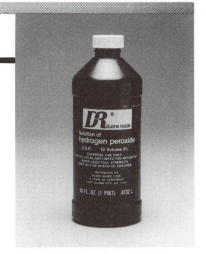

Did you ever stop to think how all of these chemical formulas that you have heard and read about were determined? How do we know their formulas are what we say they are? All chemicals have formulas! How can you possibly learn them all? Can you just look at a compound and say it necessarily has a particular formula? Who says that the formula for water is H_2O and that for limestone is $CaCO_3$? What is the formula for hydrogen peroxide (see photo)?

Formulas are determined from chemical analyses, and the analyses must be **quantitative**. Formulas are to chemists as the keyboard is to typists; in each case, letters are placed in a certain sequence by a set of rules that is understood by everyone to create a representation of an image. For example, the word "ocean" represents an image, just as the formula H_2O represents an image, but one that is quite different from that of H_2O_2.

To determine the formula for water we need to know the amount (the number of moles) of hydrogen that reacts with the available oxygen to produce a measured mass of water; or in the decomposition of a measured mass of limestone, the number of moles of calcium, carbon, and oxygen present. Thereafter, we can apply facts that we have already learned about the relative masses of the elements to determine a formula of the compound.

- To determine the chemical formulas of compounds formed from the thermal decomposition of copper bromide
- To determine the chemical formula of a compound synthesized from magnesium and hydrochloric acid

OBJECTIVES

PRINCIPLES

Quantitative: a measurement made with calibrated equipment, such as a balance or volumetric glassware.

The formula of a compound specifies a whole-number, mole ratio of atoms in a compound. Two procedures by which the formula of a compound can be determined in the laboratory are from the thermal decomposition of the compound into its respective elements or from its synthesis from the elements.

An example for determining the formula from a thermal decomposition reaction is given for mercuric oxide on the Lab Preview. Sodium chloride can be synthesized from its elements: 2.75 g of sodium react with 4.25 g of chlorine. The amounts (in moles) of the two elements are:

$$2.75 \text{ g Na} \quad \times \quad \frac{1 \text{ mol Na}}{23.0 \text{ g Na}} \quad = \quad 0.120 \text{ mol Na}$$

$$4.25 \text{ g Cl} \quad \times \quad \frac{1 \text{ mol Cl}}{35.45 \text{ g Cl}} \quad = \quad 0.120 \text{ mol Cl}$$

Therefore, the mole ratio of sodium to chlorine is 0.120 to 0.120. Since a formula is always expressed in whole numbers, the mole ratio of sodium to chlorine is 1 to 1 and the formula is Na_1Cl_1 or $NaCl$.

A formula also provides the mass ratio of the elements in the compound. The formula NaCl implies that 23.00 g (1 mole) of sodium combines with 35.45 g (1 mole) of chlorine. The mass percentages of sodium and chlorine in sodium chloride are

$$\frac{23.00 \text{ g}}{(23.00 \text{ g} + 35.45 \text{ g})} \times 100 = 39.35\% \text{ Na}$$

$$\frac{35.45 \text{ g}}{(23.00 \text{ g} + 35.45 \text{ g})} \times 100 = 60.65\% \text{ Cl}$$

The mass percent of sodium to chlorine in sodium chloride is always the same (constant and definite), a statement of the **law of definite composition**.

When hydrogen chemically combines with oxygen to form water, the mass percent of hydrogen (11.2%) and oxygen (88.8%) is constant. However, when hydrogen chemically combines with oxygen to form hydrogen peroxide, the mass percent of hydrogen (5.94%) and oxygen (94.1%) is also constant, but different from that found in water. Thus, hydrogen and oxygen combine to from more than one compound with definite compositions.

The mass ratio of oxygen to hydrogen in water is:

$$\frac{88.8 \text{ g O}}{11.2 \text{ g H}} = 7.93 \text{ g O/g H}$$

The mass ratio of oxygen to hydrogen in hydrogen peroxide is:

$$\frac{94.1 \text{ g O}}{5.94 \text{ g H}} = 15.8 \text{ g O/g H}$$

The masses of oxygen combining with a fixed mass (1 g) of hydrogen for the two compounds are a ratio of 1 to 2, again a ratio of small whole numbers.

$$\frac{7.93 \text{ g O}}{15.8 \text{ g O}} = \frac{1}{2}$$

This experiment describes procedures for determining the formulas of compounds resulting from the thermal decomposition of an anhydrous copper bromide salt and from the reaction of magnesium and hydrochloric acid. For each compound the formula and the mass percentage are determined using a series of techniques, measurements, and calculations.

At least two trials are to be completed.

1. a. Measure the mass (±0.001 g) of a clean 150 mm test tube (Figure 8.1). Add 1–2 g of the copper bromide and remeasure. Set up the apparatus shown in Figure 8.2. The test tube is inclined at about 60° from the horizontal, the funnel hood is positioned down over the mouth of the test tube, and the gas inlet tube of the trap extends below the 0.1 M NaOH solution.

    b. Have your instructor approve your apparatus.

A. Decomposition of Copper Bromide

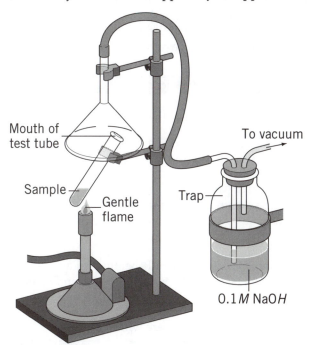

Mouth of test tube

To vacuum

Sample

Gentle flame

Trap

0.1 M NaOH

Figure 8.1
Use a small beaker to assist in measuring the mass of a test tube.

Figure 8.2
Apparatus for removing bromine vapor, a product of the thermal decomposition of copper bromide.

2. a. Turn on the aspirator until air bubbles gently pass through the 0.1 M NaOH in the trap.

    b. Slowly and gently (starting near the top level) heat the copper bromide salt. (**Caution:** *bromine is a poisonous, corrosive vapor. Do not inhale or allow skin contact.*) Gradually intensify the heat until bromine gas is no longer evolved. Periodically heat the upper part of the test tube to vaporize any condensed bromine.

3. Once bromine is no longer evolved, turn off the aspirator and allow the test tube and contents to cool. Again measure the mass of the test tube and its contents.

4. a. Dissolve the solid in the test tube with a minimum volume of 3 M HCl (**Caution:** *avoid skin contact with HCl(aq)*). Polish (with steel wool) several 2 cm strips of magnesium.

    b. Add a strip of magnesium ribbon to the solution—magnesium metal displaces hydronium ion (from the HCl(*aq*)) *and* the copper ion (from the solid) from solution. What are the products of the displacement?

    c. Continue to add (one at a time) polished strips of magnesium until no further chemical change is observed. Allow the magnesium metal to remain in the solution for about 30 minutes, occasionally breaking up the solid with a stirring rod.

    d.   Add 6 *M* HCl to remove the excess magnesium metal.

5.   Allow the reaction mixture to settle, decant and discard (as instructed by your instructor!) the supernatant solution. **Save the copper!** Wash the copper with three 10 mL portions of deionized water—discard the washings in a similar manner.

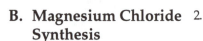

Dispose of the waste solutions in the "Waste Salts" Container.

6.   Over a cool flame, dry the solid copper in the test tube. Do not overheat to avoid the copper metal from reacting with oxygen. Measure the mass of the test tube and solid copper. Repeat the heating and cooling of the test tube and its contents and until successive mass measurements are within ±1%.

Dispose of the copper in the "Waste Metals" Container.

1.   Measure the mass (±0.001 g) of a clean, 150 mm test tube (Figure 8.1).

B. Magnesium Chloride Synthesis

2.   Cut 0.2–0.3 g of polished (with steel wool) Mg ribbon into short lengths. Place the Mg into the test tube, measure the combined mass, and record.

3.   Slowly add drops of 6 *M* HCl (**Caution**: *avoid skin contact*) as the Mg reacts. After no further reaction is apparent with the addition of the HCl, *gently* heat the reaction mixture with a cool flame to dryness—do *not* boil.

4.   Once the sample appears dry, continue gentle heating for an additional minute, but avoid excessive heating. Allow the test tube and contents to cool to room temperature, measure the combined mass, and record.

5.   Repeat Part B.4 until mass measurements are within ±1%.

Dispose of the magnesium chloride in the "Waste Salts" Container.

NOTES, OBSERVATIONS, AND CALCULATIONS

Formula of a Compound

Date _____ Name _____ Lab Sec. _____ Desk No. _____

1. Elemental oxygen was first discovered when mercuric oxide was decomposed with heat, forming mercury metal and oxygen gas. When a 1.048 g sample of mercuric oxide is heated, 0.971 g of mercury remains.

    a. Calculate the moles of mercury and oxygen in the sample.

    b. What is the formula of mercuric oxide?

2. A 2.60 g sample of titanium metal chemically combines with 7.71 g of chlorine gas. Determine the formula for the titanium chloride salt.

3. How many grams of oxygen gas combine with 0.843 g vanadium to form V_2O_5?

4. a. What is the mass percent of carbon and oxygen in carbon monoxide?

    b. What is the mass percent of carbon and oxygen in carbon dioxide?

    c. What is the mass ratio oxygen that combines with 1.0 g of carbon in the two compounds?

5. Write the equation for the displacement of hydronium ion by magnesium metal in Part B.3.

6. Explain how the mass of chlorine is determined in the synthesis of the magnesium-chlorine compound in Part B of the Procedure.

◇ Formula of a Compound

Date _____ Name _____ Lab Sec. _____ Desk No. _____

A. Decomposition of Copper Bromide

Data

	Trial 1	Trial 2
1. Mass of test tube (*g*)		
2. Mass of test tube and initial copper bromide (*g*)		
3. Mass of initial copper bromide (*g*)		
4. Mass of test tube and copper bromide after heating (*g*)		
5. Mass of copper bromide after heating (*g*)		
6. Mass of test tube and copper 1st mass measurement (*g*)		
2nd mass measurement (*g*)		
7. Mass of copper in sample (*g*)		

Calculations

	Trial 1	Trial 2
1. Mass of bromine in initial copper bromide (*g*)		
2. Moles of bromine in initial copper bromide (*mol*)	*	
3. Mass of bromine in copper bromide after heating (*g*)		
4. Moles of bromine in copper bromide after heating (*mol*)		
5. Moles of copper in each salt (*mol*)		
6. Formula of the initial copper bromide	*	
7. Formula of the copper bromide after heating		
8. Mass percent of copper in initial copper bromide (%)		
9. Mass percent of copper in copper bromide after heating (%)		
10. $\dfrac{\text{mass of bromine in initial copper bromide}}{\text{mass of bromine in copper bromide after heating}}$		

* Show sample calculation for Trial 1

B. Magnesium Chloride Synthesis

Data

	Trial 1	Trial 2
1. Mass of test tube (g)		
2. Mass of test tube and Mg (g)		
3. Mass of Mg (g)		
4. Mass of test tube and sample after reaction 1st mass measurement (g)		
2nd mass measurement (g)		
5. Mass of compound (g)		

Calculations

	Trial 1	Trial 2
1. Moles of Mg (mol)		
2. Mass of chlorine (g)		
3. Moles of chlorine (mol)	**	
4. Mole ratio of Mg to chlorine		
5. Formula of compound		
6. Percent Mg by mass in compound (%)		
7. Percent chlorine by mass in compound (%)		

* *Show sample calculation for Trial 1

QUESTIONS

1. Explain how the law of definite composition is applicable to this experiment.

2. The **law of multiple proportions** states: when two elements combine to from more than one compound, the masses of one element that combine with a fixed mass of another element are in a ratio of small whole numbers. Explain how the Law of Multiple Proportions is applicable to Part A of this experiment.

3. Suppose that in Part A.3, all of the bromine had not been removed from the initial copper bromide. Would the mass percent of copper in the copper bromide after heating be reported too high or too low? Explain.

4. If all of the copper is *not* recovered in Parts A.4–5, what will happen to the mole ratio of copper to bromine in the formula of copper bromide? Explain. What might be a representative formula of the initial copper bromide (before heating)?

5. Suppose that in the synthesis of magnesium chloride, the magnesium metal in Part B.2 is not polished.

   a.   Would the moles of magnesium that actually react be greater or less than the amount measured? Explain

   b.   How would this same error affect the reported moles of chlorine that combine with the magnesium?

6. In Part B.3, suppose all of the magnesium does not react with the HCl.

   a.   How will this affect the reported mass of chlorine in the magnesium chloride compound?

   b.   Does this affect the reported number of moles magnesium in the compound? Explain.

Experiment 9

◇ Limiting Reactant

Every process that requires two or more "things" to complete is faced with the prospect that one of the things may be consumed before the other. Think about the situation where there are 10 hotdogs in a package but only 8 hotdog buns per package. Two hotdogs will be left over unless another package of buns is purchased, but then 6 buns will be left without a hotdog, and the dilemma continues.

So it is with chemical reactions requiring two or more reactants. One reactant is necessarily depleted before the other, in which case the reaction stops and an excess of the other reactant(s) will remain. Seldom is a chemical system designed in which *exact* stoichiometric amounts of reactants are present in the system; if that situation ever did exist, it might take "forever" for the last two reactant molecules to find each other. This experiment reflects the dilemma of a limiting reactant that all scientists encounter in the "real life" of chemical research and industrial syntheses.

OBJECTIVES

- To determine the limiting reactant in the formation of a precipitate
- To determine the percent composition of a salt mixture

PRINCIPLES

Stoichiometric amounts: mole ratio of reactants and products according to a balanced equation.

Two factors that limit the yield of products in a chemical reaction are (1) the amounts of starting materials (reactants) and (2) the percent yield of the reaction. Many experimental conditions (for example, temperature and pressure) can be adjusted to increase the yield for a reaction, but because chemicals react according to fixed mole ratios (**stoichiometric amounts**), only a limited amount of product(s) can form from given amounts of starting material. The reactant restricting the amount of product that can be produced in a chemical reaction is called the **limiting reactant**.

To better understand the limiting reactant concept, let's look at the reaction studied in this experiment. The molecular form of the equation for the reaction system is

$$BaCl_2(aq) + Na_2SO_4(aq) \rightarrow BaSO_4(s) + 2\,NaCl(aq)$$

Since $BaCl_2$ (available as the salt, $BaCl_2 \cdot 2H_2O$) and Na_2SO_4 are soluble salts and since $BaSO_4$ is insoluble, the ionic equation is

$$Ba^{2+}(aq) + 2\,Cl^-(aq) + 2\,Na^+(aq) + SO_4^{2-}(aq) \rightarrow$$
$$BaSO_4(s) + 2\,Na^+(aq) + 2\,Cl^-(aq)$$

Spectator ions: ions that do not participate in a chemical reaction.

Disregarding the presence of the **spectator ions** in the chemical system, the final equation, called the **net ionic equation**, is:

$$Ba^{2+}(aq) + SO_4^{2-}(aq) \rightarrow BaSO_4(s)$$

Precipitate: the formation on an insoluble ionic compound, generally in water.

The equation states: one mole of Ba^{2+} [from 1 mole of $BaCl_2 \cdot 2H_2O$ (244.2 g/mol)] reacts with 1 mole of SO_4^{2-} [from 1 mole of Na_2SO_4 (142.1 g/mol)] to produce 1 mole of $BaSO_4$ **precipitate** (233.4 g/mol), *if the* reaction proceeds to completion (100% yield). This equation assumes no by-products in the reaction and, therefore, represents the theoretical yield of product.

Suppose, however, only 2.00 g of $BaCl_2 \cdot 2H_2O$ and 1.20 g of Na_2SO_4 are mixed in aqueous solution. How many moles and grams of $BaSO_4$ can *theoretically* precipitate? Since the equation reads in moles, not grams, the amount (moles or millimoles) of each reactant must be determined. Therefore,

1 mmol = 1 x 10^{-3} mol

$$2.00 \text{ g } BaCl_2 \cdot 2H_2O \times \frac{1 \text{ mol } BaCl_2 \cdot 2H_2O}{244.2 \text{ g } BaCl_2 \cdot 2H_2O} = 0.00819 \text{ mol } BaCl_2 \cdot 2H_2O$$

$$= 8.19 \text{ mmol } Ba^{2+}$$

$$1.20 \text{ g } Na_2SO_4 \times \frac{1 \text{ mol } Na_2SO_4}{142.1 \text{ g } Na_2SO_4} = 0.00844 \text{ mol } Na_2SO_4 = 8.44 \text{ mmol } SO_4^{2-}$$

The reaction mixture contains 8.19 mmol Ba^{2+} and 8.44 mmol SO_4^{2-}. Since 1.0 mol Ba^{2+} reacts with only 1.0 mol SO_4^{2-}, then the 8.19 mmol Ba^{2+} in the reaction vessel can only react with 8.19 mmol SO_4^{2-} (of the 8.44 mmol present) producing 8.19 mmol $BaSO_4$ *theoretically* . This consumes all of the Ba^{2+} in the vessel (Ba^{2+} is the limiting reactant) and leaves an excess of (8.44 − 8.19 =) 0.25 mmol of SO_4^{2-} (SO_4^{2-} is called the **excess reactant**).

Since the limiting reactant (Ba^{2+}) is now known, the theoretical yield of product ($BaSO_4$) is calculated from the balanced equation. Ba^{2+}, the limiting reactant, controls the moles and mass of $BaSO_4$ produced.

From the balanced equation, 1.0 mol Ba^{2+} produces 1.0 mol $BaSO_4$ *or* 8.19 mmol Ba^{2+} produces 8.19 mmol $BaSO_4$; therefore, 0.00819 mol

$$BaSO_4 \text{ produces } \left(0.00819 \text{ mol } BaSO_4 \times \frac{233.4 \text{ g } BaSO_4}{1 \text{ mol } BaSO_4} = \right) 1.91 \text{ g } BaSO_4$$

Thus, 1.91 g $BaSO_4$ is the theoretical yield (when the reaction is 100% complete).

In this experiment an unknown solid mixture of Na_2SO_4 and $BaCl_2 \cdot 2H_2O$ is added to water and $BaSO_4$ precipitates. We will measure the mass and calculate the moles of $BaSO_4$ that precipitate; from the balanced equation we can calculate the moles and masses of $BaCl_2 \cdot 2H_2O$ and Na_2SO_4 in the original unknown mixture.

From a series of tests, the limiting and excess reactants are determined. The difference between the mass of the original salt mixture, m_{sm}, and the mass of the limiting reactant, m_{lr} allows us to calculate the mass of excess reactant, m_{xr}, in the salt mixture

$$m_{xr} = m_{sm} - m_{lr}$$

The mass percents of $BaCl_2 \cdot 2H_2O$ and Na_2SO_4 in the original salt mixture are also determined.

PROCEDURE

Two trials are recommended for this experiment. To hasten the analyses, determine the mass of duplicate unknown salt mixtures, and simultaneously follow the Procedure for each. Label 200 mm test tubes accordingly for Trial 1 and Trial 2 to avoid intermixing samples and solutions.

1. Measure 0.6–0.8 g (±0.001 g) of the unknown salt mixture on weighing paper. Transfer the mixture to a 200 mm test tube and add 50 mL (±0.2 mL) of deionized water. Add about 10 drops of conc HCl (**Caution**: *conc HCl is a severe skin irritant. Flush the affected area with a large amount of water*). Stir the aqueous mixture with the stirring rod for about 1 minute and then allow the precipitate to settle.

2. a. Set up a hot water bath (70°–90°C) using a 600 mL beaker, three-fourths filled with water—do *not* boil. Place the test tube in the hot water bath and maintain the solution at this temperature for 40–50 minutes (Figure 9.1).[1] Proceed to Part A.4 during this procedure.

   b. Remove the test tube from the hot water bath and allow the precipitate to settle.

3. Decant two 10 mL (±0.2 mL) volumes of the supernatant liquid into separate 150 mm test tubes, labeled tt1 and tt2; save for Part B.

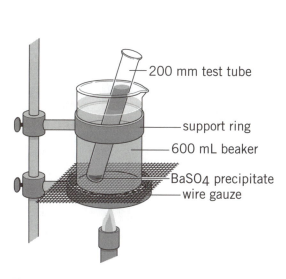

Figure 9.1
Warm the precipitate to digest the BaSO₄ precipitate

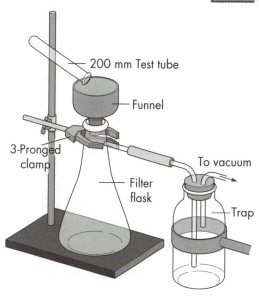

Figure 9.2
Transfer of the precipitate to the filter funnel

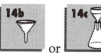

4. The BaSO₄ precipitate may either be gravity filtered or vacuum filtered. In either case, use fine porosity filter paper, such as Whatman No. 42 or Fisher*brand* Q2. Premeasure the mass (±0.001 g) of the dry filter paper and seal it into the funnel with a few milliliters of deionized water. Have your instructor approve your filtering apparatus.

5. While the solution is still warm, quantitatively transfer the precipitate to the funnel (Figure 9.2). Rinse any precipitate from the test tube wall with warm deionized water, swirl, and transfer the precipitate onto the filter. Wash the precipitate on the filter paper with two 5 mL portions of warm deionized water.

---

[1]The purpose of this step is to digest the precipitate; that is, by allowing the precipitate to maintain an equilibrium with its ions in solution for an extended period of time, larger aggregates of particles will form, making the filtration of the precipitate more efficient.

6. Dry the precipitate on the filter paper; either dry in a constant temperature drying oven set at 110°C or until the next laboratory period. Meanwhile, proceed to Part B.

Measure and record the mass of the filter paper and precipitate.

B. Determination of the Excess (and the Limiting) Reactant

The limiting reactant in the salt mixture is determined in the following tests. Use the two 10 mL volumes collected in Part A.3. See Figure 9.3.

1. **Testing for excess Ba^{2+}.** Add 2 drops of 0.5 M SO_4^{2-} ion (from 0.5 M Na_2SO_4) to the 10 mL of solution in tt1. If a precipitate forms, Ba^{2+} is in excess and SO_4^{2-} is the limiting reactant in the original salt mixture.

2. **Testing for excess SO_4^{2-}.** Add 2 drops of 0.5 M Ba^{2+} ion (from 0.5 M $BaCl_2$) to the other 10 mL of solution in tt2. If a precipitate forms, SO_4^{2-} is in excess and Ba^{2+} is the limiting reactant in the original salt mixture.

Dispose of the $BaSO_4$ precipitate in the "Solid Salts" container and the test solutions in the "Waste Salts" container.

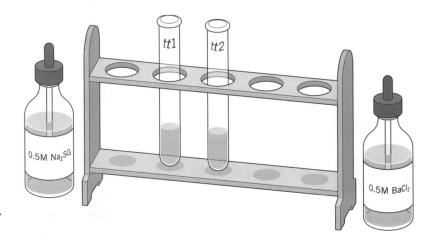

Figure 9.3
Testing for the excess (and the limiting) reactant

NOTES, OBSERVATIONS, AND CALCULATIONS

◇ Limiting Reactant

Date _____Name _____ Lab Sec. _____Desk No. _____

1. If 0.468 g Na_2SO_4 is mixed with an excess of $BaCl_2 \cdot 2H_2O$ to form 50 mL of solution, what mass of $BaSO_4$ can form?

2. A 0.268 g sample of Na_2SO_4 is mixed with 0.402 g $BaCl_2 \cdot 2H_2O$ to form 50 mL of solution.

    a. Calculate the moles of $BaCl_2 \cdot 2H_2O$ and the moles of Na_2SO_4 in the mixture.

    b. What is the limiting reactant for the precipitation of $BaSO_4$?

    c. What mass of $BaSO_4$ can theoretically form?

3. How is the test for the presence of excess $BaCl_2 \cdot 2H_2O$ in your unknown salt mixture conducted in today's experiment?

4. Describe the technique for transferring a precipitate from a beaker to a funnel.

5. A 0.782 g sample of a $BaCl_2 \cdot 2H_2O/Na_2SO_4$ salt mixture, mixed with water, filtered, and dried, produces 0.503 g of $BaSO_4$. The Na_2SO_4 salt is determined to be the limiting reactant.

    a. Calculate the mass of Na_2SO_4 in the mixture.

    b. Calculate the mass of $BaCl_2 \cdot 2H_2O$ in the mixture.

    c. What is the percentage of Na_2SO_4 and $BaCl_2 \cdot 2H_2O$ in the salt mixture?

◇ Limiting Reactant

Date _____ Name _____ Lab Sec. _____ Desk No. _____

A. Precipitation of $BaSO_4$

Unknown Number _____

	Trial 1	Trial 2
1. Mass of unknown salt mixture (g)		
2. Mass of filter paper (g)		
3. Instructor's approval of filtering apparatus		
4. Mass of filter paper and precipitate (g)		
5. Mass of $BaSO_4$ precipitate (g)		

B. Determination of the Excess (and the Limiting) Reactant

1. Limiting reactant in original salt mixture _____

2. Excess reactant in original salt mixture _____

Calculations

	Trial 1	Trial 2
1. Moles of $BaSO_4$ precipitate (mol)		
2. Moles of $BaCl_2 \cdot 2H_2O$ reacted (mol)		
3. Mass of $BaCl_2 \cdot 2H_2O$ reacted (g)		
4. Moles of Na_2SO_4 reacted (mol)		
5. Mass of Na_2SO_4 reacted (g)		
6. Mass of salt mixture (from A.1) (g)		
7. Mass of excess reactant (g)		
8. Percent $BaCl_2 \cdot 2H_2O$ in mixture		
9. Percent Na_2SO_4 in mixture		

QUESTIONS

1. Because $BaSO_4$ is a very finely divided precipitate, some is lost in the filtering process. If coarse filter paper is used instead of one with fine porosity, will the reported percent of limiting reactant in the original salt mixture be high or low? Explain.

2. The solubility of $BaSO_4$ at 25°C is 9.04 mg/L. How many grams and moles of $BaSO_4$ dissolve in the 50 mL of solution?

3. How do excessive quantities of wash water affect the amount of $BaSO_4$ collected on the filter (see Part A.5)? Explain.

*4. a. Sodium carbonate, Na_2CO_3, is an inherent contaminant of Na_2SO_4. How does its presence affect the expected mass of $BaSO_4$ reported? $BaCO_3$ is insoluble, as is $BaSO_4$.

    b. Will this cause the reported mass percent of limiting reactant to be high or low? Explain.

Experiment 10

◇ Calorimetry

Nearly all chemical and physical changes that occur in the laboratory or in nature are accompanied by a change in energy, most often in the form of thermal energy or heat. Energy changes accompany the changing colors of the leaves in the Fall, the growing of a child to an adult, the blinking of an eye, and the corrosion of a metal; these examples illustrate energy changes that accompany *chemical* reactions. The melting of ice, the release of a ball toward the Earth, and the cooling of magma (see photo) are energy changes accompanying *physical* changes. Energy and energy changes, then, are very important for understanding of natural phenomena. Early scientists considered energy to be one of the four basic elements of nature, the other three being Earth, air, and water. Energy was needed to make changes from one basic element to another.

- To determine the specific heat capacity of a metal
- To measure the heat of neutralization for a strong acid–strong base reaction
- To measure the heat evolved or absorbed for the dissolution of a salt

OBJECTIVES

PRINCIPLES

Chemical and physical changes which evolve energy are exothermic and have negative ΔH values (**enthalpy change**) whereas changes that absorb energy are endothermic and have positive ΔH values. Heat changes (q) equal enthalpy changes when the chemical and physical changes occur in a constant pressure **calorimeter**. When heat is evolved by the system (exothermic) ΔH *and* q are negative; when heat is absorbed by the system (endothermic) ΔH and q are positive. The calorimeter and the systems of study therein for this experiment are at constant pressure.

Enthalpy change: the energy change accompanying a chemical or physical change at constant pressure.

Calorimeter: an apparatus used to measure heat changes for chemical and physical changes.

Three calorimetric measurements are made in this experiment. Each requires the measurement of temperature changes, masses of materials, and the plotting of the data.

The energy (or heat, q, in joules) required to change the temperature of 1 g of a substance by 1°C is the **specific heat capacity** (commonly, specific heat) of that substance. In the laboratory the specific heat capacity for a substance can be determined from the data of heat, mass, and temperature change by the equation

Specific Heat Capacity of a Metal

$$\text{specific heat capacity } (c_p) = \frac{\text{heat(joules)}}{\text{mass(grams) } \times \Delta T(°C)}$$

A rearrangement of the equation, and substituting the appropriate notations for heat (q), specific heat capacity (c_p), and mass (m), is

$$q = c_p \times m \times \Delta T$$

The temperature change of the substance, ΔT, equals the temperature *after* the change minus the temperature *before* the change, $T_{final} - T_{initial}$. The specific heat capacity for most substances changes only slightly with temperature; we will assume that it remains constant over the temperature range used in this experiment. The specific heat capacity of H_2O is 4.18 J/(g•°C).

The specific heat capacity of a metal (that is nonreactive with water) is measured by first heating a known mass of the metal to a measured temperature. The metal is then placed into a calorimeter containing a known mass of (cooler) water also at a measured temperature. The heat from the metal is transferred to the water until the metal and water reach the same temperature (called thermal equilibrium). This final equilibrium temperature is recorded. Expressed in equation form

$$-q \text{ [heat(J) lost by metal]} = +q \text{ [heat(J) gained by } H_2O]$$

A substitution of the specific heat capacity equation given above for heat in this equation gives

$$-c_{p, \text{metal}} \times m_{\text{metal}} \times \Delta T_{\text{metal}} = c_{p, H_2O} \times m_{H_2O} \times \Delta T_{H_2O}$$

The specific heat capacity of the metal is calculated from a rearrangement of the equation,

$$c_{p, \text{metal}} = -\frac{c_{p H_2O} \times m_{H_2O} \times \Delta T_{H_2O}}{m_{\text{metal}} \times \Delta T_{\text{metal}}}$$

Each equation assumes no heat loss to the calorimeter.

Heat (Enthalpy) of Neutralization for an Acid-Base Reaction

The reaction of a strong acid with a strong base produces water and heat as products:

$$H^+(aq) + OH^-(aq) \rightarrow H_2O(l) + \text{heat}$$

The heat of neutralization, q_n, is calculated by (a) assuming the densities and specific heat capacities of the acidic and basic solutions are the same as that for water and (b) measuring the temperature change when the two solutions are mixed.

$$q_n = -c_{p H_2O} \times m_{\text{acid + base}} \times \Delta T_{\text{solution}}$$

The enthalpy of neutralization, ΔH_n, and q_n are generally expressed in units of kJ/mol of acid (or base) reacted.

Heat (Enthalpy) of Solution for a Salt

The dissolving of a salt may be an endothermic or exothermic process depending upon two factors often considered in the dissolution of the salt: the lattice energy of the salt and the hydration energy of the ions. The **lattice energy** is the energy required (endothermic process, a $+\Delta H_{LE}$) to vaporize 1 mole of the solid salt into its gaseous ions; the **hydration energy** is the energy released (exothermic quantity, a $-\Delta H_{hyd}$) when the gaseous ions are attracted to and surrounded by water molecules. The enthalpy of solution of a salt, ΔH_s, is the sum of these two factors (Figure 10.1 for NaCl).

The heat of solution, q_{solution}, is determined by adding the heat losses for the water and salt.

$$q_{\text{solution}} = -q_{H_2O} + -q_{\text{salt}}$$

$$q_{\text{solution}} = -c_{p, H_2O} \times m_{H_2O} \times \Delta T_{H_2O} + -c_{p, \text{salt}} \times m_{\text{salt}} \times \Delta T_{\text{salt}}$$

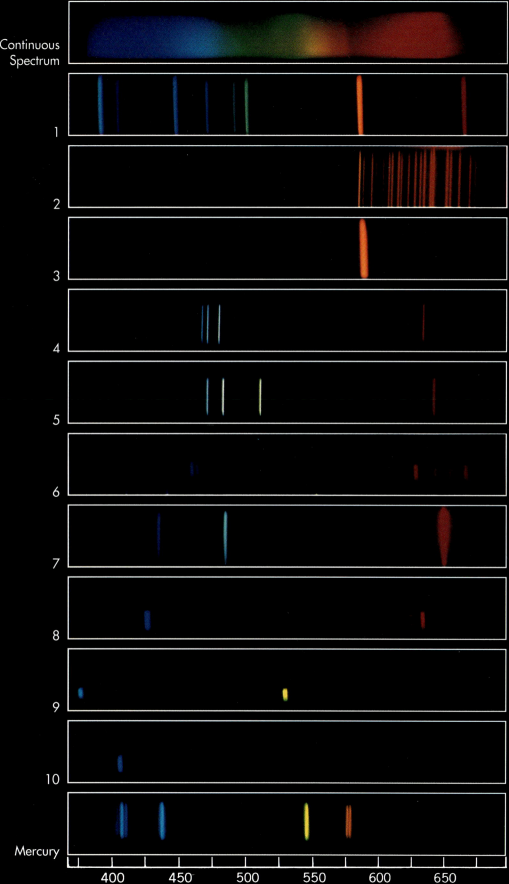

Continuous Spectrum

1

2

3

4

5

6

7

8

9

10

Mercury

400   450   500   550   600   650

The color of universal indicator varies with pH

A color wheel. A sample that absorbs one color transmits the complementary color (across the color wheel).

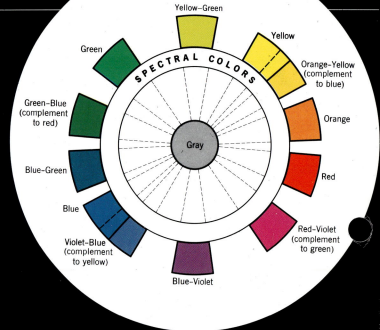

SPECTRAL COLORS

Yellow–Green
Yellow
Green
Orange-Yellow (complement to blue)
Green-Blue (complement to red)
Orange
Gray
Blue-Green
Red
Blue
Red–Violet (complement to green)
Violet-Blue (complement to yellow)
Blue-Violet

The phenolphthalein indicator is commonly used in acid-base titrations.

Phenolphthalein indicator at pH 8.2 (left) and pH 10.0 (right).

Methyl Orange indicator at pH 3.2 (left) and pH 4.40 (right).

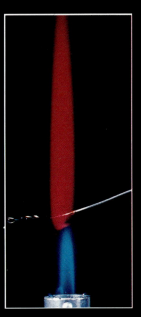
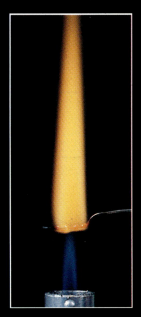

The colors of (a) lithium, (b) sodium, and (c) potassium salts in a Bunsen burner flame.

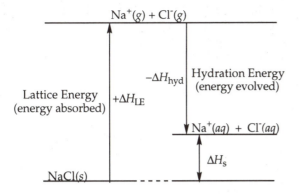

Figure 10.1
Energy changes for the dissolving of NaCl in water

Because the heat of solution, q_s, is generally expressed in units of joules per gram salt (J/g), then

$$q_s = \frac{q_{solution}}{m_{salt}}$$

The specific heat capacities of some common salts are listed in Table 10.1.

Table 10.1. Specific Heat Capacities of Some Salts

Salt	Formula	Specific Heat Capacities (J/g•°C)
ammonium chloride	NH_4Cl	1.57
ammonium nitrate	NH_4NO_3	1.74
ammonium sulfate	$(NH_4)_2SO_4$	1.409
sodium hydroxide	$NaOH$	1.49
sodium sulfate	Na_2SO_4	0.903
sodium thiosulfate pentahydrate	$Na_2S_2O_3 \cdot 5H_2O$	1.45
potassium bromide	KBr	0.435
potassium chloride	KCl	0.688
potassium hydroxide	KOH	1.16
potassium nitrate	KNO_3	0.95

PROCEDURE

Ask your instructor which parts of the experiment you are to complete. You and a partner are to complete two trials for each part assigned. Obtain two styrofoam cups, a lid, and a 110°C thermometer will serve as your calorimeter. In this experiment do *not* use the thermometer as a stirring rod!! Instead, use a coffee stirrer or copper wire as the stirrer.

1. a. Ask the instructor for a ≈10 g sample of an unknown metal. Measure its mass (±0.01 g) in a *dry*, previously determined (±0.01 g) 200 mm test tube.

    b. Place the 200 mm test tube in a 400 mL beaker filled with water well above the level of the metal sample in the test tube (Figure 10.2).

    c. Heat the water to boiling and maintain this temperature for at least 5 minutes in order for the metal to reach **thermal equilibrium** with the boiling water. Measure the temperature of the water (±0.1°C).

A. Specific Heat Capacity of a Metal

Thermal equilibrium: the temperature of two objects in contact is the same.

2. a. Set up the calorimeter (Figure 10.3). Using a graduated cylinder, add approximately 25.0 mL (±0.2 mL) of water to the calorimeter. Secure the thermometer with a small three-pronged clamp. Be certain the thermometer bulb is below the water level.

b. Read and record the temperature (±0.2°C) of the water in the calorimeter.

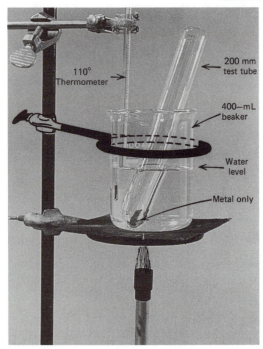

Figure 10.2
The placement of the metal below the level of the water in the beaker

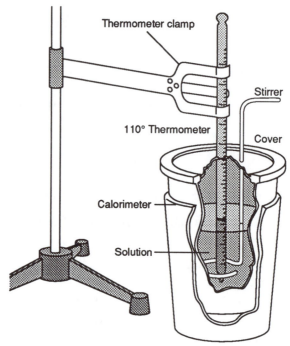

Figure 10.3
The setup of a calorimeter

3. a. Once thermal equilibrium has been reached in Parts A.1 and A.2, remove the test tube from the boiling water; quickly transfer the metal to the calorimeter. Be careful not to break the thermometer and not to splash water from the calorimeter. Replace the lid and swirl gently.

b. Read and record the temperature (±0.2°C) of the mixture in the calorimeter at timed intervals (every 10–20 s) on the table on the Data Sheet.

4. Repeat the experiment for Trial 2.

Return your metal to the appropriately marked container for the unknown metals.

5. Plot on linear graph paper, temperature (ordinate) *vs.* time (abscissa) (Figure 10.4). The maximum temperature reached by the mixture occurs at the intersection of two lines: a straight line drawn perpendicular to the initial temperature/time line at the time when the metal is added to the calorimeter and the best straight line drawn through the points after the recorded maximum temperature is reached.[1] Have the instructor approve your graph.

---

[1]The maximum temperature is never actually measured because heat is always being transferred to or from the wall of the calorimeter.

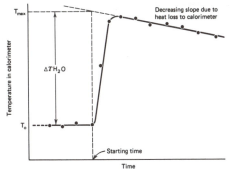

1. a. Clean and dry the calorimeter. Use a graduated cylinder to transfer 50.0 mL (±0.2 mL) of standard[2] 1.0 M NaOH solution into the calorimeter, secure and insert the thermometer through the lid into the solution.

    b. Record the temperature (±0.2°C) of the NaOH solution. Record the exact concentration of the NaOH solution on the Data Sheet.

2. a. Measure 50.0 mL (±0.2 mL) of 1.1 M HCl in a clean, graduated cylinder. Rinse the thermometer to remove any NaOH solution and then measure the temperature of the HCl solution. The NaOH and the HCl solutions should be at the same temperature.

    b. Carefully but quickly, add the acid to the base, replace the lid, and swirl gently.

3. Record the solution temperature (±0.2°C) at timed intervals on the Data Sheet. Plot the data on linear graph paper and interpret the maximum temperature, as obtained in Part A.4. Have the instructor approve your graph.

4. Repeat the experiment for a second trial.

1. Clean and rinse your calorimeter with deionized water. Add approximately 25.0 mL (±0.2 mL) of deionized water to the calorimeter and record its temperature (±0.2°C).

2. Measure about 5.0 g (±0.01 g) of your assigned salt on weighing paper. Add the salt to the calorimeter, replace the lid, and swirl gently. Measure and record the temperature at timed intervals on the Data Sheet.

3. Plot the data on linear graph paper and interpret the maximum temperature *change* (Part A.4). Have the instructor approve your graph.

4. Repeat the experiment for a second trial.

B. Heat (Enthalpy) of Neutralization for an Acid-Base Reaction

C. Heat (Enthalpy) of Solution for a Salt

---

[2] A standard solution is one in which the concentration of the solute has been very accurately determined.

NOTES,
OBSERVATIONS,
AND
CALCULATIONS

This is a plain jacket bomb calorimeter that can be used for measuring the heat of combustion (at constant volume), ΔH_{comb}, for any solid or liquid fuel and the caloric value of the various foodstuffs found in diet books.

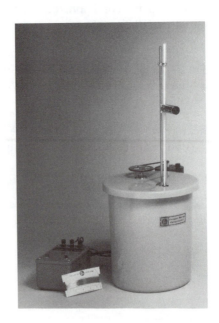

◇ Calorimetry

Date _____Name _____ Lab Sec. _____Desk No. _____

1. A metal of mass 13.11 g and at 81.0°C is transferred to a calorimeter containing 25.0 mL of deionized
    water at 25.0°C. The final equilibrium temperature is 30.0°C. What is the specific heat capacity of the
    metal? The specific heat capacity of H_2O is 4.18 J/(g•°C). Assume the density of water to 1.00 g/mL.

2. A 4.50 g sample of a salt dissolves in 30.0 mL of water initially at 25.0°C. The final equilibrium
    temperature is 18.0°C. What is the enthalpy of solution (J/g) of salt? The specific heat of the salt is
    0.692 J/(g•°C).

3. In Part B, excess moles of HCl are added to the NaOH solution for the neutralization reaction. Why is
    this procedure preferred rather than adding a 1:1 mole ratio according to the balanced equation?

4. Will the recorded temperature change for an exothermic reaction performed in a glass calorimeter be
    greater or less than that for today's styrofoam "coffee cup" calorimeter? Assume glass to be a better
    conductor of heat than styrofoam.

5. a. How can "bumping" be avoided when water is being heated in a beaker?

    b. When determining the volume of solution in a graduated cylinder, you should always read the

        _____ of the meniscus.

    c. A balance with _____ g sensitivity is used in today's experiment.

    d. The margin of error in reading the thermometer in today's experiment is ± _____ °C.

    e. In today's experiment, you are *not* to use a thermometer as a stirring rod. Explain.

◇ Calorimetry

Date _____ Name _____ Lab Sec. _____ Desk No. _____

A. Specific Heat Capacity of a Metal

Unknown number _____

		Trial 1	Trial 2
1.	a. Mass of test tube + metal (*g*)		
	b. Mass of test tube (*g*)		
	c. Mass of metal (*g*)		
	d. Temperature of metal (°C)		
2.	a. Volume of water in calorimeter (*mL*)		
	b. Mass of water (*g*) Assume density of water is 1.0 g/mL		
	c. Temperature of water (°C)		
3.	a. Instructor's approval of graph		
	b. Maximum temperature of metal and water from graph (°C)		

Calculations

		Trial 1	Trial 2
1.	Temperature change of water, ΔT (°C)		
2.	Heat gained by water (*J*)		
3.	Temperature change of metal, ΔT (°C)		
4.	Specific heat capacity of metal (*J/g*°C)		
5.	Average specific heat capacity of metal (*J/g*°C)		

B. Heat (Enthalpy) of Neutralization for an Acid-Base Reaction

1.	a. *Exact* volume of NaOH solution (*mL*)		
	b. Initial temperature of NaOH solution (°C)		
	c. Concentration of NaOH solution (*mol/L*)		
2.	a. Volume of HCl solution (*mL*)		
	b. Initial temperature of HCl solution (°C)		
3.	a. Instructor's approval of graph		
	b. Maximum final temperature of mixture from graph (°C)		

Calculations

	Trial 1	Trial 2
1. Average of initial temperatures (NaOH + HCl) ($°C$)		
2. Temperature change, ΔT ($°C$)		
3. Mass of final mixture (g) Assume density of mixture is 1.0 g/mL		
4. Specific heat capacity of solution	4.18 J/g•°C	
5. Heat evolved (J)		
6. Moles of OH^- reacted (limit. reactant) (mol)		
7. Moles of H_2O formed (mol)		
8. Heat evolved per mole of water ($kJ/mol\ H_2O$)		
9. Average q_n ($kJ/mol\ H_2O$)		

C. Heat (Enthalpy) of Solution for a Salt

Salt _____

	Trial 1	Trial 2
1. a. Volume of water (mL)		
b. Mass of water (g). Assume density of H_2O is 1.0 g/mL		
c. Initial temperature of water ($°C$)		
2. Mass of salt (g)		
3. a. Instructor's approval of graph		
b. Maximum (or minimum) temperature of solution from graph ($°C$)		

Calculations

	Trial 1	Trial 2
1. Change in temperature of solution, ΔT ($°C$)		
2. Heat change of water, q_{H_2O} (J)		
3. Heat change of salt, q_{salt} (J). See Table 10.1		
4. Heat of solution, $q_{solution}$ (J)		
5. q_s of salt (J/g)		
6. Average q_s of salt (J/g)		

Specific Heat Capacity of a Metal				Heat (Enthalpy) of Neutralization for an Acid-Base Reaction				Heat (Enthalpy) of Solution for a Salt			
Trial 1		Trial 2		Trial 1		Trial 2		Trial 1		Trial 2	
Temp	Time	Temp	Time	Temp	Time	Temp	Time	Temp	Time	Temp	Time

QUESTIONS

1. The coffee cup calorimeter, although a good insulator, absorbs some heat when the chemical system is above room temperature.

    a. How does this heat loss affect the experimental value of the specific heat capacity of the metal?

b. How does this heat loss affect the reported q_n value in the acid-base reaction?

2. Suppose that when the metal was transferred to the calorimeter in Part A, some of the water splashed from the calorimeter. How does this "sloppiness" affect the reported specific heat capacity of the metal? Explain.

3. The specific heat capacity of styrofoam is 1.34 J/g•°C. If we assume that the entire inner cup of the calorimeter reaches thermal equilibrium with the solution, how much heat is lost to the calorimeter in Part B? Assume the inner cup has a mass of 2.35 g.

4. If the maximum *recorded* temperature in Part B is used to determine ΔT rather than the maximum temperature extrapolated from the plotted data, will the reported q_n be greater or less than the actual q_n for this reaction? Explain.

Experiment 11

◇ Chemical Literature

What is the boiling point of nitrogen? What is the (bond) distance between the hydrogen atom and the oxygen atom in water? Tritium, ^3H, is a radioactive isotope of hydrogen—what is its half-life? What other isotopes of hydrogen exist? Where can I learn more about molecular structure? Phenolphthalein is an acid-base indicator—how is the solution prepared? Where can I find the most recent information about acid rain? These, and many, many others, are common questions encountered by scientists. The information is available somewhere, but, specifically, where?

Research data has been recorded in the chemical literature at an ever increasing rate since the advent of the printing press and with the realization that the use of accumulated data can expedite the interpretation of current research or experimentation and that this data can provide insight to further research endeavors. For example, a cryogenic chemist need *not* measure the boiling point of oxygen, nitrogen, and other air components every time air is to be liquefied for an experiment. Rather the boiling point data of each principal air component can be retrieved from the chemical literature and then the air sample can knowingly be cooled to a lower temperature for liquefaction.

OBJECTIVES

- To become familiar with common sources of chemical data
- To search the chemical literature to obtain specific chemical data
- To search the chemical literature for general information

PRINCIPLES

Anyone who plans to become a professional chemist or plans to use chemistry in an allied field, such as physics, biology, agriculture, engineering, or a health profession, needs to be familiar with the sources of chemical literature, what is available as chemical data, and how to extract the information in a most efficient manner.

Several handbooks of data and an assortment of chemical journals are known to all practicing chemists. But as your interests in chemistry and science increase and as your need for up-to-date research information and data become more specialized, additional sources of information will be needed and used.

Most libraries use the Library of Congress (LC) classification system for cataloging library materials. In the LC system all subject areas are divided into disciplines according to a letter of the alphabet. Those of major significance to chemists, scientists, and engineers are:

Q	Science	S	Agriculture	Z	Library Science: Information
R	Medicine	T	Technology		and retrieval systems

A second letter is then added to further subdivide the discipline. In the "Q" series, the second letter designations indicate:

QA	Mathematics		QK	Botany
QB	Astronomy		QL	Zoology
QC	Physics		QM	Human Anatomy
QD	**Chemistry**		QP	Physiology
QE	Geology		QR	Bacteriology
QH	Natural History (incl. Biology)			

Some libraries retain the Dewey Decimal Classification (DDC)while others retain the Universal Decimal Classification (UDC) to catalog their library holdings. Table 11.1 shows the correlation of these two systems of cataloging library holdings with the LC system.

Table 11.1. Correlation of the Library of Congress (LC), Dewey Decimal (DDC), and Universal Decimal (UDC) Classification Systems of Library Materials

LC		DDC	UDC
Q	Pure Science	500	50
QA	Mathematics	510	51
QB	Astronomy	520	52
QC	Physics	530	53
QD	**Chemistry**	**540**	**54**
QE	Geology	550, 560	55, 56
QH	Natural History (incl. Biology	570	57
QK	Botany	580	58
QL	Zoology	590	59
QM	Human Anatomy	610	
QP	Physiology	612	
QR	Bacteriology	616, 636	

Sources of Chemical Information

Chemical information and data can be obtained from primary, secondary, or tertiary sources. Primary sources of chemical information and data are obtained from original journal articles published by the researcher. Primary sources are often the most recent data; they are also subjected to critical peer review and the conditions for the securing of the data are detailed. Professional chemical societies and special interest groups publish these journals which present articles that are on the "cutting-edge" of research in the scientist's field of expertise.

Chemical Abstracts is the most comprehensive source in finding information on a chemistry topic or author. Concise abstracts of not only original journal articles but also conference proceedings, theses, reports and books, and patents from over 14,000 publications (about 250 of which are abstracted cover to cover) from more than 150 countries are listed and summarized weekly by the Chemical Abstracts Service of the American Chemical Society (ACS). The American Chemical Society alone publishes more journals than any other professional society in the world. The list of ACS publications are in Table 11.2.

Secondary sources are compilations of data that are published throughout the scientific literature and are recorded in **handbooks** of various names and purpose. Perhaps the most common of the handbooks are the Chemical Rubber Company's (CRC) *Handbook of Chemistry and Physics* and the *Merck Index*. Secondary sources may not report the best available (or most recent) data and do not report the detailed experimental conditions by which the data were obtained. However, for most purposes, secondary sources are

reliable, especially if an effort is made by the publisher to continue to update the handbook of data.

Table 11.2. Professional Journals of the American Chemical Society

Accounts of Chemical Research	Journal of the American Chemical Society
Analytical Chemistry	
Biochemistry	Journal of Chemical and Engineering Data
Bioconjugate Chemistry	
Biotechnology Progress	Journal of Chemical Information & Computer Science
Chemical Research in Toxicology	
Chemical Reviews	Journal Of Medicinal Chemistry
Chemistry of Materials	Journal of Organic Chemistry
ChemTech	Journal of Pharmaceutical Sciences
Energy & Fuels	Journal of Physical Chemistry
Environmental Science & Technology	Langmuir
Industrial & Engineering Chemical Research	Macromolecules
	Organometallics
Inorganic Chemistry	Journal of Chemical Education
Journal of Agriculture & Food Chemistry	

Tertiary sources are the least specific with regard to date of reported data, the experimental conditions used for securing the data, and the peer review process. Textbooks, general science books, dictionaries, and encyclopedias are considered tertiary sources. Encyclopedic sources include:

- Encyclopedia of Chemical Technology, John Wiley & Sons, Inc., New York
- McGraw-Hill Encyclopedia of Science and Technology, McGraw-Hill, New York
- Scientific Encyclopedia, Van Nostrand Reinhold, New York

On-Line Sources

Computerized, on-line searches of the chemical literature are currently the most expeditious method to obtain the scientific information for your needs. A computer search system with the appropriate software is required to access a telecommunications network in order to conduct an on-line search. Nearly 100 databases of interest to chemists are accessible through the network—the most valuable and comprehensive is CAS (Chemical Abstracts Service) Online. Most databases are available through vendors (which charge relatively large fees), others can only be accessed directly from the producer, such as CAS Online from the Chemical Abstracts Service. Generally, a science library has the hardware and software to access the various databases, including CAS Online.

But, like using any computer system or computer network, it is no smarter than the operator. To intelligently and effectively use the telecommunications network to access databases and "cutting-edge" chemical and scientific research information, it is important that you know the specifics of your search so as to minimize on-line time—generally charges are assessed for access to the database and for the "connect time" to the database. If you are interested in knowing more about on-line searches of the scientific literature, consult with your science librarian.

PROCEDURE

The information required for the completion of this experiment is found in your campus library either in the main library or in the chemistry/science library. The procedure is described as if your library had adopted the LC

System for classifying library materials. Your instructor will make appropriate assignments from the suggestions in each exercise.

A. Primary Sources

1. Find and write instructions for locating the "Q" and "QD" sections of your library. Where are the *current* issues of chemical journals located? Most likely these issues are *not* hard bound or "in the stacks."

2. Write the LC number for the journal (below) to which your assigned. Complete the corresponding requested information on the Data Sheet.
   a. Analytical Chemistry
   b. Biochemistry
   c. Canadian Journal of Chemistry
   d. Environmental Science & Technology
   e. Inorganic Chemistry
   f. Journal of the American Chemical Society
   g. Journal of Chemical Education
   h. Journal of Chemical Information & Computer Science
   i. Journal of the Chemical Society
   j. Journal of Chemical Physics
   k. Journal of the Electrochemical Society
   l. Journal of Organic Chemistry

B. Secondary Sources

1. Use the CRC *Handbook of Chemistry and Physics* to obtain the information requested on the Data Sheet for your compound (below):

   a. aluminum bromide
   b. aluminum orthophosphate
   c. ammonium iodate
   d. ammonium nitrate
   e. boric acid, ortho-
   f. cadmium iodide
   g. calcium hydroxide
   h. calcium sulfate dihydrate
   i. copper(I) bromide
   j. gold(I) chloride
   k. nitrous oxide
   l. sodium chloride

2. Use the CRC *Handbook of Chemistry and Physics* to obtain the information requested, as assigned by your instructor:
   a. Origin of the names for the elements beryllium, holmium, iodine, and copper
   b. Formulas and the solubility product constants for cadmium iodate, lead carbonate, mercury(I) chloride, and zinc sulfide
   c. Percent natural abundance and half-life of ^{40}K
   d. Formula, molar mass, physical state, and melting point of hexamethylenediamine (used in the manufacture of nylon)
   e. Thermodynamic value of $\Delta H_f°$ at 25°C for the assigned compound in B.1 above
   f. Human exposure limit to nitrogen dioxide
   g. Metal compounds that appear scarlet in a flame test
   h. Carbonate salts that are *in*soluble in water, but soluble in acids
   i. Approximate pH range of dill pickles
   j. Standard reduction potential for $Np^{4+} + e^- \rightarrow Np^{3+}$
   k. Density of garnet in g/cm^3 and lb/ft^3
   l. Definition of a poise. Where is it used?
   m. What is the S–S bond length and bond strength in S_2?
   n. What is the S–O bond length and bond angle in SO_2?
   o. Describe the procedure for preparing a phenolphthalein solution.

3. Use a recent edition of the *Merck Index* to provide the information requested on the Data Sheet for your assigned compound, called Merck monograph numbers for the listed compounds:

a. 2	e. 1717	i. 5886
b. 281	f. 2251	j. 7692
c. 901	g. 3389	k. 8881
d. 1089	h. 4090	l. 9311

C. Literature Search and Report

A search of the current literature for information related to a general topic in chemistry, science, and engineering can begin with a science/engineering index, much like the *Reader's Guide to the Periodical Literature* is used to search current events. Several such publications are available, most common of which are:

- *General Science Index*, H. W. Wilson Co., New York
- *Applied Science and Technology Index*, H. W. Wilson Co., New York

Both of these indices are published on a regular basis and cite current articles related to most any area of scientific interest.

You will be assigned or you may be asked to select a topic of personal interest. The degree to which you will use the scientific literature will depend upon the assignment of your instructor, anything from a partial bibliography of the topic to a complete research paper having a designated length.

A research paper that includes cutting-edge chemical research information may require the use of Chemical Abstracts to search the chemical literature or the CAS Online services in your library. If either of these sources are required, you will need to receive additional instructions from your instructor or science librarian.

NOTES

Chemical Literature

Date _____ Name _____ Lab Sec. _____ Desk No. _____

1. Where is the chemical literature located on your campus? Be specific as to the building, floor, wing, stack, etc.

2. Browse your library. Does your library use the LC, DDC, or UDC system for cataloging library materials?

3. Use your dictionary to define the following chemical terms:

   a. chemistry

   b. atom

   c. salt

   d. phosphate

   e. boil

◇ Chemical Literature

Date _____ Name _____ Lab Sec. _____ Desk No. _____

A. Primary Sources

1. Instructions for locating the "Q" and "QD" sections of your library.

    Location of most recent (current, unbound) journals.

2. Name of journal assigned: _____ LC No. _____

    Publisher: _____

    Country in which the journal is published: _____

    Frequency of publication: _____

    Volume number of first issue in your library _____

    List the title, author and affiliation of the first journal article in the most recent issue of the journal:

B. Secondary Sources

1. Write the bibliographical data and LC number for the CRC *Handbook of Chemistry and Physics* that your are using.

Compound(s) Assigned	Formula	Molar Mass	Crystalline Form	Melting Point	Solubility in Cold Water

2. Assigned information: _____
   Data:

3. Write the bibliographical data and LC number for the *Merck Index* that your are using.

Merck Monograph No.   _____

Listed Name   _____

Molecular Formula: _____

Molar mass   _____

Property (list one)   _____

Use or therap cat (list one):   _____

C. Literature Search and Report

Present an overview of your assignment that requires the use of the chemical literature.

Experiment 12

◆ Spectroscopy

The interaction of light with matter is a phenomenon that is intriguing to most everyone. Think about the effects of public lighting on the colors of skin tones, clothes, and automobiles. Mercury vapor lights tend to make objects appear blue, whereas the sodium vapor lights (see photo) emit a yellow-orange hue. What this does to affect other visible colors is considered "strange!" For example, a red automobile appears gray in the light of a low pressure sodium vapor lamp.

Sunlight is considered "white light" in that all colors of the visible spectrum are emitted. When sunlight strikes an object that absorbs none of the visible light, the object appears white, but if the object absorbs all of the visible light, then the object appears black. If the object absorbs only the blue light from the sunlight, it then appears yellow, the complimentary color of blue in the color wheel (see color plate). How and why atoms or molecules of an object absorb or do not absorb sunlight is the basis of this experiment.

- To account for the appearance of line spectra
- To observe the flame emission colors from excited metal atoms
- To study the hydrogen spectrum
- To identify an element from its spectrum

OBJECTIVES

PRINCIPLES

The current model of the atom includes a nucleus, comprised of protons and neutrons, surrounded by electrons. From experimental observations of atomic absorption and emission spectra, and from a quantum mechanical interpretation of spectra, the energy of an electron in an atom is assumed to be **quantized**; that is, an electron in an atom has only discrete energies.

Photon: a quantity of energy having mass-like properties.

When an atom absorbs energy from a flame or electric discharge, it absorbs the energy necessary to excite one of its electrons to a higher energy state— the atom is now in an excited state. When the electron returns to its original (ground) energy state, it emits that absorbed energy in the form of one or more **photons**. The excited atom decreases in energy by ΔE_{atom} when the electron returns from the higher, E_h, to a lower, E_l, energy state. The energy change of the atom from this one electron de-excitation equals the energy of one photon.

$$\Delta E_{atom} = E_h - E_l = E_{photon}$$

The photon energy, E_{photon}, equal to the energy difference between the two energy states for the electron, is inversely proportional to its wavelength, λ, by the equation,

$$E_{photon} = h \frac{c}{\lambda} = h\nu$$

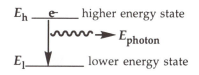

c is the speed of light, 3.00×10^8 m/s, h is Planck's constant, 6.63×10^{-34} J•s/photon, ν is the frequency of the photon in s^{-1} (or Hertz, Hz), and λ is the wavelength of the photon in meters.

Many possible energy states exist for an electron in an atom. For example, suppose that the same electron in a large number of atoms of an element is excited to the same energy state. As the electrons of the various atoms cascade to the ground state (the lowest energy state), each according to its own pathway, photons of different energies are emitted; the pathway for the de-excitation of the electron depends on the stability of the intermittent energy states of the atom. When these photons pass through a prism, they produce a **line spectrum** (Figure 12.1); each line results from a particular electron transition (and with a particular wavelength and energy) in a large number of atoms.

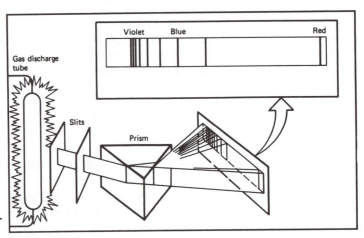

Figure 12.1
A line spectrum

Each element has its own line spectrum because the energy states for electrons in its atoms are unique. Sodium having 11 electrons has a different set of electronic energy states, and thus exhibits a different line spectrum, than does calcium (with 20 electrons) or mercury (with 80 electrons). The most prevalent mode of electron de-excitation produces the most intense line in the spectrum, producing the color characteristic of that element. For example, many of the electrons in excited sodium atoms commonly de-excite to produce a photon with a wavelength in the orange region of the visible spectrum. We, therefore, associate excited sodium atoms as being orange which explains the yellow-orange hue emitted by sodium vapor lamps.[1] By the same account, we associate blue light with mercury vapor lamps (a de-excitation of electrons in excited mercury atoms emit photons with a wavelength in the blue region of the spectrum) and green with excited barium atoms. The characteristic colors associated with a number of excited atoms are observed in their flame tests of Part A of this experiment.

In Part B of this experiment, we will determine the wavelengths of the emission lines in the spectra appearing on the color plate. The bottom spectrum of the color plate is that of mercury, known to show emission lines at the wavelengths listed in Table 12.1. A linear wavelength axis will be established for the mercury spectrum to adapt to its known emission lines. The wavelength axis, extended across the color plate, will then establish the position of the emission lines for the remaining emission spectra on the color plate.

---

[1]High and low pressure sodium vapor lamps are now common for lighting streets and highways in public areas.

Table 12.1. The Wavelengths of the Visible Lines in the Mercury Spectrum

Violet	404.7 nm
Violet	407.8 nm
Blue	435.8 nm
Green	546.1 nm
Yellow	577.0 nm
Yellow	579.1 nm

For the hydrogen spectrum, in particular, the wavelengths of the emitted photons will be estimated and the energy changes of the atom associated with the photon emissions calculated.

In Part C, an element is identified by determining the position of its emission lines in the line spectrum relative to the known wavelengths from the mercury line spectrum, Table 12.1. A match of the unknown's spectrum with the known spectral data (Table 12.1) for several elements is made for the identification of the element.

PROCEDURE

A. Flame Tests

1. Dip the end of a platinum or nichrome wire into conc HCl (**Caution**: *Avoid skin contact. Flush affected area with large amounts of water*). Heat the wire in the hottest region of the flame (Figure 2.3) until no visible color (Figure 12.2) appears. Repeat this cleaning procedure as necessary.

2. In a watch glass (or well of a 24 well plate), add a few crystals of $CaCl_2$ to 2–3 drops of conc HCl. Stir. Dip the clean wire into the solution and return it to the flame. Note the color of the flame (view it through a spectroscope, if available). Correlate the color of the flame with its dominate wavelength, using the chart on the Data Sheet. Check the appropriate section.

3. Repeat Steps A.1 and A.2 with $CuCl_2$, NaCl, $BaCl_2$, LiCl, $SrCl_2$, and KCl salts. For KCl view the flame through a cobalt glass plate.[2]

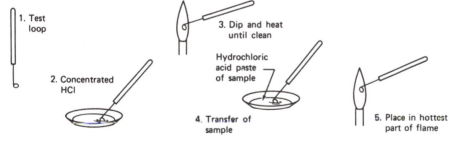

Figure 12.2
The procedure for performing a flame test

B. Hydrogen Spectrum

1. Notice the various emission line spectra on the color plate. A continuous spectrum appears at the top, the line spectra for various elements appear in the middle, and the Hg-spectrum appears at the bottom.

2. a. Calibrate the color plate with the wavelengths of the Hg-spectrum. Establish a wavelength axis across the bottom of the color plate such that the known wavelengths of the emission lines of mercury (Table 12.1) appear at the appropriate position along the axis. A wax marker or

---

[2]A cobalt glass plate filters all wavelengths *except* those emitted by excited K^+ ions.

"permanent" felt tip pen may be required for marking the wavelength scale along the axis.

b. Extend the wavelength markings upward across the color plate.

3. The spectrum marked "7" is the hydrogen spectrum. From your calibrated color plate estimate the wavelengths of its emission lines. Calculate the energy change for an electron of the hydrogen atom associated with the photon emission at each wavelength. Determine the frequency of the photon at each wavelength.

C. Unknown Spectrum

Your instructor will assign one of the remaining emission spectra. Mark and record the wavelengths for the emission lines in that spectrum. Using Table 12.2, identify the element having the shown spectrum.

Table 12.2 Wavelengths and Relative Intensities of the Emission Spectra for Several Elements

	Wavelength (nm)	Relative Intensity		Wavelength (nm)	Relative Intensity		Wavelength (nm)	Relative Intensity		Wavelength (nm)	Relative Intensity
Argon	451.1	100	**Cadmium**	467.8	200	**Neon**	585.2	500	**Rubidium**	420.2	1000
	560.7	35		478.0	300		587.2	100		421.6	500
	591.2	50		508.6	1000		588.2	100		536.3	40
	603.2	70		610.0	300		594.5	100		543.2	75
	604.3	35		643.8	2000		596.5	100		572.4	60
	641.6	70					597.4	100		607.1	75
	667.8	100	**Cesium**	455.5	1000		597.6	120		620.6	75
	675.2	150		459.3	460		603.0	100		630.0	120
	696.5	10000		546.6	60		607.4	100			
	703.0	150		566.4	210		614.3	100	**Sodium**	466.5	120
	706.7	10000		584.5	300		616.4	120		466.9	200
	706.9	100		601.0	640		618.2	250		497.9	200
				621.3	1000		621.7	150		498.3	400
Barium	435.0	80		635.5	320		626.6	150		568.2	280
	553.5	1000		658.7	490		633.4	100		568.8	560
	580.0	100		672.3	3300		638.3	120		589.0	80000
	582.6	150					640.2	200		589.6	40000
	601.9	100	**Helium**	388.9	500		650.7	150		616.1	240
	606.3	200		396.5	20		660.0	150			
	611.1	300		402.6	50				**Thallium**	377.6	12000
	648.3	150		412.1	12	**Potassium**	404.4	18		436.0	2
	649.9	300		438.8	10		404.7	17		535.0	18000
	652.7	150		447.1	200		536.0	14		655.0	16
	659.5	3000		468.6	30		578.2	16		671.4	6
	665.4	150		471.3	30		580.1	17			
				492.2	20		580.2	15	**Zinc**	468.0	300
				501.5	100		583.2	17		472.2	400
				587.5	500		691.1	19		481.1	400
				587.6	100					507.0	15
				667.8	100					518.2	200
										577.7	10
										623.8	8
										636.2	1000
										647.9	10
										692.8	15

NOTES, OBSERVATIONS, AND CALCULATIONS

Spectroscopy

Date _____ Name _____ Lab Sec. _____ Desk No. _____

1. Distinguish between an absorption spectrum and an emission spectrum.

2. The radiation emitted from the de-excitation of an SrCl* molecule, formed in a starburst of a fireworks display, has a wavelength range of 620–750 nm.
   a. Calculate the energy range of these photons.

   b. What color range do these photons have?

3. The wavelength for an electron transition in the hydrogen atom appears at 97.0 nm.
   a. Where in the electromagnetic spectrum (visible, ultraviolet, or infrared) does this line appear?

   b. Calculate the energy of the photon resulting from this electron transition.

c. Determine the frequency of the photon resulting from this electron transition.

4. A large number of hydrogen atoms have electrons excited to the $n_h = 4$ energy state. How many possible spectral lines can appear in the emission spectrum as a result of electrons reaching the ground state ($n_l = 1$)? Remember a spectral line appears when an electron de-excites from a higher to a lower energy state. Diagram all possible pathways for the de-excitation of an electron from $n_h = 4$ to $n_l = 1$.

$n = 4$ _____

$n = 3$ _____

$n = 2$ _____

$n = 1$ _____

5. What dominant color appears for the mercury street bulb? the sodium street lamp? an incandescent bulb?

◇ Spectroscopy

Date _____ Name _____ Lab Sec. _____ Desk No. _____

A. Flame Tests

$\lambda(nm)$	400	450	500	550	600	650	700	750
$CaCl_2$								
$CuCl_2$								
NaCl								
$BaCl_2$								
LiCl								
$SrCl_2$								
KCl								

ultraviolet | violet | blue | green | yellow | orange | red | infrared

B. Hydrogen Spectrum

$\lambda(nm)$	color	$\Delta E_{atom} = E_{photon}$	$\nu(s^{-1})$

Show sample calculation of E_{photon} and ν

C. Unknown Spectrum (spectrum no.) _____

Lines in the spectrum (nm) _____, _____, _____, _____,

_____, _____, _____, _____.

Unknown element: _____.

QUESTIONS

1. Why does the mercury light from public lighting appear blue, even though yellow and green lines appear in the spectrum?

2. Explain why the color for ions in the flame tests differ.

3. The Cl—Cl bond energy is 242 kJ/mol. What wavelength of electromagnetic radiation is necessary to break the bond? In what region of the spectrum does this wavelength appear? Hint: Planck's constant, h, equals 6.63×10^{-34} J•s/$photon$.

Experiment 13

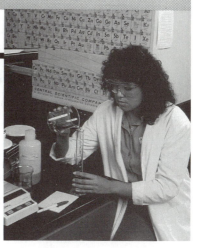

◇ Chemical Periodicity

Nearly every chemistry lecture auditorium and laboratory, as well as nearly every chemist's office, has on the wall at least one seemingly necessary article of decor. This piece of art to a layman is "the table" which seemingly answers many questions posed to a chemist, and is clearly a very integral part of the profession. What is so "magic" or what is so significant about this table? Is it really that important?

This table, called the **periodic table** of the elements, is an organized set of data stemming from years of research. The organization of the data is used to predict the chemical and physical properties of existing elements and can be used to predict the properties of elements of very low natural abundance or of synthesized elements. The early data were organized in 1869 by the Russian Dimitri Mendeleev (see photo), a professor of chemistry at the University of St. Petersburg. In organizing his lectures of inorganic chemistry, he found it apparent that inorganic chemistry did not have the system and order that was evident in organic chemistry at that time. In his attempt to organize the elements, he used the accumulated data of their known chemical properties and placed them in columns and rows whereby adjacent elements were similar. Recognizing that such an organization was also related to the atomic masses of the elements led to the current organization of the periodic table.

Mendeleev was outspoken on his views over academic freedom. Mendeleev, in an outward expression of his independent and somewhat radical ideas, chose to cut his hair only once each year, in the Spring. Because he remained friendly with the Czar and his colleagues in spite of his views, the dates for many of the scientific meetings in Russia were held in the springtime.

OBJECTIVES

- To observe the physical appearance of thirteen common elements
- To observe the chemical reactivity of several elements in Period 3 and Group 7A
- To predict the physical and chemical properties of other elements

PRINCIPLES

Malleable: the property of being formed into very thin sheets.

Amorphous: the property of a solid having no orderly arrangement of atoms, molecules, or ions.

Many chemical and physical properties of an element can be predicted from its location in the periodic table. For example, elements at the left of the table are metals—they are shiny, conduct electricity, and are **malleable**; elements at the right of the table are nonmetals—they are **amorphous**, nonconductors of electricity, and may even be gases. The metals at the left form salts with the nonmetals at the right. Elements of a **group** (vertical column) have similar chemical and physical properties with each successive element showing a gradual difference. Successive elements in a **period** (horizontal row) show more dramatic differences in properties, progressing from those with characteristic metallic properties at the left of the period to those with nonmetallic properties at the right.

In this experiment, we will look at the similarities and differences in the chemical and physical properties of several elements of Period 3 and Group 7A (the halogens). These elements and some of their parameters are listed in Table 13.1.

Table 13.1. Measured Parameters of Some Elements

Properties/Groups	1A	2A	3A	4A	5A	6A	7A
	Li	Be	B	C	N	O	F
Atomic Radius (*pm*)	145	105	85	70	65	60	50
Ionization Energy (*kJ/mol*)	519	900	799	1090	1400	1310	1680
Electronegativity	1.0	1.5	2.0	2.5	3.0	3.5	4.0
Density (*g/cm^3*)	0.53	1.85	2.47	2.25[1]	1.25g/L	1.43g/L	1.70g/L
Melting Point (°C)	181	1285	2030	3370s	-210	-218	-220
Boiling Point (°C)	1347	2470	3700	—	-196	-183	-188
	Na	Mg	Al	Si	P	S	Cl
Atomic Radius (*pm*)	180	150	125	110	100	100	100
Ionization Energy (*kJ/mol*)	494	736	577	786	1060	1000	1260
Electronegativity	0.9	1.2	1.5	1.8	2.1	2.5	3.0
Density (*g/cm^3*)	0.97	1.74	2.70	2.33	1.82	2.09	3.21g/L
Melting Point (°C)	98	650	660	1410	44[2]	115	-101
Boiling Point (°C)	883	1100	2350	2620	280	445	-34
	K	Ca	Ga	Ge	As	Se	Br
Atomic Radius (*pm*)	220	180	130	125	125	115	115
Ionization Energy (*kJ/mol*)	418	590	577	762	966	941	1140
Electronegativity	0.8	1.0	1.6	1.8	2.0	2.4	2.8
Density (*g/cm^3*)	0,86	1.53	5.91	5.32	5.78	4.81	3.12
Melting Point (°C)	64	840	30	937	613s	220	-7
Boiling Point (°C)	774	1490	2070	2830	—	685	59
	Rb	Sr	In	Sn	Sb	Te	I
Atomic Radius (*pm*)	235	200	155	145	145	140	140
Ionization Energy (*kJ/mol*)	402	548	556	707	833	870	1010
Electronegativity	0.8	1.0	1.7	1.8	1.9	2.1	2.5
Density (*g/cm^3*)	1.53	2.58	7.29	7.29[3]	6.69	6.25	4.95
Melting Point (°C)	39	770	157	232	631	450	114
Boiling Point (°C)	688	1380	2050	2720	1750	990	184
	Cs	Ba	Tl	Pb	Bi		
Atomic Radius (*pm*)	266	215	190	180	160		
Ionization Energy (*kJ/mol*)	376	502	812	920	1040		
Electronegativity	0.7	0.9	1.8	1.8	1.9		
Density (*g/cm^3*)	1.87	3.59	11.87	11.34	8.90		
Melting Point (°C)	28	710	304	328	271		
Boiling Point (°C)	678	1640	1460	1760	1650		

[1]Graphite   [2]White phosphorus   [3]White tin

Definitions for the parameters in Table 13.1 are
- **Atomic radius:** the radius of an atom, expressed in picometers where $1 \text{ pm} = 1 \times 10^{-12} \text{ m}$

- **Ionization Energy:** energy required to remove one mole of electrons from a mole of the gaseous element, expressed in kJ/mol
- **Electronegativity:** an atom's relative attraction of the electrons used for bonding to another atom, expressed on a scale relative to fluorine being assigned a number of 4.0
- **Density:** the ratio of the mass of substance to its volume, expressed as g/cm^3 for solids and liquids and g/L at STP[1] for gases
- **Melting point:** the temperature at which the solid and liquid phases of a substance coexist at a specific temperature
- **Boiling point:** the temperature at which the liquid and gaseous phases of a substance coexist at a specific temperature[2]

Chemical properties are dependent upon the chemical environment of the element or compound. If a substance "reacts" with another substance, then it is said to be **reactive**. An element is cited as being reactive if it prefers its ionic (or charged) state to its elemental state. For example, sodium metal is "reactive" because it exists in common compounds as the Na^+ ion rather than as Na; elemental chlorine, Cl_2, is reactive because it prefers the Cl^- ion as its most stable state.

Reactive: the ease with which a substance undergoes a chemical reaction in the presence of another substance.

PROCEDURE

A. Physical Properties

Allotrope: more than one combination of atoms of the same element.

1. Samples of Na, Mg, Al, Si, P, and S are on the reagent table. Note that Na metal is stored under a nonaqueous liquid because of its reactivity with water and the oxygen in air (Part B.1). Your lab instructor will cut a piece of Na; quickly notice its luster and other metallic characteristics. Use steel wool to polish the pieces of Mg and Al for better viewing. Two allotropic forms of phosphorus exist: white P_4 is so reactive with O_2 that it ignites in air; therefore white P_4 is stored under water. When stored under water for a long time, white P_4 slowly changes to the more stable red allotrope of phosphorus, which does not ignite in air. Record data and your observations for the elements on the Data Sheet.

2. **Preparation of Cl_2.** Place $1/2$ mL (10 drops) of 5% NaClO (household laundry bleach) in a 75 mm test tube and add 5 drops of toluene (**Caution**: *do not inhale*). Which layer is toluene and what is its color? Add 5 drops of 6 M HCl (**Caution**: 6 M HCl *is corrosive; wash it immediately from skin and clothes*). Agitate the solution by holding the upper part of the test tube with your thumb and index finger and tapping the lower part with your "pinky" finger (Figure 13.1). Observe the color of the toluene layer. What is the color of Cl_2 in the toluene layer?

3. **Preparation of Br_2 and I_2.**
   a. Mix equal pea-size portions of the solids KBr and MnO_2; transfer a portion of the mixture to a dry, clean 150 mm test tube until a depth of 3 mm ($\approx 1/8$ inch) is reached.[3] Add 3–5 drops of *conc* H_2SO_4 (**Caution**: *don't let it touch your skin*). *Gently* and very *carefully* warm the mixture over a low flame to initiate the reaction. What evidence of a reaction has occurred? Allow the test tube to cool, add 10 drops of water, and 5 drops of toluene, and agitate (see Figure 13.1). What is the color of Br_2 in the toluene layer?

---

[1]STP (**s**tandard **t**emperature and **p**ressure) conditions are 0°C (273K) and 1 atm (101.325 kPa).

[2]More specific definitions of melting and boiling points will be given later in the course.

[3]Share the unused portion of the mixture with your neighbor—don't waste it!

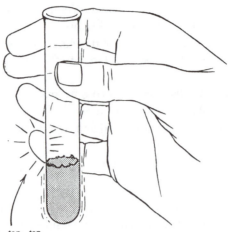

Figure 13.1

Agitating a solution in a test tube with the "pinky" finger

Tap – tap – tap
w/pinky

b. Repeat the procedure in Part A.3, substituting KI for the KBr. What is the color of I_2 in the toluene layer?

Dispose of the test chemicals from Parts A.2, 3 in the "Waste Halogens" container.

4. Plot on graph paper the atomic radius (ordinate) *vs.* atomic number (abscissa) and on the *same* graph, the ionization energy *vs.* atomic number for the elements of Period 3. Label each axis and title the graph. Connect the data points with straight lines. Have your instructor approve your graph.

B. Chemical Properties

Gas evolution: the escape of gas from a system.

1. Na/H_2O. **Demonstration Only**. (**Caution**: *never allow Na to touch the skin; it causes a severe skin burn*)

a. Wrap a *pea-sized* (no larger!) piece of freshly cut Na metal in aluminum foil. Fill a 200 mm test tube with water and invert it into an 800 mL beaker $^3/_4$ filled with water; test the water with litmus. Punch 5 pin-sized holes in the aluminum foil, grasp it with crucible tongs, and place it beneath the mouth of the water-filled test tube (Figure 13.2).

b. After **gas evolution** has ceased, remove the test tube from the beaker, keeping it inverted. Place the mouth of the test tube over a Bunsen flame. What is the evolved gas? Don't be so alarmed as to drop the test tube—it may cost you 70¢ to replace it! Perform the litmus test on the water in the beaker. Is the water now more/less acidic or basic? Based upon the properties of the gas and the acidity (or basicity) of the solution in the beaker, write a balanced equation for the reaction of Na with H_2O.

c. **Demonstration Only**. Na/H_2O and CH_3OH (methanol). Set up a 150 mL beaker containing 20 mL of methanol, CH_3OH, and a 150 mL beaker containing 20 mL of water behind a safety shield (Yes, *behind a safety shield*—if you don't have one, omit this step and proceed to Part B.2). Cut two "BB"-sized (no larger!) pieces of Na metal and, with tongs or tweezers, place one into each beaker. *Immediately* cover each beaker with a watch glass. Describe the reaction. How are water and CH_3OH similar? How do they differ?

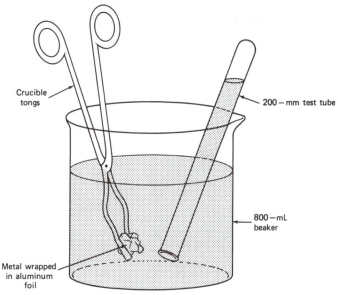

Figure 13.2
Collection of the H$_2$(g) evolved
from the reaction of Na with H$_2$O

2. Mg and Al/H$_2$O and HCl

 a. Place 5 mL of deionized water into separate 150 mm test tubes.
 Place these test tubes in a half-filled 250 mL beaker of water and heat to
 boiling.

 b. Polish, with steel wool or sand paper, 2 cm strips of Mg and Al to
 remove their oxide coatings. Quickly add each metal to a separate test
 tube of hot water. Maintain the temperature at near boiling for
 10 minutes. What is observed?

 c. Remove the test tubes from the hot water bath, containing any
 remaining Mg and Al samples. Perform a litmus test on the water in
 each test tube. Add 2 mL of 6 M HCl (**Caution**: *avoid skin contact*) to each
 test tube. Record your observations. What is the evolved gas?

 d. How does the reactivity of Na, Mg, and Al change in proceeding
 across Period 3?

*Pinch: a volume no greater than a
grain of rice.*

3. Cl$_2$, Br$^-$, and I$^-$

 a. Refer again to Part A.2: prepare Cl$_2$ using the 5% NaClO and 6 M
 HCl solutions with the added toluene. Add a pinch of KBr to the
 Cl$_2$/toluene/H$_2$O mixture and agitate. Account for your observation.

 b. Repeat Part B.3a substituting a pinch of KI (for the KBr) to the
 Cl$_2$/toluene/H$_2$O mixture and agitate. What chemical reaction has
 occurred? Is chlorine more or less reactive than bromine? than iodine?

4. Cl$^-$, Br$_2$, and I$^-$

 a. Dissolve a pinch of KCl in $^1/_2$ mL of water in a 75 mm test tube; add
 5 drops of toluene. In the fume hood, add 5 drops of 2% Br$_2$/H$_2$O
 (**Caution**: Br$_2$ *is corrosive and causes severe skin burns*) to the test tube and
 agitate. What happens? Does the color of the Br$_2$ in the toluene layer
 disappear?

 b. Repeat Part B.4a, substituting KI for the KCl. Is the occurrence of a
 chemical reaction evident? Write a balanced equation to represent your
 observation.

Dispose of the test chemicals from Parts B.3, 4 in the "Waste Halogens" container.

NOTES AND OBSERVATIONS

Chemical Periodicity

Date _____Name _____ Lab Sec. _____ Desk No. _____

1. a.   What is a group of elements?

    b.   What is a period of elements?

2. Refer to Table 13.1 to answer the following.

    a.   Which element in Period 3 is most dense?_____ least dense?_____

    b.   Which element in Group 7A has the highest electronegativity?_____ the lowest
    electronegativity?_____

    c.   Which element in Group 1A has the highest melting point?_____ the lowest boiling
    point?_____

    d.   In general, the densities of the elements in a group _____ as the atomic number increases.

    e.   In general, the ionization energies of the elements in a period _____ as the atomic number
    increases.

    f.   The electronegativity of the elements in a group _____ and in a period _____ as the atomic
    number increases.

3. The relative chemical reactivity for a number of elements is studied in today's experiment.

    a.   List the elements.

    b.   List the element whose reactivity will be demonstrated by the instructor.

4. a. The following reactions of metals occur spontaneously:

$$Zn(s) + Cu^{2+}(aq) \rightarrow Zn^{2+}(aq) + Cu(s)$$
$$Mg(s) + Ni^{2+}(aq) \rightarrow Mg^{2+}(aq) + Ni(s)$$
or generically,
$$Q(s) + M^{n+}(aq) \rightarrow Q^{n+}(aq) + M(s)$$

Is copper or zinc more reactive?

Is magnesium or nickel more reactive?

Is M or Q more reactive?

b. Consider a generic equation for the reaction of the halogens:
$$X_2(g) + 2Y^-(aq) \rightarrow 2X^-(aq) + Y_2(g)$$

Is X_2 or Y_2 the more reactive halogen? Explain.

5. What commercially available compound is used to generate Cl_2 in the experiment?_____

◇ Chemical Periodicity

Date _____ Name _____ Lab Sec. _____ Desk No. _____

A. Physical Properties

Element	Symbol	Atomic Number	Atomic Mass	Physical State (g, l, s)	Color	Comments
sodium						
magnesium						
aluminum						
silicon						
phosphorus						
sulfur						

2. **Preparation of Cl_2.** Which is the toluene layer?_____ What color is

   toluene?_____ What is the color of Cl_2 in the toluene layer?_____

3. **Preparation of Br_2 and I_2.**
   Evidence for the preparation of Br_2.

   Evidence for the preparation of I_2.

   Color of Br_2 _____ Color of I_2 _____

4. Instructor's approval of graph._____

   From the graph, what general statement can you make about the relationship between atomic radii and ionization energies for a period of elements?

   Is the same statement valid for a group of elements?_____ If not, what statement holds true?

B. Chemical Properties

1. Na /H_2O
   a. Gas evolved. _____

   b. Litmus test, acidic or basic, before: _____ ; after: _____

   Write a balanced equation for the reaction of Na with H_2O.

   c. Na/H_2O and CH_3OH (methanol)
      What similarities exist between the reactions of Na with H_2O and CH_3OH?

      Is Na more reactive in H_2O or CH_3OH?_____

2. Mg and Al/H_2O and HCl

	Observation	Gas evolved	Litmus test
Mg/H_2O			
Al/H_2O			
Mg/HCl			
Al/HCl			

   Compare the relative chemical reactivity of Na, Mg, and Al with H_2O and with HCl. Referring to the physical properties of the elements in Table 13.1, speculate as to the trend.

3. Cl_2, Br^-, and I^-

	Observation	Balanced equation for the reaction
Cl_2 + Br^-		
Cl_2 + I^-		

Is Cl_2 more or less reactive than Br_2? (i.e., Does Cl_2 react with Br^- to form a less reactive Br_2?) than I_2?

4. Cl^-, Br_2, and I^-

	Observation	Balanced equation for the reaction
$Cl^- + Br_2$		
$Br_2 + I^-$		

List the halogens in order of decreasing reactivity, the most active halogen listed first.

_____ > _____ > _____ Explain your listing.

QUESTIONS

1. What tool did your lab instructor use to cut the Na metal?_____ What property does Na exhibit, one that we don't normally associate with a metal, that allowed the lab instructor to use that tool?

2. a. Metallic oxides, like Na_2O and BaO, dissolve in water and turn red litmus blue. What chemical property does this indicate about metallic oxides?

   b. Predict the effect on litmus when *non*metallic oxides, such as SO_2 and CO_2, dissolve in water. Explain.

3. Cl_2 is used extensively as a bleaching agent and as a disinfectant. Without regard to any adverse effects, would Br_2 be more or less effective as a bleaching agent and disinfectant? Explain.

4. Predict the reactivity of cesium in methanol relative to that of sodium. Explain.

5. Predict the reactivity of silicon in water relative to that of sodium, magnesium, and aluminum. Explain.

6. Predict the chemical reactivity of fluorine gas, F_2, relative to other halogens of Group 7A. Explain.

Experiment 14

◇ Spectrophotometric Iron Analysis

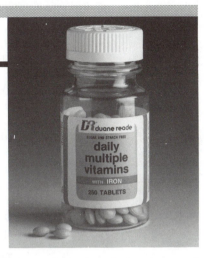

The exchange of oxygen with carbon dioxide during respiration could not occur without iron being a part of the protein hemoglobin. A healthy adult body contains about 3 g of iron with a loss of about 1 mg/day through sweat, feces, and hair. A deficiency of iron leads to a condition called anemia, resulting in symptoms of complacency and tiredness. Iron supplements (see photo) are often used to replenish the loss.

Scientists use a number of laboratory techniques to analyze for the amount of iron in a sample. Previously in this laboratory manual we used the balance to assist in our analyses. Many substances have color and the intensity of that color relates to the amount of substance present. For example, the more drops of blue ink that are added to water, the higher is its concentration in the water and the more intense is the blue color. Quantitative and qualitative measurements based upon the property of a substance to absorb visible light are used extensively in chemical laboratories. An iron analysis is one of those applications.

OBJECTIVES

- To determine an unknown iron(II) ion concentration in solution
- To develop techniques for the use and operation of a spectrophotometer
- To determine the amount of iron in a vitamin tablet

PRINCIPLES

The principle underlying a spectrophotometric method of analysis involves the interaction of electromagnetic (EM) radiation with matter. The ultraviolet, visible, and infrared regions of the EM spectrum are the most common used in analyses; in this experiment the visible region is used. The wavelength range for the visible spectrum is from about 400 nm to 700 nm; the 400 nm radiation approximates a violet color while the 700 nm region has a red color.

The principles of spectroscopy are also discussed in Experiment 12.

Every chemical species (atoms, molecule, or ion) possesses a characteristic set of electronic, vibrational, and rotational energy states. Because they are characteristic of a chemical species, energy transitions between these states are often used to identify their presence and/or concentration in a mixture. This unique set of energy states for a chemical specie is therefore analogous to the unique set of fingerprints possessed by each person—both can be used for characteristic identifications.

The absorption of EM radiation from the visible spectrum is a result of the excitation of an electron from a lower to a higher state in the chemical specie. The energy of the radiation that is absorbed is equal to the difference between these two energy states. The species (atom, molecule, or ion) that absorbs the radiation is in an **excited state**. The EM radiation that is not absorbed, and therefore passes through the sample, is detected by an EM detector (either our own eye or an instrument). The absorbed energy, *E*, is related to the wavelength, λ, of the EM radiation by the equation,

$$E = h\frac{c}{\lambda}$$

where h is Planck's constant and c is the speed of light.

When our eye is the detector, the color that we see is the EM radiation which the sample does *not* absorb. The appearance of the sample is that of the **complementary color** of the absorbed radiation (see color wheel on the color plate). For example, if our sample absorbs EM radiation from the yellow region of the visible spectrum, then the remaining EM radiation is transmitted to our eye and the sample appears violet. The greater the concentration of the yellow absorbing specie, the darker is its violet appearance. Concord grapes having a violet appearance absorb yellow light from the EM spectrum. Table 14.1 lists the wavelengths of the visible spectrum.

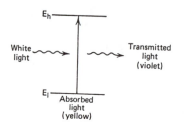

Table 14.1. Color and Wavelengths in the Visible Region of the Electromagnetic Spectrum

Color	Wavelength (nm*)	Color Transmitted
red	750-610	blue-green
orange	610-595	blue
yellow	595-580	violet
green	580-500	purple
blue	500-435	orange
violet	435-380	yellow

*1 nanometer = 1×10^{-9} meter

In this experiment the absorption of visible radiation is used to determine the concentration of the iron(II), Fe^{2+}, ion in an aqueous solution. The wavelength at which the maximum absorption of visible radiation occurs is set on the **spectrophotometer** (Figure 14.1).

Several factors affect the amount of visible radiation that the Fe^{2+} absorbs.

- the concentration of the Fe^{2+} in solution
- the thickness of the solution through which the visible radiation passes (this is determined by the diameter of the **cuvet**)
- the extent to which the Fe^{2+} absorbs the radiation at a particular wavelength (this is called its molar absorptivity). This factor is constant for a chemical specie at a set wavelength.

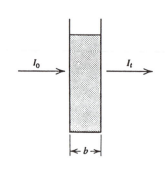

The ratio of the intensities of the transmitted visible radiation, I_t, through the cuvet to the incident visible radiation, I_o, is the sample's **transmittance, T**, or expressed as a percent, %T

$$\%T = \frac{I_t}{I_o} \times 100$$

Frequently a chemist is more interested in knowing the amount of radiation that the Fe^{2+} absorbs rather than the amount that it transmits, the absorption being directly proportional to the Fe^{2+} concentration. The **absorbance, A,** of the radiation is related to the percent transmittance by the equation

$$A = -\log\frac{I_t}{I_o} \times 100 = \log\frac{100}{\%T}$$

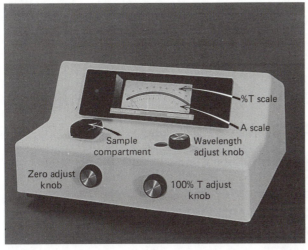
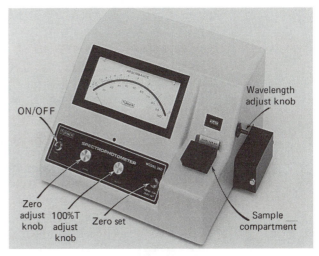

The absorbance of radiation is directly proportional to the molar concentration of the Fe^{2+} by the equation

$$A = a \cdot b \cdot [Fe^{2+}]$$

a is the molar absorptivity for Fe^{2+} and b is the thickness of the solution—both of which are constants at a given wavelength in a given cuvet. This equation is commonly referred to as Beer's Law, the important relationship being that $A \propto [Fe^{2+}]$ (Figure 14.2).

Figure 14.1

Common laboratory visible spectrophotometers

α: means "proportional to".

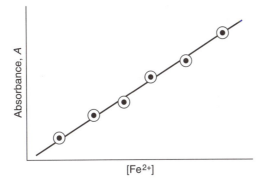

Figure 14.2

A plot of absorbance, A, vs. $[Fe^{2+}]$

In this experiment an iron sample is dissolved in solution and all of the dissolved iron ions are reduced to Fe^{2+}. The Fe^{2+} forms a red-orange **complex ion** with 1,10-phenanthroline (also called ortho-phenanthroline and abbreviated o-phen), $[Fe(o\text{-phen})_3]^{2+}$. The absorbance, A, of this ion is determined at a wavelength where the complex ion has a maximum molar absorptivity, called λ_{max}. The λ_{max} for $[Fe(o\text{-phen})_3]^{2+}$ is determined in Part A from a plot of A vs. λ for a standard solution of $[Fe(o\text{-phen})_3]^{2+}$.

The iron(III) ion, Fe^{3+}, does *not* form the red-orange complex ion with o-phen; therefore, in the experiment all iron(III) ion is reduced to iron(II) to ensure that the total iron concentration is being determined. The reducing agent is hydroxylamine hydrochloride, $NH_3OH^+Cl^-$.

$$2\,Fe^{3+}(aq) + 2\,NH_3OH^+Cl^-(aq) \rightarrow$$
$$2\,Fe^{2+}(aq) + N_2(g) + 2\,H_2O + 4\,H^+(aq) + 2\,Cl^-(aq)$$

After the total reduction of all iron ions to iron(II) ion is complete, 1,10-phenanthroline is added to the solution to form the red-orange complex ion with Fe^{2+}.

Measuring the Fe^{2+} Concentration

Complex ion: in this experiment, $[Fe(o\text{-phen})_3]^{2+}$.

$$Fe^{2+} + 3 \quad \rightleftharpoons \quad Fe$$

$$Fe^{2+}(aq) + 3\ o\text{-phen}(aq) \rightarrow [Fe(o\text{-phen})_3]^{2+}(aq)$$

In addition, since the $[Fe(o\text{-phen})_3]^{2+}$ ion is most stable in the pH range from 2 to 9, sodium acetate, $NaCH_3CO_2$, is added, reacting with the H^+ (formed as the product) from the $NH_3OH^+Cl^-$ reaction to form an $CH_3CO_2^-/CH_3COOH$ combination that maintains the pH of the solution between 4 and 6.

$$H^+(aq) + CH_3CO_2^-(aq) \rightarrow CH_3COOH(aq)$$

PROCEDURE

A. Setting the λ_{max} on the Spectrophotometer

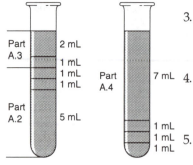

You and a partner are using a visible spectrophotometer, an expensive instrument, in today's experiment. Follow your instructor's suggestions on its safe operation. You will need three 1 mL pipets and two 10 mL graduated (Mohr) pipets from the stockroom. Read ahead and be prepared.

1. Obtain (and thoroughly clean with soap and water) three 1 mL pipets and two 10 mL pipets, graduated at 1 mL increments (a Mohr pipet), and five 150 mm (6-inch) test tubes.[1] Use the graduated 10 mL pipet for the standard Fe^{3+} solution (Part A.2b).

2. a. When using pipets, rinse the pipet twice with the reagent and then discard the reagent before preparing the test solution. This technique is especially important when using the 10 mL pipet in this experiment.

Dispose of the test solutions in the "Waste Iron Salts" container.

   b. Pipet 5.00 mL of the standard Fe^{3+} solution ($\approx 10\ \mu g\ Fe^{3+}/mL$) into a clean 150 mm test tube. Using 1 mL pipets, add 1 mL of 10% $NaCH_3CO_2$ and 1 mL of 10% $NH_3OH^+Cl^-$. Agitate the solution periodically for at least 10 minutes to complete the reduction of Fe^{3+} to Fe^{2+}. In the meantime continue to Parts A.4 and B.2 for the preparation of the subsequent solutions for testing.

3. Use the third 1 mL pipet to add 1 mL of 0.1% o-phen to the solution; with the second 10 mL pipet add 2.0 mL of water. Stir, with a stirring rod, the solution for several minutes. This is test solution #4 in Table 14.2.

4. To prepare a blank solution,[2] pipet 1 mL of 10% $NaCH_3CO_2$, 1 mL of 10% $NH_3OH^+Cl^-$, 1 mL of 0.1% o-phen and 7 mL of water into a 150 mm test tube and stir. This is test solution #1 in Table 14.2.

5. Prepare two cuvets: rinse one cuvet twice with the blank solution and then fill to the $^3/_4$ level; rinse a second cuvet (also twice) with the $[Fe(o\text{-phen})_3]^{2+}$ solution (from Parts A.2 and A.3) and similarly fill.

---

[1] 10 mL volumetric flasks may be substituted for the 150 mm test tubes.

[2] A **blank** solution is used to calibrate the spectrophotometer; the solution corrects for all substances that absorb EM radiation *except* the one of interest, in this case, $[Fe(o\text{-phen})_3]^{2+}$.

Carefully dry the outside of each cuvet with a clean Kimwipe to remove fingerprints and water droplets. Thereafter, handle only the lip of the cuvets.[3]

6. Set the wavelength, λ, on the spectrophotometer at 400 nm. Insert the blank solution (from Part A.4) into the sample holder and set the meter at 100%T. Remove the blank and (without any cuvet in the sample holder) set the meter to read 0%T. Repeat until no further adjustments are necessary.

7. Place the cuvet containing the $[Fe(o\text{-phen})_3]^{2+}$ solution (from Part A.3) in the sample holder, read the meter, and record the %T.

8. Repeat Part A.7 at 20 nm intervals between 400 nm and 600 nm. Make additional %T measurements at 5 nm intervals in the region of λ_{max}. Periodically repeat Part A.6.
Save these solutions for Part B of the experiment.

9. Convert all %T readings to absorbance values. Plot the data as A (ordinate) vs. λ (abscissa) on linear graph paper. Draw the best smooth curve through the data points. Have your instructor approve the graph.

B. Constructing the Standard Curve

1. Set the spectrophotometer at the λ_{max} determined from the graph in Part A.9.

2. Prepare the solutions in Table 14.2 in the same manner as described in Parts A.2, 3. Be sure to add the 0.1% o-phen *after* the Fe^{3+} has been reduced to Fe^{2+} (again, allow at least 10 min for the reduction reaction). Use the 10 mL pipet, calibrated at 1 mL intervals, for transferring the standard Fe^{3+} solution. Be sure to rinse the pipet twice before preparing each test solution. Use the second 10 mL graduated pipet for the dilution with water.

3. Use the blank solution (from Part A.4, Test solution 1) to repeat the calibration of the spectrophotometer (see Part A.6).

4. Read and record the %T of the five solutions at the λ_{max}. Follow the same techniques for handling the cuvets as indicated in Part A.5. Calculate the absorbance for each solution.

Table 14.2. Test Solutions for the Construction of the Standard Curve

Test sol'n	10 µg Fe^{3+}/mL sol'n (mL)	10% $NaCH_3CO_2$ sol'n (mL)	10% $NH_3OH^+Cl^-$ sol'n (mL)	0.1% o-phen sol'n (mL)	Volume of water (mL)
1*	0.0	1.0	1.0	1.0	7.0
2	1.0	1.0	1.0	1.0	6.0
3	3.0	1.0	1.0	1.0	4.0
4*	5.0	1.0	1.0	1.0	2.0
5	7.0	1.0	1.0	1.0	0.0

*Prepared in Parts A.2–4

5. Prepare a standard curve by plotting on linear graph paper A (ordinate) vs. Fe^{2+} concentration (µg/mL) (abscissa). Draw the best straight line

---

[3]Foreign material on the cuvet affects the intensity of the transmitted EM radiation.

through the five points on the graph. Have your instructor approve your graph.

C. Preparation of a Water Sample for Iron Analysis

1. Obtain at least 20 mL of a water sample known to contain dissolved iron. It may be an unknown that has already been prepared for the experiment or it may be obtained from a drinking water supply, reservoir, or river. If there is any evidence of cloudiness, filter the sample.

2. Pipet 5.00 mL of the water sample into a 100 mL volumetric flask and dilute to the "mark" with water. Pipet 1.0 mL of this solution into a 150 mm test tube, add 1.0 mL each of the $NaCH_3CO_2$, $NH_3OH^+Cl^-$, and (wait ≈10 minutes) o-phen solutions and 6.0 mL of water as in Parts A and B. Proceed to Part E.

D. Preparation of a Vitamin Tablet for Iron Analysis

1. Crush a vitamin tablet with a mortar and pestle. Transfer the powder to a clean, dry, 100 mL beaker of known mass (±0.01 g). Measure the combined mass of the beaker and crushed tablet. Add 25 mL of 6 M HCl into the beaker and stir to dissolve (≈10 to 20 min).[4]

2. Quantitatively transfer the solution to a 250 mL volumetric flask. Wash the solid remaining in the beaker with several portions of deionized water and add the washings to the flask. Dilute to the mark with deionized water and agitate for several minutes; if cloudiness persists, filter the solution.

3. Pipet 1.0 mL of the (filtered) solution into a 150 mm test tube and add 1.0 mL each of the $NaCH_3CO_2$, $NH_3OH^+Cl^-$, and (wait ≈10 minutes) o-phen solutions and 6.0 mL of water. Proceed to Part E.

4. Rinse clean each pipet with several volumes of deionized water.

E. Determination of the Fe²⁺ Concentration in the Sample

1. Visually compare the color intensity of this prepared solution with the standard solutions prepared in Part B; if the red-orange color is within the range of the standards, OK; if not, discard this solution and prepare a second sample solution (quantitatively) in which the second dilution has a color intensity that lies within that of the standards. You will need to make some judgment on the volume of sample to use.

2. Record the %T of the solution on the spectrophotometer and calculate its absorbance. It may be advisable at this point to again check the calibration of the spectrophotometer with the blank solution; if it needs recalibration, the %T of the solution will need to be repeated.

3. Use the standard curve prepared from Part B to determine the Fe^{2+} concentration (μg/mL) in the prepared sample. Calculate the quantity of iron in the original sample; be sure to account for the dilution of the original sample.

Dispose of the test solutions in the "Waste Iron Salts" container.

---

[4]Some of the tablet's binder may not dissolve; heat the solution gently in the hood to dissolve the tablet, but do not boil.

Spectrophotometric Iron Analysis

Date _____ Name _____ Lab Sec. _____ Desk No. _____

1. List the purpose for each of these substances in the iron analysis.

    a. 1,10-phenanthroline (*o*-phen)

    b. $NaCH_3CO_2$

    c. $NH_3OH^+Cl^-$. Write a balanced equation for its reaction with Fe^{3+}. The oxidation half-reaction of NH_3OH^+ is

$$2\,NH_3OH^+(aq) \rightarrow N_2(g) + 2\,H_2O(l) + 4\,H^+(aq) + 2\,e^-$$

2. A concentration of 2 ppm Fe^{2+} means 2 g Fe^{2+} in 10^6 g solution. Assuming the density of the solution is 1.0 g/mL, express 2 ppm Fe^{2+} in

    a. μg Fe^{2+}/mL solution

    b. mg Fe^{2+}/L solution

3. What is the color of a solution that absorbs 600 nm EM radiation?

4. If [Fe(o-phen)$_3$]$^{2+}$ appears red-orange, what is the approximate wavelength for its maximum absorption?

5. A sample containing 18.0 mg of iron dissolves in 250 mL of solution. One milliliter is then withdrawn and diluted to 10 mL. What is the iron concentration in the diluted sample? Express your answer in µg Fe/mL *and* in ppm Fe.

6. A test solution is prepared by pipetting 2.0 mL of an original sample into a 250 mL volumetric flask and then diluting to the mark. The concentration of iron in the test solution is determined (from a calibration curve) to be 1.66 µg/mL or 1.66 ppm. What is the iron concentration in the original sample? Express your answer in µg Fe/mL *and* in ppm Fe.

◇ # Spectrophotometric Iron Analysis

Date _____Name _____ Lab Sec. _____Desk No. _____

A. Setting the λ_{max} on the Spectrophotometer

$\lambda(nm)$	%T	A (calc)	$\lambda(nm)$	%T	A (calc)	$\lambda(nm)$	%T	A (calc)

Plot the data, A (ordinate) vs. λ(abscissa) on linear graph paper.

Instructor's approval of graph._____

B. Constructing the Standard Curve

Test sol'n	Fe^{2+} concentration (total volume = 10 mL)		%T	A (calc)
	$\mu g\ Fe^{2+}/mL$	ppm Fe^{2+}		
1				
2	*			
3				
4				
5				

* Sample calculation for Fe^{2+} ($\mu g\ Fe^{2+}/mL$) for test solution 2 (show work here).

Plot the data, A (ordinate) vs. Fe^{2+} concentration ($\mu g\ Fe^{2+}/mL$) (abscissa) on linear graph paper.

C/E. Iron Concentration in Water Sample

	Trial 1	Trial 2
1. %T		
2. Absorbance		
3. Iron concentration in diluted sample from calibration curve ($\mu g\ Fe/mL$)		
4. Iron concentration in original sample ($\mu g\ Fe/mL$)		
($ppm\ Fe$)		

Show sample calculation.

D/E. Iron Concentration in Vitamin Tablet

	Trial 1	Trial 2
1. Mass of 100 mL beaker (g)		
2. Mass of 100 mL beaker + crushed tablet		
3. Mass of tablet (g)		
4. %T		
5. Absorbance		
6. Iron concentration in diluted sample from calibration curve ($\mu g\ Fe/mL$)		
7. Iron concentration in original sample ($\mu g\ Fe/mL$)		
8. mg Fe/gram tablet		
9. %Fe in tablet		

Show sample calculation.

QUESTIONS

1. This experiment measures the iron concentration in the 0.1 ppm to 10 ppm range. Explain how a sample solution having a higher iron concentration can be analyzed, and still remain within the limits of this sensitivity range.

2. How does an unfiltered sample affect the absorbance value for a solution?

3. If not enough time is allowed for the following reactions, how will it affect the reported iron concentration in the sample?

   a. the reduction of the Fe^{3+}.

   b. the formation of $[Fe(o\text{-phen})_3]^{2+}$.

4. State the purpose for the blank solution in measuring the $\%T$ of the $[Fe(o\text{-phen})_3]^{2+}$ solutions. Why is deionized water not a suitable blank?

5. a. If a 0.9 cm cuvet is mistakenly substituted for a 0.8 cm cuvet in one of the measurements, will the $\%T$ reading be higher or lower than it should be for that solution? Explain.

   b. Will the reported iron concentration be high or low for that solution?

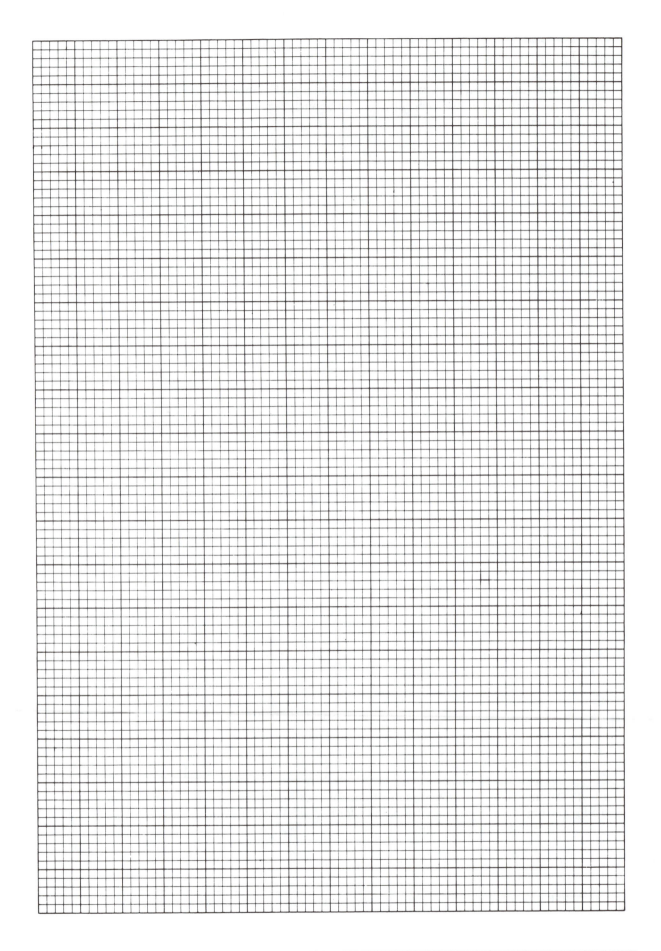

Experiment 15

◇ Molecular Geometry

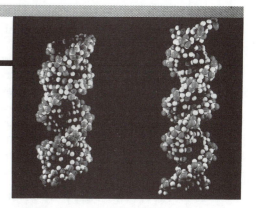

Atoms bond. Molecules and ion-pairs form. Our bodies are three-dimensional (3-D) and, since our bodies are made of compounds, the molecules and ion-pairs that form the compounds must also be 3-D. For that matter "everything" is 3-D, so an understanding of the structure of molecules is very important to the scientist. Certainly the structure of the DNA molecule (see photo: DNA-A (left) and DNA-B) is vital to our existence—any change in its structure changes its chemical activity and alters the genetic information used in subsequent cell replication. These changes can result in cancerous cells, birth defects, and terminal illnesses.

Much of the early explanations and interpretations of chemical bonding is attributed to G. N. Lewis, a native of Massachusetts and later, a professor of chemistry at the Massachusetts Institute of Technology and the University of California, Berkeley. Lewis advanced the concept of the electron pair in ionic and molecular bonding. What is amazing about Lewis' theories of bonding (first proposed in 1916) is that they preceded the **quantum theory** of the atoms. It is from this Lewis concept of the chemical bond that the VSEPR (valence shell electron pair repulsion) theory of molecular geometry was proposed.

OBJECTIVES

- To construct models for molecular compounds and polyatomic ions
- To apply the Lewis theory of bonding for predicting the three-dimensional shapes of molecules and polyatomic ions
- To predict the polarity of molecules

PRINCIPLES

The structure of a molecule is the basis for explaining its chemical and physical properties. For example, the facts that water is a liquid at room temperature, dissolves innumerable salts and sugars, is more dense than ice, boils at a relatively high temperature, and has a low vapor pressure can be explained through an understanding of its bonding and the bent arrangement of its atoms in the molecule.

Quantum theory: recognizes the wave-like properties of electrons in atoms.

Lewis proposed that for an atom to be stable it must gain, lose, or share **valence electrons** with other atoms until each electron is paired. In addition, he proposed that atoms tend to be most stable in a compound when they achieve a valence number of electrons equal to that of a nearby noble gas in the Periodic Table. With the exception of helium, this number of electrons is eight and is called the **Octet Rule**. The bond formed is ionic or covalent depending upon whether the electrons are transferred or shared (respectively). Lewis' Octet Rule is quite effective in explaining bonding, especially for bonding between the representative (main group) elements.

Valence electrons: electrons in the highest energy level (outer shell) of the atom.

For water, the Lewis structure shows that when hydrogen and oxygen *share* valence electrons; each atom is **isoelectronic** with a noble gas—hydrogen with helium and oxygen with neon.

Isoelectronic: "the same electron configuration".

The Lewis structure accounts for the bonding of each atom, but not the 3-D structure of the molecule or polyatomic ion. Various bonding theories predict the 3-D structure of a molecule; these theories account for some of the chemical and physical properties as well.

Valence Shell Electron Pair Repulsion (VSEPR) Theory

The valence shell electron pair repulsion, VSEPR, theory proposes that the geometry of a molecule or polyatomic ion is a result of a repulsive interaction of electron pairs in the valence shell of an atom; the most significant atom for determining its 3-D structure is the *central* atom. The orientation of the atoms is such that there is minimal electrostatic interaction between the valence shell electron pairs; this, in turn, maximizes the distance between the electron pairs and therefore also between the atoms in the molecule.

Table 15.1. The VSEPR Structures for Molecules and Ions

Valence Electron Pairs	Bonding Electron Pairs	Nonbonding Electron Pairs	VSEPR Formula	Approx. Bond Angles	3-D Structure	Examples
2	2	0	AX_2	180°	**linear**	$HgCl_2$, $BeCl_2$
3	3	0	AX_3	120°	**trigonal planar**	BF_3, $In(CH_3)_3$
	2	1	AX_2E	<120°	bent or angular	$SnCl_2$, $PbBr_2$
4	4	0	AX_4	109.5°	**tetrahedral**	CH_4, $SnCl_4$
	3	1	AX_3E	<109.5°	trigonal pyramidal	NH_3, PCl_3, H_3O^+
	2	2	AX_2E_2	<109.5°	bent or angular	H_2O, OF_2, SCl_2
5	5	0	AX_5	90°/120°	**trigonal bipyramidal**	PCl_5, $NbCl_5$
	4	1	AX_4E	>90° <120°	irregular tetrahedral	SF_4, $TeCl_4$
	3	2	AX_3E_2	<90°	T–shaped	ClF_3
	2	3	AX_2E_3	180°	linear	ICl_2^-, XeF_2
6	6	0	AX_6	90°	**octahedral**	SF_6
	5	1	AX_5E	>90°	square pyramidal	BrF_5
	4	2	AX_4E_2	90°	square planar	ICl_4^-, XeF_4

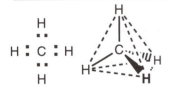

In methane, CH_4, the four valence electron pairs of the carbon atom (the central atom of the molecule) repel; this positions the electron pairs (and, for CH_4, the four hydrogen atoms) at the corners of a tetrahedron and the carbon atom at the middle in a 3-D structure. This positioning of electron pairs can be generalized to include all molecular systems having four valence shell electron pairs on the central atom.

Since the oxygen atom in water also has four valence shell electron pairs, they, too, are arranged tetrahedrally—two electron pairs bond the hydrogen atoms to the oxygen and two electron pairs are nonbonding. This causes water to have a "bent" arrangement of the H–O–H atoms in the molecule.

Methane has the VSEPR formula, AX_4, where A represents the central atom and X represents an attached atom; water has the VSEPR formula, AX_2E_2, where E represents a nonbonding electron pair. The H–C–H bond angles in the tetrahedral arrangement of the four electron pairs in methane is 109.5°; therefore, we can predict that all AX_4 molecules have similar bond angles and that molecules with AX_3E and AX_2E_2 formulas (also four electron pairs in the valence shell of the central atom), such as ammonia and water respectively, have bond angles slightly less than 109.5°.

Methane: the simplest of the hydrocarbons, better known as the major component of natural gas.

A multiple bond (more than one electron pair being shared between two atoms in the molecule), as in an $X=AX_2$ or $X\equiv AX$ molecule, holds the multiple-bonded atom in the same position as a single bond. Therefore, the multiple bond is treated as a single bond for predicting the structure of a molecule. For example, carbon dioxide, which has the Lewis structure $:\ddot{O}::C::\ddot{O}:$, has the VSEPR formula AX_2 and is a linear molecule.

In summary, the arrangement of the valence shell electron pairs, both bonding and nonbonding pairs, on the central atom gives rise to the corresponding 3-D structures of molecules and polyatomic ions. Predicting the structure for a molecule with multiple bonds is the same as that for a molecule with single bonds. The 3-D structures and bond angles for molecules and polyatomic ions with a central atom having from two to six valence shell electron pairs are listed in Table 15.1.

Dipoles and Polarity

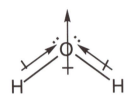

Figure 15.1
The bond polarity of the OH bonds contribute to the resultant polarity (↑) of the H_2O molecule

Once the 3-D structure of a molecule is known, its polarity can be ascertained. A **polar molecule** has a nonuniform distribution of electrons (and therefore, negative and positive charge) over its entire structure. This nonuniform distribution of charge results both from atoms in the molecule having different electronegativities and from their relative positions in space. Atoms that are highly electronegative, such as fluorine, tend to strongly attract electrons, creating a center of negative charge and, therefore, a center of positive charge elsewhere in the molecule. The separation of opposite charges by a distance is called a molecular **dipole**, designated by a vector pointed in the direction of the negative region of the molecule. For example, H–Cl is a polar molecule because chlorine, being more electronegative than hydrogen (electronegativities: Cl = 3.0, H = 2.1), has a slightly higher concentration of negative charge than hydrogen in the molecule; H–F is even more polar because fluorine (electronegativity: F = 4.0) is more electronegative than chlorine.

For molecules with more than one bond, one needs to look at its 3-D structure to determine the relative orientation of each polar bond and how it contributes to the overall polarity of the molecule. For H_2O, each O–H bond is polar and their relative orientations determine the polarity of the molecule. If the bond angle were 180°, the bond polarities would cancel and the molecule would be nonpolar, but since the bond angle is 104.5°, the bond polarities do not cancel, but rather enhance each other to give the (overall) molecule a resultant dipole (Figure 15.1). A look at the 3-D structure of a molecule will enable us to predict whether a molecule is polar or nonpolar.

PROCEDURE

Check out a set of molecular models. Your instructor will assign you a number of molecules or polyatomic ions from the lists on the Data Sheet. For each molecule or polyatomic ion, you will be required to do the following:

- Write its Lewis structure
- Determine the number of bonding electron pairs on the central atom
- Determine the number of nonbonding electron pairs on the central atom
- Predict the approximate bond angle of the atoms in the molecule or polyatomic ion
- Predict the 3-D structure of the molecule or polyatomic ion
- Predict if the molecule is polar or nonpolar

NOTES AND OBSERVATIONS

◇ Molecular Geometry

Date _____ Name _____ Lab Sec. _____ Desk No. _____

1. Describe the features of the Octet Rule.

2. Nitrogen triiodide is a shock sensitive compound—when dry, it detonates with the touch of a feather. Its

   3-D sketch is

   Nitrogen triiodide has a VSEPR formula of _____ . The approximate I–N–I bond

   angle is _____°. Predict if nitrogen triiodide is polar or nonpolar._____

3. a. A molecule that has four bonding pairs and no nonbonding pairs of electrons has a

      _____ structure. Its approximate bond angle is _____°.

   b. A molecule that has three bonding pairs and two nonbonding pairs of electrons has a VSEPR formula

   of _____ . Its approximate bond angle is _____°.

4. a. A molecule that has four bonding pairs and two nonbonding pair of electrons has a

      _____ structure. Its approximate bond angle is _____°.

   b. Sketch the 3-D shape of the molecule described in 4a.

5. The essential amino acid, phenylalanine, the amino acid in NutraSweet®, has the structure

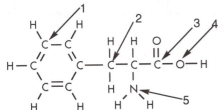

Identify the VSEPR formula and approximate bond angles at the five designated atoms.

6. a. Complete the statement, "A molecule, such as water, is polar if . . . "

   b. How is it determined if a molecule is polar or nonpolar?

7. a. The number of nonbonding electron pairs in SF_4 is _____.

   b. The VSEPR formula for SF_4 is _____.

   c. Sketch the 3-D structure of SF_4.

   d. The 3-D structure of SF_4 is _____.

   e. Is SF_4 predicted to be a polar or nonpolar molecule? Explain.

8. Explain why CO_2 is a nonpolar molecule, but CO is a polar molecule.

◇ Molecular Geometry

Date _____Name _____ Lab Sec. _____Desk No. _____

Your instructor will assign molecules and/or ions from the following lists for you to characterize according to the format shown below. The central atom of the molecule/ion is italicized and in bold face. The central atom of the molecules/ions marked with an asterisk (*) do *not* obey the octet rule.

Molecule or Ion	Lewis Structure	Bonding electron pairs	Nonbonding electron pairs	Approx. bond angle (°)	3-D Sketch	3-D Structure	Polar or nonpolar
CH_4	H $H:C:H$ H	4	0	109.5°		tetrahedral	nonpolar

1. Complete the table (as outlined above) for the following molecules/ions, all of which obey the octet rule.

   a. CF_3Cl d. H_2O g. H_3O^+ j. OF_2

   b. NH_3 e. SbH_3 h. ClO_2^- k. ICl_2^+

   c. NH_4^+ f. AsF_3 i. BF_4^- l. SiH_4

2. Complete the table (as outlined above) for the following molecules/ions.

   a. SF_2 d. SF_5^{+*} g. BrF_3^* j. BrF_5^*

   b. SF_4^* e. BrF_2^+ h. BrF_4^{-*} k. $BeCl_2^*$

   c. SF_6^* f. BrF_2^{-*} ·   i. BrF_4^{+*} l. XeF_2^*

3. Complete the table (as outlined above) for the following molecules/ions.

   a. GaI_3^* d. PF_3 g. SbF_6^{-*} j. XeF_4^*

   b. CH_3^- e. PF_5^* h. SnF_4 k. SnF_6^{2-*}

   c. CH_3^{+*} f. PF_4^+ i. SnF_2^* l. SO_4^{2-}

4. Complete the table (as outlined above) for the following molecules/ions. For molecules or ions with two or more atoms considered as central atoms, consider each atom separately in the analysis according to the table.

   a. $OPCl_3$ d. Cl_3CCF_3 g. $COCl_2$

   b. H_2CCH_2 e. Cl_2O h. $OCCCO$

   c. $CH_3NH_3^+$ f. $ClCN$

QUESTIONS

1. A double bond does not affect the 3-D structure of a molecule. Sketch the 3-D shapes of SO_2 and $CO_3{}^{2-}$ and predict their 3-D structures.

2. Match the molecules/polyatomic ions with the 3-D sketches on the right. The sketches only include the bonded atoms (bonding electron pairs).

   a. SCl_2

   b. SiF_4

   c. $IF_4{}^+$

   d. IF_5

   e. NH_3

   f. XeF_4

   1 
   2 
   3 
   4 
   5 
   6

Experiment 16

◇ Molar Mass Determination

To make ice cream at home, the ice cream mix is bathed and turned in a bucket containing a mixture of salt, ice, and water. Salt is used in the bath because it lowers the temperature of the ice-water slurry below 0°C. At the lower temperature the ice cream freezes much more quickly. Salts are also applied to sidewalks and highways in the wintertime (see photo) to lower the freezing temperature of the accumulated ice and snow. Could sugar be substituted for the salt in each case?

Antifreeze, or ethylene glycol, is added to cooling systems in automobiles to protect against freezing temperatures. While water freezes at 0°C, and could be used exclusively in cooling systems, the addition of the ethylene glycol appreciably lowers the freezing temperature.

In these instances the addition of a substance to water lowers its melting temperature; the substance also increases the boiling temperature of water as well. Try the experiment of adding salt to boiling water; the water stops boiling momentarily, *not* because the salt is cooler than the water, but because a higher temperature is needed for the boiling of a solution relative to a pure solvent.

OBJECTIVE

- To determine the molar mass of a nonvolatile solute by observing the difference between the freezing points of a solvent and a solution

PRINCIPLES

Colligative properties: properties of a solvent that result from the presence of the number of solute particles in the solution, and not their chemical composition.

The addition of a nonvolatile solute to a solvent produces a decrease in the freezing point and the increase in the boiling point of the solvent. These changes are called **colligative properties** of a solution, where the colligative property is due to the number of moles of solute particles dissolved in the solvent and not on the kind of solute. This means that one mole of sugar, $C_{12}H_{22}O_{11}$, and one mole of ethylene glycol have the same effect on the freezing point and boiling point of water. It is important to note that two moles of solute, whether they be molecules or ions, have nearly twice the effect on the melting and boiling points as does one mole; for example, two moles of sugar and one mole of salt, NaCl, (2 moles of solute particles, $NaCl(aq) \rightarrow Na^+(aq) + Cl^-(aq)$) have approximately the same effect. Other colligative properties are the lowering of the vapor pressure of a solvent and the phenomenon of **osmosis**.

Osmosis: the passage of a solvent through a semipermeable membrane from a solution of lower concentration to one of higher concentration.

Molality: an expression of the concentration of a solute dissolved in a solvent, equal to
$$\frac{moles\ of\ solute}{kg\ of\ solvent}$$

The freezing point change, ΔT_f, and boiling point change, ΔT_b, between the pure solvent and the solvent in solution are proportional to the **molality**, m, of the solute in solution. The proportionality is made an equality by inserting a constant; k_f and k_b are called the molal freezing and boiling point constants for the solvent. Constants for several solvents are listed in Table 16.1. The temperature changes are related to the molality, m, moles, and molar mass of the solute by the equations

$$\Delta T_f = k_f m = k_f \left[\frac{\text{mol solute}}{\text{kg solvent}} \right] = k_f \left[\frac{\frac{\text{g}}{\text{g/mol}}\ \text{solute}}{\text{kg solvent}} \right]$$

$$\Delta T_b = k_b m = k_b \left[\frac{\text{mol solute}}{\text{kg solvent}} \right] = k_b \left[\frac{\frac{\text{g}}{\text{g/mol}} \text{ solute}}{\text{kg solvent}} \right]$$

Table 16.1. Molal Freezing Point and Boiling Point Constants for Solvents

Substance	Freezing Point (°C)	k_f (°C•kg/mol)	Boiling Point (°C)	k_b (°C•kg/mol)
water	0.0	1.86	100.0	0.512
t-butanol	25.5	9.1	...	...
cyclohexane	*	20.0	80.7	2.79
naphthalene	80.2	6.9	....	....
acetic acid	16.6	3.90	118.3	3.07
camphor	178.4	37.7	....	....

* The freezing point of cyclohexane is measured in this experiment

Cyclohexane

In this experiment you will measure the freezing point change between pure cyclohexane[1] and a cyclohexane solution to determine the molar mass of a nonvolatile solute. A mass of the unknown solute, added to a measured mass of cyclohexane, causes a freezing point change, ΔT_f, which is measured. Since ΔT_f is proportional to the moles of solute added, the molar mass of the unknown is calculated.

$$\text{molar mass (g/mol)} = \frac{\text{mass of solute (g)}}{\text{moles of solute (mol)}}$$

PROCEDURE

You and a partner should complete at least two trials of Parts A and B of the experiment. A 110°C thermometer and a wire stirrer are needed from the stockroom. One of you should fill a 400 mL beaker with ice.

A. Freezing Point of Cyclohexane (Solvent)

1. Place a clean, dry 200 mm test tube in a 250 mL beaker and determine their combined mass (±0.01 g). Add approximately 15 g (15–20 mL) of cyclohexane to the test tube and again measure the mass, using the same balance (Figure 16.1).

2. a. Prepare about 300 mL of an ice-water slurry[2] in a 400 mL beaker. Maintain the ice-water slurry through Parts A.4 and B.

   b. Place the test tube containing the cyclohexane (the freezing point apparatus) in the ice-water slurry. Clamp a thermometer with a small three-pronged clamp and insert its bulb into the cyclohexane to measure its temperature (Figure 16.2).

3. While stirring the cyclohexane with the wire stirrer record the temperature (±0.1°C) *and* the time at regular intervals (10 s or 30 s) on your Data Sheet. The temperature remains nearly constant at the freezing point until the solidification is complete. Continue recording until the temperature begins to drop again.

---

[1]*t*-Butanol is a suitable substitute for cyclohexane in this experiment.

[2]Rock salt may be added to the ice-water slurry to keep the temperature of the slurry at a low temperature.

4. To repeat the measurement for the melting point of cyclohexane, remove the freezing point apparatus from the ice-water slurry until its temperature is near room temperature. Return the apparatus to the ice-water slurry and repeat Part A.3.

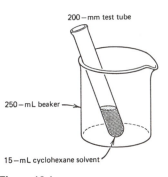

Figure 16.1

Determine the mass of cyclohexane in a beaker

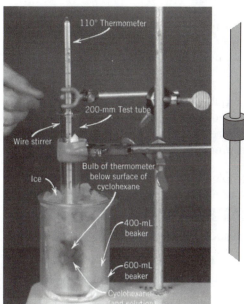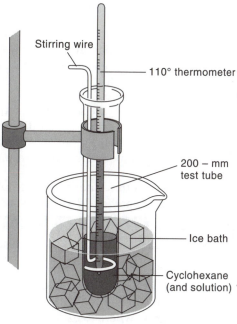

Figure 16.2

Freezing point apparatus for measuring the freezing point of cyclohexane and a cyclohexane solution

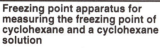

5. On linear graph paper, plot the temperature (°C, ordinate) *vs.* time (seconds, abscissa) to obtain the "cooling curve" for cyclohexane for both measurements. (See the solid line in Figure in 16.3.) Determine the freezing point of cyclohexane from the plots. Obtain your instructor's approval for your graph.

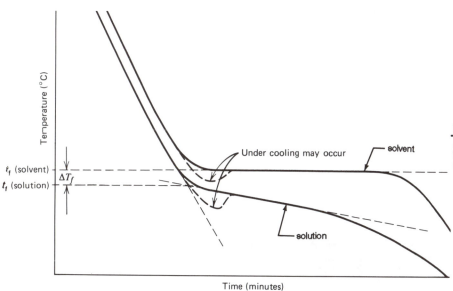

Figure 16.3

A sketch for the cooling curve for the cyclohexane and for the cyclohexane solution

B. Freezing Point of the Solution and Molar Mass of an Unknown Solute

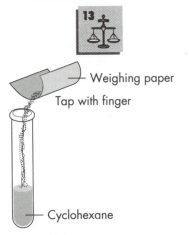

Weighing paper

Tap with finger

Cyclohexane

Figure 16.4
Curl the weighing paper and tap with the finger to transfer the sample to the solvent

1. Dry the outside of the test tube containing the cyclohexane and again measure its mass in the same 250 mL beaker with the same balance as in Part A.1. Obtain an unknown from your instructor. Measure approximately 0.2–0.4 g of the unknown (ask your instructor for the approximate mass to use).[3] Quantitatively transfer the unknown[4] to the cyclohexane in the 200 mm test tube (Figure 16.4).

2. Restore the ice-water slurry in the beaker. Determine the freezing point of the (now) solution in the same way as that for cyclohexane in Part A.3. When the solution nears its freezing point, record the temperature at more frequent intervals (15–20 s).

3. For Trial 2, warm the solution to about room temperature. Restore the ice-water slurry; add rock salt if suggested by your instructor. Measure an additional 0.2–0.3 g solute (see Part B.1) and transfer it to the solution. Repeat the temperature-time measurements in Part B.2.

4. (Optional) A third trial can be completed by repeating the procedure and adding another 0.2–0.3 g of solute to the solution in Part B.3.

Discard the test solution into a jar marked "Waste Cyclohexane Solution".

5. Plot the temperature *vs.* time data for the two (three) trials on the *same* graph as in Part A.5 to obtain the cooling curve for the solution. (See the broken line in Figure 16.3.) The curve will show a "break" at the temperature where freezing begins; this is only an approximate freezing point and is not as well defined as that for pure cyclohexane.

6. To determine the freezing point of the cyclohexane solution:
   • draw a line tangent to the curve *prior* to the freezing point.
   • draw a line tangent to the curve *after* the freezing point is reached.
   • at the intersection point of the two drawn tangents, draw a line parallel to the abscissa (time axis) until it intersects the ordinate (temperature axis)—this temperature value is the freezing point of the solution.

   Again obtain your instructor's approval for your graphs.

NOTES, OBSERVATIONS, AND CALCULATIONS

---

[3]If the unknown is a **solid**, measure its mass on weighing paper. If the unknown is a **liquid**, transfer about 3 mL from a 10 mL graduated cylinder into the cyclohexane; again measure the combined masses, now containing the liquid unknown. Calculate the mass difference to determine the mass of the liquid unknown.

[4]In the transfer, be certain that *none* of the solute adheres to the wall of the test tube. If some does, roll the test tube so that the cyclohexane contacts and dissolves the solute.

Molar Mass Determination

Date _____Name _____ Lab Sec. _____Desk No. _____

1. Define a colligative property.

2. What is the freezing point of a 1 m sugar (aqueous) solution?

3. A 0.442 g sample of a nonvolatile solute is dissolved in 15.0 g of t-butanol. The solution freezes at 23.9°C.
    a. What is the molality of the nonvolatile solute in solution?

    b. What is the molar mass of the solute?

    c. What would be the freezing point change, ΔT_f, of a solution containing 0.442 g of the solute dissolved in 15.0 g of naphthalene instead of the 15.0 g of t-butanol?

4. a. What mass of *p*-nitrotoluene must be dissolved in 15.0 g of *t*-butanol to lower its freezing point by 2.0°C? The molar mass of *p*-nitrotoluene is 137 g/mol.

    b. How many grams of *p*-nitrotoluene must be dissolved in 15.0 g of camphor to lower its freezing point by 2.0°C?

5. A pure solvent has a constant freezing point, but a solution does not. Why is this so?

6. a. In plotting data, which axis is the abscissa? _____

    b. In a plot of temperature *vs.* time for a boiling point measurement, which values are customarily plotted along the y-axis?

◇ Molar Mass Determination

Date _____ Name _____ Lab Sec. _____ Desk No. _____

A. Freezing Point of Cyclohexane (Solvent)

	Trial 1	Trial 2
1. Mass of beaker, test tube and cyclohexane (g)		
2. Mass of beaker and test tube (g)		
3. Mass of cyclohexane (g)		
4. Freezing point, from cooling curve (°C)		
5. Instructor's approval of graph		

B. Freezing Point of the Solution and Molar Mass of an Unknown Solute

	Trial 1	Trial 2
1. Mass of beaker, test tube, and cyclohexane (g)		
2. Mass of cyclohexane in solution (g)		
3. Mass of unknown solute in solution (g)		
4. Freezing point of solution, from cooling curve (°C)		
5. Instructor's approval of graph		
6. k_f for cyclohexane	20.0°C kg/mol	
7. Freezing point change, ΔT_f (°C)		
8. Molar mass of unknown solute (g/mol)	*	
9. Average molar mass (g/mol)		

* Show calculations for Trial 1.

QUESTIONS

1. a. If the freezing point of the cyclohexane solution is recorded 0.2°C lower than its actual freezing point, will the molar mass of the unknown be reported too high or too low? Explain.

   b. If the freezing point of pure cyclohexane is recorded 0.2°C lower than its actual freezing point, will the molar mass of the unknown be reported too high or too low? Explain.

2. If a thermometer is miscalibrated to read 0.5°C higher than the actual temperature over its entire scale, how will it affect the *reported* molar mass of the solute? Explain.

3. Cyclohexane is a volatile compound. If some of the cyclohexane evaporates during the experiment, how will its loss affect
   a. the ΔT_f in the experiment?

   b. the reported molar mass of the unknown solute? Explain.

4. Suppose the "pure" cyclohexane in today's experiment was initially contaminated with a nonvolatile solute.
   a. How would the ΔT_f measurement have been affected?

   b. Would the molar mass of the unknown have been reported as being too high, too low, or unchanged? Explain.

Experiment 17

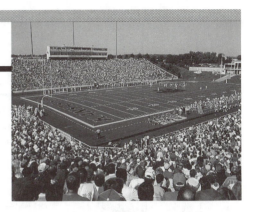

◇ Molar Mass of a Volatile Liquid

Gas molecules move at incredible velocities; for example, helium atoms travel at 1930 m/s while CO_2 molecules travel at 410 m/s at 25°C—the latter being approximately four lengths of a football field (see photo from Kansas State University) per second! The greater velocity of the helium atoms is attributed to its smaller mass. Even further, the molecules of a cologne travel from its source to the nose in a time span that is different from those of an orange blossom. All molecules have mass, but how do chemists measure the mass of a molecule? One modern instrumental technique employs the use of a mass spectrometer, an instrument that vaporizes the molecules at a high vacuum, ionizes them, and then subjects the molecular ions to electric and magnetic field for mass selection. This technique requires the molecule to be stable while in the gaseous state.

In this experiment we also take advantage of the volatility of a liquid, but instead use the basic principles of gas phase behavior to measure its molar mass.

OBJECTIVE

* To determine the molar mass of a volatile liquid and the density of its vapor

PRINCIPLES

Many compounds that we normally consider as liquids at room temperature and pressure may also be relatively volatile, that is, their intermolecular forces are so weak that the molecules readily escape into the vapor state. Gasoline, for example, is considered "volatile." As molecules in the vapor state, the compound exhibits all of the properties of a gas, and, within experimental error, follows ideal gas behavior.

The Dumas method (John Dumas, 1800–1884) uses the ideal gas law equation, $PV = nRT$, for determining the molar mass of a volatile liquid.

In this experiment a liquid is vaporized at a measured temperature, T, into a measured volume, V, of an Erlenmeyer flask. Once the barometric **pressure**, P, is read from the laboratory barometer, the moles of vapor, n, are calculated from the ideal gas law equation. When pressure data are in atmospheres, volume in liters, and temperature in kelvins, then the gas constant, R, equals 0.0821 L atm/(mol K).

Pressure: the force exerted by a gas over a surface area. Common laboratory units are atmospheres and Torr. The SI unit for pressure is the pascal.

The mass of the vapor, m, is measured from the difference between an empty Erlenmeyer flask and the vapor-filled flask. The molar mass of the volatile liquid is calculated from the equation,

Molar mass units: grams/mole (g/mol).

$$\text{molar mass} = \frac{m}{n}$$

The density of the vapor for the liquid is generally recorded at STP conditions.[1] A correction of the measured volume of the vapor in the

---

[1]STP (standard temperature and pressure) conditions are 273 K and 1 atm pressure.

Erlenmeyer flask to STP, V_{corr}, is made before the density of the vapor is reported.

$$\text{density of vapor (STP)} = \frac{m}{V_{corr}}$$

PROCEDURE

You are to complete two trials in today's experiment. Obtain no more than 15 mL of an unknown volatile liquid from your laboratory instructor and check out a 110°C thermometer from the stockroom. Record the sample number of your unknown.

1. a.   Set up the apparatus shown in Figure 17.1. Clean a 125 mL Erlenmeyer flask with soap and water, rinse thoroughly with deionized water, and *dry* either in a drying oven or by allowing it to air-dry.

    b.   Fit the mouth of the flask with aluminum foil and place a rubber band around the foil and the neck of the flask. The aluminum foil should be no larger than what is necessary to cover the mouth of the flask and secure it.[2] Measure the mass (±0.001 g) of the *dry* flask and aluminum foil/rubber band.

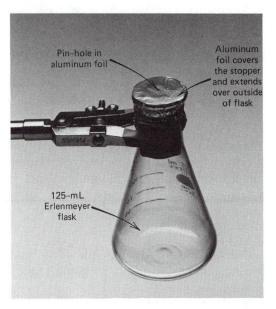

Figure 17.1

Foil-covered flask to contain the volatile liquid

2. a.   Half-fill a 600 mL beaker with tap water, add 2 or 3 drops of 6 M HCl,[3] and a boiling stone. Support the beaker of water on a wire gauze and an upper support ring. Heat the water to boiling.

    b.   While waiting for the water to boil, add approximately 6 mL of your unknown volatile liquid to the flask and cover with the foil/rubber band.

---

[2]If the aluminum covering is too large, water will become entrapped beneath the foil in Part 5a and will contribute to the mass of the volatile unknown liquid.

[3]The HCl prevents the buildup of mineral deposits on the Erlenmeyer flask and the beaker as the water evaporates.

3. a. Remove the heat from the hot water bath. Place the covered Erlenmeyer flask, containing your unknown volatile liquid, in the hot water bath and secure it with a utility clamp; be certain the flask does not touch the beaker wall (Figure 17.2). Adjust the water level high on the neck of the flask; you may need to add or remove water from the bath. Punch 2–4 pin-sized holes in the aluminum foil.

b. Resume heating of the hot water bath. (**Caution**: *most of the volatile liquid unknowns are flammable; use a moderate flame for heating.*)

4. As the unknown volatile liquid is heated, some of its vapor escapes through the pinholes of the aluminum foil. When vapors are no longer visible,[4] continue heating for another 5 minutes. The flask should now be filled with the vapor of your unknown—no liquid should remain in the flask. Record the temperature (±0.1°C) of the boiling water.

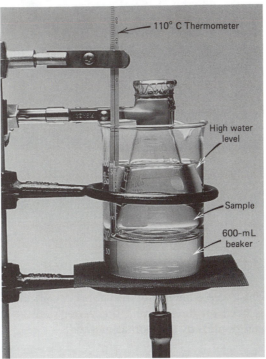

110° C Thermometer

High water level

Sample

600-mL beaker

Figure 17.2
Setup for the molar mass determination of a volatile liquid

5. a. Remove the flask from the hot water bath. Allow to cool to room temperature. Sometimes the vapor remaining inside the flask condenses; that's OK. Dry the outside of the flask and the aluminum foil/rubber band assembly.

b. Measure the combined mass of the apparatus, which contains some liquid and vapor of the unknown volatile liquid, using the same balance that was used earlier.

6. To repeat the procedure, add 5 mL of the unknown volatile liquid to the Erlenmeyer flask and repeat Parts 3 through 5.

Discard the liquid unknown in the "Waste Organic Liquids" container.

---

[4]To see vapors escaping through the pinholes, look across the top of the aluminum foil toward a lighted area. The vapor causes a diffraction of the lighted area.

7. Fill the empty 125 mL Erlenmeyer flask to the brim with tap water. Measure the volume (±0.1 mL) of the flask by transferring portions of the water to a 50 mL graduated cylinder until it is all transferred. Sum the volumes of water and record the total volume.

8. Read the laboratory **barometer** and record the barometric pressure.

NOTES, OBSERVATIONS, AND CALCULATIONS

This is a commercially available borosilicate glass bulb with a drawn tip that is used for determining the vapor density and the molar mass of volatile compounds by the Dumas method.

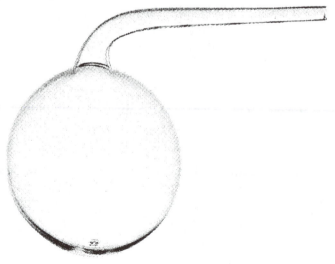

Molar Mass of a Volatile Liquid

Date _____Name _____ Lab Sec. _____Desk No. _____

1. The vapor from an unknown volatile liquid occupies a 269 mL Erlenmeyer flask at 98.7°C and 748 Torr.
    The mass of the vapor is 0.791 g.
    a. How many moles of vapor are present?

    b. What is the molar mass of the volatile liquid?

    c. What is the density of the vapor at STP conditions?

2. a. If the atmospheric pressure is mistakenly recorded as 760 Torr in Question #1, what would be the
        reported molar mass of the volatile liquid?

    b. What would be the percent error in the reported molar mass caused by this assumed pressure?

3. Explain how the mass of the vapor for the unknown volatile liquid is measured in today's experiment.

4. What is the purpose of placing a boiling stone into the beaker of a hot water bath?

◇ Molar Mass of a Volatile Liquid

Date _____Name _____ Lab Sec. _____ Desk No. _____

	Trial 1	Trial 2
Unknown Number _____		
1. Mass of dry flask + Al foil/rubber band (*g*)		
2. Mass of dry flask + Al foil/rubber band, and volatile liquid (*g*)		
3. Mass of volatile liquid, **m** (*g*)		
4. Temperature of boiling water, **T** (°C)		
5. Volume of 125 mL flask _____+_____+_____ , **V**(*mL*)		
6. Barometric pressure, **P** (*atm*)		
7. Moles of volatile liquid, **n** (*mol*)		
8. Molar mass of volatile liquid, **m/n** (*g/mol*)	*	
9. Average molar mass of volatile liquid (*g/mol*)		
10. Density of vapor for volatile liquid at STP (*g/L*)		

*Show your calculations for Trial 1.

QUESTIONS

1. If the mass of the flask is measured after the liquid has been vaporized (in Part 5a) but before the outside of the flask is dried (or if water has become entrapped between the aluminum foil and the flask), will the molar mass of the unknown volatile liquid be too high or too low? Explain.

2. a. Suppose the barometric pressure during the experiment is recorded as 780 Torr instead of an actual value of 750 Torr. Would the molar mass of the unknown volatile liquid be reported too high or too low? Explain.

   b. What would be the percent error for the molar mass if the error had been made?

3. If the volume of the vapor is assumed to be 125 mL instead of the measured volume in today's experiment, what would be the percent error for the molar mass of the unknown volatile liquid? Show your work.

4. If all of the unknown volatile liquid does not vaporize into the 125 mL Erlenmeyer flask in Part 4, will the reported molar mass be too high or too low? Explain.

Experiment 18

◈ Acids, Bases, and Salts

Fresh water lakes are acidic, salt water bays and estuaries (see photo of salt marsh near Ipsovich, Massachuestts) are basic. The foods and fluids that we consume are acidic (as are our stomachs) but our intestinal tract and blood are basic. Acids are corrosive, but bases remove layers of skin. The extent of such effects depend upon the strength and the concentration of the acid or base. All aqueous solutions are generally either acidic or basic, few are "neutral". Because aqueous solutions are so important to our livelihood, then it becomes obvious as to why we should become familiar with the sources and properties of "home" and laboratory acids and bases, many of which exist because of a particular ion of a salt dissolved in water.

OBJECTIVE

- To become familiar with the chemical and physical properties of acids, bases, and salts

PRINCIPLES

Acids are substances that, when dissolved in water, produce hydronium ion, H_3O^+. For example, pure HCl is a gas at room temperature and pressure, but dissolved in water, it releases H^+ to a water molecule, forming the hydronium ion, H_3O^+.

$$HCl(g) + H_2O(l) \rightarrow H_3O^+(aq) + Cl^-(aq)$$

Bases dissolved in water produce hydroxide ion, OH^-. Barium hydroxide, $Ba(OH)_2$, is a solid at room temperature and pressure, but in water, it dissociates into Ba^{2+} and OH^- ions.

$$Ba(OH)_2(s) \xrightarrow{-H_2O} Ba^{2+}(aq) + 2\,OH^-(aq)$$

When aqueous solutions of an acid and a base are mixed, a reaction occurs producing water and a salt as products. This is a **neutralization reaction**. For example, a mixture of HCl and $Ba(OH)_2$ produce water and the salt barium chloride, $BaCl_2$, as products.

$$2\,HCl(g) + Ba(OH)_2(s) \xrightarrow{-H_2O} 2\,H_2O(l) + BaCl_2(aq)$$

A **salt** therefore is any **ionic compound** that is a neutralization product of an acid-base reaction.

Ionic compound: a compound that consists of a ratio of cations and anions that results in a neutral compound.

But in water, hydrogen chloride actually exists as H_3O^+ and Cl^- and barium hydroxide as Ba^{2+} and OH^-; barium chloride is a soluble, ionic compound—in an aqueous solution it dissociates into Ba^{2+} and Cl^-. Therefore, a better representation of the reaction between HCl and $Ba(OH)_2$ in aqueous solution is with the **ionic equation**,

Ionic equation: an equation that represents the species (ions, molecules, etc.) in the form of their actual existence in an aqueous solution.

$$2\,H_3O^+(aq) + 2\,Cl^-(aq) + Ba^{2+}(aq) + 2\,OH^-(aq) \rightarrow$$
$$4\,H_2O(l) + Ba^{2+}(aq) + 2\,Cl^-(aq)$$

Net ionic equation: an equation that presents only the species (ions, molecules, etc.) that participate in a measurable/observable chemical reaction.

Considering only the ions participating in the neutralization reaction, written as a **net ionic equation**, the equation becomes

$$H_3O^+(aq) + OH^-(aq) \rightarrow 2\,H_2O(l)$$

Electrolyte: a substance, usually a solute dissolved in an aqueous solution, that conducts an electric current because of the movement of ions.

Electrolytes are substances that dissociate in water to produce ions and conduct an electrical current. Some compounds are **strong electrolytes** in water—these substances completely dissociate ($\approx$100%) into ions (Figure 18.1, left). Hydrochloric acid, HCl, barium hydroxide, $Ba(OH)_2$, and barium chloride, $BaCl_2$, are strong electrolytes. **Nonelectrolytes** are substances that dissolve in water but do not form ions ($\approx$0% ionization). Sugar and alcohol are nonelectrolytes. A large number of compounds only partially dissociate in an aqueous solution; these are called **weak electrolytes** (Figure 18.1, right). Acetic acid is a weak electrolyte.

Figure 18.1
Strong electrolytes are excellent conductors of electrical current (left); weak conductors are not.

Acids and bases may be strong or weak, concentrated or dilute. "Strong and weak" refer to the degree of ion formation. Strong acids/bases completely dissociate ($\approx$100%) into ions; weak acids/bases only partially dissociate—often less that 5% of the molecules dissociate to form ions. "Concentrated and dilute" refer to the amount of acid/base dissolved in aqueous solution. Concentrated acids/bases have a relatively large amount dissolved, dilute acids/bases have only a relatively small amount dissolved.

Acids

Nonmetallic oxides: also called acidic **anhydrides**. "Anhydride" means without water.

Acidic solutions result from the action of water on (1) nonmetallic hydrides, such as HCl and HBr, (2) **nonmetallic oxides**, such as CO_2 and SO_3, or (3) compounds of hydrogen, oxygen, and one other element (usually a nonmetal), such as H_2SO_4 and HNO_3.

$$HCl(g) + H_2O(l) \rightarrow H_3O^+(aq) + Cl^-(aq)$$
$$CO_2(g) + H_2O(l) \rightarrow H_3O^+(aq) + HCO_3^-(aq)$$
$$H_2SO_4(l) + H_2O(l) \rightarrow H_3O^+(aq) + HSO_4^-(aq)$$

Sulfuric acid, also called oil of vitriol, is perhaps the most versatile of all inorganic industrial chemicals. There is hardly a chemical industry that does not use sulfuric acid in its manufacturing process. In 1994, when it ranked number one in chemical usage, over 89 billion pounds of H_2SO_4 were produced in the United States, nearly twice that of the second ranked chemical. Other acids that ranked among the "Top 50" chemicals in production were phosphoric acid, H_3PO_4, nitric acid, HNO_3, and hydrochloric acid, HCl (called muriatic acid). See Table 18.1.

acetic acid

In addition, many organic acids that rank in the "Top 50", such as acetic, adipic, and oleic acids, are useful and important to the chemical industry.

Bases

Metallic oxides: also called basic anhydrides.

Three sources of basic solutions are: (1) metallic hydroxides, such as $Ba(OH)_2$ and $NaOH$, (2) **metallic oxides**, such as Na_2O, or (3) a select number of polyatomic anions, such as PO_4^{3-} and CO_3^{2-}.

$$Ba(OH)_2(s) \xrightarrow{-H_2O} Ba^{2+}(aq) + 2\,OH^-(aq)$$
$$Na_2O(s) + H_2O(l) \rightarrow 2\,Na^+(aq) + 2\,OH^-(aq)$$
$$PO_4^{3-}(aq) + H_2O(l) \rightarrow HPO_4^{2-}(aq) + OH^-(aq)$$

The strong bases such as $NaOH$ (called caustic soda or lye) and KOH (called caustic potash) are also known as **alkalis**.

Table 18.1. Acids and Bases Ranked Among the "Top 50" in Production for 1994 in the United States from *Chemical and Engineering News* **, April 10, 1995**

Rank	Chemical	Formula	Billions of Pounds (1994)
1	sulfuric acid	H_2SO_4	89.20
5	lime	CaO	38.35
6	ammonia	NH_3	37.93
8	sodium hydroxide	$NaOH$	25.83
9	phosphoric acid	H_3PO_4	25.26
11	sodium carbonate	Na_2CO_3	20.56
13	nitric acid	HNO_3	17.65
28	hydrochloric acid	HCl	6.71
34	acetic acid	CH_3COOH	3.82
46	adipic acid	$C_4H_8(COOH)_2$	1.80

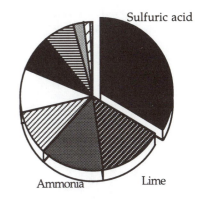

The most notable bases ranked in the "Top 50" are ammonia, NH_3, sodium hydroxide, calcium oxide, CaO (also called quicklime or, more simply, lime), and sodium carbonate, Na_2CO_3 (also known as soda ash). Again, refer to Table 18.1. The pie diagram summarizes the data of Table 18.1—sulfuric acid contributed 33.4% of the total amount of acids and bases produced in the United States in 1994.

A number or organic bases are also important to the chemical industry, most notable are hydroxylamine and aniline.

During this experiment, **STOP** at each numbered superscript (i.e., [1]) and record your observation(s) or conclusion(s) on the Data Sheet.

PROCEDURE

Caution: *Dilute and concentrated (conc) acids and bases cause severe skin burns and irritation to mucous membranes. Be very careful in handling these chemicals. Clean up all spills immediately with excess water, followed by a covering of baking soda, $NaHCO_3$. Notify your instructor if a spill occurs. Read the "Laboratory Safety" section of Experiment 1.*

A. Conductivity

The apparatus shown in Figure 18.2 is used to determine the strength, or extent of dissociation, of various electrolytes. When the two electrodes are submerged in a solution with a high concentration of ions (a strong electrolyte), the circuit is completed and the bulb shines brightly; if the solution is a weak electrolyte, the bulb burns dimly; or if a nonelectrolyte, not at all. Therefore, the brightness of the bulb is a qualitative measure of the degree of dissociation of a substance into ions.

1. Connect the apparatus to an electrical outlet. Half-fill a 100 mL beaker with deionized water and place it in contact with the electrodes. Does the bulb glow?[1] Add about 0.5 g of NaCl to the water and stir to dissolve. What happens to the conductivity of the solution?[2] Is NaCl a strong, weak, or nonelectrolyte?[3] Remove the solution and rinse the electrodes with deionized water from your wash bottle.

2. Place ≈20 mL of conc acetic acid (read the **Caution** for handling acids) into a clean 100 mL beaker . Place the acid in contact with the electrodes. Classify its strength as an electrolyte.[4] While keeping the acetic acid in contact with the electrodes, add deionized water from your wash bottle and swirl the solution. Observe. Continue to add water (up to ≈75 mL), swirl, and observe the glow of the bulb. What happens to the

conductivity as the acetic solution is diluted?[#5] Rinse the electrodes with deionized water and discard the rinse.

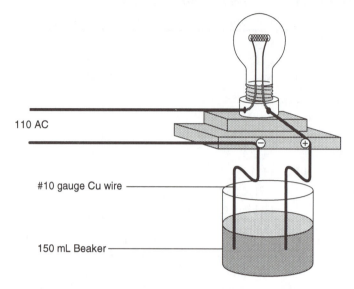

Figure 18.2

Apparatus for testing the conductivity of a solution

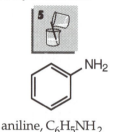

aniline, $C_6H_5NH_2$

110 AC

#10 gauge Cu wire

150 mL Beaker

3. Test the conductivity of the following solutions.* Be sure to rinse the electrodes after each test to avoid contamination of the solutions.

   0.1 M HCl, 0.1 M NH$_3$, 0.1 M NaOH, 0.1 M NaNO$_3$, 0.1 M NH$_4$Cl, 0.1 M $C_{12}H_{22}O_{11}$, and 0.1 M $C_6H_5NH_2$ or other test solutions as suggested by your instructor. Record your observations on the Data Sheet. Classify the solutions as strong, weak, or nonelectrolytes.[#6]

4. Place 20 mL of 0.1 M H$_2$SO$_4$ in a 100 mL beaker; submerge the electrodes.[#7] Obtain about 15 mL of 0.2 M Ba(OH)$_2$. Maintain the electrodes in contact with the solution and slowly add the 0.2 M Ba(OH)$_2$; keep the mixture stirred. What happens to the conductivity of the mixture? Why?[#8] Continue adding 0.2 M Ba(OH)$_2$ until the glow disappears. Now add additional Ba(OH)$_2(aq)$. Explain your observations.[#9]

5. Disconnect the apparatus from the electrical outlet.

B. Concentrated Acids

1. Several bottles of concentrated acids (purchased from the chemical supplies distributor) are on the reagent shelf. Select one of the acids and closely read the label. Answer the questions on the Data Sheet.

C. Acids: Sulfuric, Nitric, Hydrochloric, and Acetic Acids

1. Place 10 drops of water in wells A1–A4 in a 24 well plate. Place a finger flush underneath well A1 and add, while swirling, 5 drops of conc HCl; is noticeable heat produced?[#10] Test the solution with litmus paper.[#11] Repeat the tests in wells A2–A4 with 5 drops (*slowly*) of conc H$_2$SO$_4$, conc HNO$_3$, and conc CH$_3$COOH (**Caution**: *do not allow the concentrated acids to touch the skin*).[#12]

---

*
 To avoid waste, the test solutions can be shared with other students, especially if care is taken to avoid contamination.

2. Some acids can be prepared by the action of another acid on a salt. H_2SO_4 and H_3PO_4 are often used as the acids for the preparation. Place about 0.1 g of NaCl in a well C1. Slowly, while swirling, add 3–4 drops of conc H_2SO_4 to the solid salt and hold both red and blue moistened litmus papers over the reaction. Explain the effect on litmus.[13] Write a balanced equation for the reaction.[14]

3. Place a small piece of polished (with steel wool) Mg, Zn, Fe, and Cu in a wells B1–B4. *Slowly* add 10 drops of 6 *M* HCl to each metal. After each drop and after several minutes, observe any chemical reaction that occurs in the mix.[15] Draw off the HCl(*aq*) solution with a Beral pipet and discard in a waste acids beaker. Repeat the test with 6 *M* HNO_3.[16]

4. **Instructor Demonstration.** Test the dehydrating effects of conc H_2SO_4 by placing a few drops on a wood splint and, in an evaporating dish, on a small amount (≈1 g) of sugar. Repeat the test with conc HCl and conc HNO_3. Record your observations for each acid on the Data Sheet.[17-19]

5. Pipet 1 mL of 0.1 *M* NaOH into each wells D1–D3 and add 1 drop of phenolphthalein indicator. Add drops of 0.5 *M* HCl to well D1 until a color change (from pink to colorless) occurs (swirl after each drop). Record the number of drops.[20]

Repeat the determination, substituting 0.5 *M* H_2SO_4 and 0.5 *M* HNO_3 for the 0.5 *M* HCl.[21,22] What can you conclude about the available acidity of the three acids?[23] Write balanced equations for the reactions.[24]

Discard the acid test solutions of the 24 well plate into the "Waste Acids" container.

1. In wells A1–A4 of a 24 well plate, place in succession a small BB-sized piece of NaOH , a sample of oven cleaner or solid drain cleaner, a fresh sample of CaO, and a sample of Na_2CO_3. Place the tip of your finger flush underneath the well and add 1 mL of water to the first well A1. Note the heat generated in the dissolving of the NaOH.[25] Repeat the test with the other samples. Test each solution with litmus.[26]

D. Bases: Sodium Hydroxide, Calcium Oxide, and Sodium Carbonate, and Ammonia

2. Set up the apparatus shown in Figure 18.3. Place 3 g (±0.01 g) of a 2:1 mixture (by volume of solid) of NH_4Cl and $Ca(OH)_2$ into the 200 mm test tube. Heat the mixture gently; begin at the mouth of the test tube and gradually extend the heat over the entire test tube. As soon as NH_3 gas evolves freely and the apparatus is free of air, collect *two* bottles of the NH_3 by air displacement.[§] Write balanced equations for the reaction of NH_4Cl with $Ca(OH)_2$ and for the reaction of HCl with NH_3 (see footnote).[27] Note the color and odor of the NH_3.[28]

[§]To test if a bottle is filled with NH_3, suspend a drop of conc HCl from the tip of a glass rod at the mouth of the bottle; the conc HCl fumes strongly when ammonia is present. Consult the instructor on this technique.

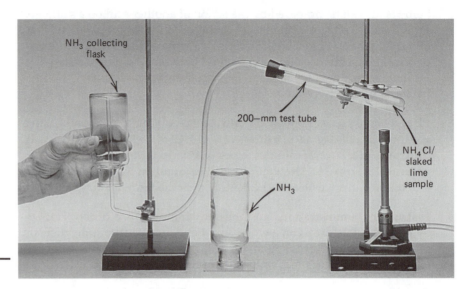

3. To test for the solubility of NH_3 in water, place about 300 mL of water in an 800 mL beaker. Add ≈5 drops of **phenolphthalein** to the water. Place the mouth of a bottle of NH_3 under the surface of the water (Figure 18.4). Allow it to remain there for several minutes; be sure that the mouth of the bottle stays submerged. What happens to the water level in the bottle?[29] What happens to the color of the solution? Explain. [30]

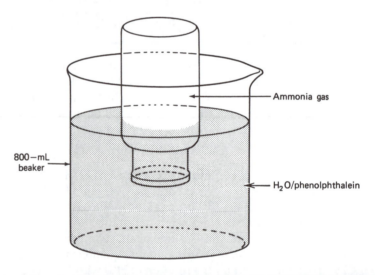

Figure 18.4

A setup for testing the solubility of NH₃(g) in water

Dispose of the base test solutions in the "Waste Bases" container.

Acids, Bases, and Salts

Date _____Name _____ Lab Sec. _____ Desk No. _____

1. a. What is an acid?

   b. What is a base?

   c. What is a salt?

2. a. Distinguish between a strong electrolyte, a weak electrolyte, and a nonelectrolyte.

   b. Distinguish between a concentrated solution and a dilute solution.

3. a. Sketch at right a (molecular/ionic) "picture" of a
   hydrochloric acid solution, showing the species present.

   b. How is the hydrochloric acid picture different from
   that of an acetic acid solution? Sketch at right the appearance
   of an acetic acid solution.

4. a. What is a neutralization reaction?

   b. Write a net ionic equation to represent a neutralization reaction?

5. a. What is the color of litmus paper in an acidic solution? _____

   b. What is the color of litmus paper in a basic solution? _____

   c. What is the color of phenolphthalein in an acidic solution? _____

   d. What is the color of phenolphthalein in a basic solution? _____

6. Describe the test to determine if a flask is filled with $NH_3(g)$.

7. List the common (or commercial) names and the correct chemical names associated with these compounds.

	Common (commercial) Names	Chemical Names
a. Na_2CO_3		
b. HCl		
c. H_2SO_4		
d. NaOH		
e. CaO		
f. $Ca(OH)_2$		
g. $CaCO_3$		
h. NH_3		
i. $NaHCO_3$		

8. How should acid spills be cleaned up in the laboratory?

◇ Acids, Bases, and Salts

Date _____ Name _____ Lab Sec. _____ Desk No. _____

A. Conductivity

	Compound	Observation	Electrolyte		
			Strong	Weak	Non–
#1	H_2O				
#2,3	NaCl				
#4	conc CH_3COOH				
#5	dil CH_3COOH				
#6	0.1 M HCl				
	0.1 M NH_3				
	0.1 M NaOH				
	0.1 M $NaNO_3$				
	0.1 M NH_4Cl				
	0.1 M $C_{12}H_{22}O_{11}$				
	0.1 M $C_6H_5NH_2$				
#7	0.1 M H_2SO_4				

#8,9   Describe and account for the changes in conductivity of the H_2SO_4 solution as the $Ba(OH)_2$ solution is being added.

B. Concentrated Acids

1.   What is the percent (range) by mass of the acid in the bottle? _____

     What is its major impurity? _____

2.   List two **Danger** warnings printed on the label.

     a.   _____

     b.   _____

3. List two **First Aid** remedies for the acid.

a. _____

b. _____

C. Acids

	Acid	Heat Change	Litmus Test
#10,11	HCl	_____	_____
#12	H_2SO_4	_____	_____
	HNO_3	_____	_____
	CH_3COOH	_____	_____

#13   Litmus test on vapors _____

#14   Balanced equation: $NaCl + H_2SO_4 \rightarrow$ _____

Acid/Metal	Mg	Zn	Fe	Cu
#15 HCl	_____	_____	_____	_____
#16 HNO_3	_____	_____	_____	_____

Acid	conc H_2SO_4	conc HCl	conc HNO_3
#17-19 wood splint	_____	_____	_____
sugar	_____	_____	_____

Acid	HCl	H_2SO_4	HNO_3
#20-22 Drops of NaOH	_____	_____	_____

#23   Conclusion on available acidity

#24   Balanced equations: $NaOH + HCl \rightarrow$ _____

                                   $NaOH + H_2SO_4 \rightarrow$ _____

                                   $NaOH + HNO_3 \rightarrow$ _____

D. Bases

	Heat Change	Litmus Test
#25 NaOH	_____	_____
#26 Oven/Drain Cleaner	_____	_____
CaO	_____	_____
Na_2CO_3	_____	_____

#27   Balanced equations: $NH_4Cl + Ca(OH)_2 \rightarrow$ _____

                                   $HCl + NH_3 \rightarrow$ _____

#28   Color of NH_3 _____ . Does NH_3 have an odor? _____

#29   Is NH_3 soluble in water? _____

#30   Why does the color of the solution change?

Experiment 19

LeChatelier's Principle

The flow of traffic across a bridge seems to never cease (see photo of George Washington Bridge, NYC, at night). In the morning hours the traffic moves preferentially in one direction, but in the evening the flow is reversed. Although construction and accidents often inhibit and alter its flow, traffic never seems to stop; even at midday and at midnight, the traffic continues to flow. It is evident that traffic flow is continuous and dynamic, its direction and magnitude is dependent upon the time of day and upon the conditions of the highway. The "stresses" to continuous and dynamic traffic flow are varied and many.

The movement of molecules in physical systems or in chemical reaction systems is also continuous and dynamic, and is influenced by many stresses. Suppose that two tanks (#1 and #2) of water are connected by the glass capillary tubing as shown in the diagram. A stopcock is connected to each tank. If we focus on the water molecules that are moving from one tank through the glass tubing to the other, we can visualize an equal flow of water molecules. This condition is called a **dynamic equilibrium** (as contrasted to a static equilibrium) in a physical system. Now if we open a stopcock in tank #2 to upset this equilibrium condition, what can we predict?

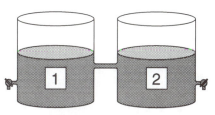

OBJECTIVES

- To study the effects of concentration and temperature changes on the equilibrium position in various chemical systems
- To study the pH effect of strong acid and strong base addition on buffered and unbuffered systems

PRINCIPLES

A chemical system reaches a dynamic equilibrium when reactants combine to form products at the same rate at which products combine to reform reactants, a condition that exists in reversible reactions. For the reaction,

$$2\,SO_2(g) \; + \; O_2(g) \; \rightleftarrows \; 2\,SO_3(g) \; + \; 197\,kJ$$

chemical equilibrium is reached when the reaction rate for SO_2 and O_2 forming SO_3 is the same as that for SO_3 molecules reforming SO_2 and O_2.

If the amount of one of the species in the *equilibrium* system changes or the temperature changes, then the equilibrium position of the reaction system tends to shift compensating for the change. For example, when more SO_2 is added it reacts (stoichiometrically) with the O_2, increasing the amount of SO_3 until a new dynamic equilibrium is established. The new equilibrium has more SO_2 (because it was added) and SO_3 (because it formed from the reaction), but less moles of O_2 (because it reacted with the SO_2). Therefore, the addition of the SO_2 resulted in a *shift* in the position of the original equilibrium to the *right*, a shift to compensate for the added SO_2.

Figure 19.1
Henri Louis LeChatelier (1850–1936)

A general statement governing all chemical systems at dynamic equilibrium is:

> *if an external stress (change in concentration, temperature, ...) is applied to a chemical system at equilibrium, the equilibrium shifts in the direction which minimizes the effect of that stress*—this is **LeChatelier's Principle,**

proposed by Henri Louis LeChatelier in 1888 (Figure 19.1). For instance, if more bleach is added to the laundry, more stains are removed (more oxidation occurs) as the bleach is consumed.

Often the concentrations of the species in a chemical system at equilibrium can be quantitatively determined. From this, an equilibrium constant is calculated; its magnitude then indicates the position of the equilibrium. Equilibrium constants are determined in Experiments 22 and 23.

Changes in Concentration

Many salts are only slightly soluble in water. The Ag^+ ion forms slightly soluble salts with many anions. This experiment studies several equilibria involving the relative solubility of the CO_3^{2-}, Cl^-, I^-, and S^{2-} silver salts; an explanation of the sequential testing of the equilibria follows.

Silver carbonate equilibrium. Ag_2CO_3 precipitates in the presence of excess Ag^+ and CO_3^{2-} ions in aqueous solution to establish a dynamic equilibrium between the solid and the ions in solution.

$$Ag_2CO_3(s) \rightleftarrows 2\,Ag^+(aq) + CO_3^{2-}(aq)$$

Figure 19.2
"Insoluble" silver chloride quickly forms in a mixture of silver ions and chloride ions

Addition of HNO_3 causes Ag_2CO_3 to dissolve: the H^+ ions from the HNO_3 react with the CO_3^{2-} ions, removing the CO_3^{2-} from the equilibrium. The system shifts *right* to compensate for the removal of the CO_3^{2-} ions, causing the Ag_2CO_3 to dissolve. The H^+ and CO_3^{2-} combine to form H_2CO_3 which, because of its instability at room temperature and pressure, decomposes to CO_2 and H_2O.

$$CO_3^{2-}(aq) + 2\,H^+(aq) \text{ from } HNO_3 \rightarrow H_2CO_3(aq) \rightarrow H_2O(l) + CO_2(g)$$

The Ag^+ and NO_3^- remain in solution.

Silver chloride equilibrium. The addition of Cl^- ions to the Ag^+ ions remaining in solution causes solid, white $AgCl$ to form (Figure 19.2).

$$Ag^+(aq) + Cl^-(aq) \rightleftarrows AgCl(s)$$

When $NH_3(aq)$ is added to this $AgCl$ equilibrium, a complex ion of Ag^+ and NH_3, the diamminesilver ion, $[Ag(NH_3)_2]^+$, forms

$$Ag^+(aq) + 2\,NH_3(aq) \rightarrow [Ag(NH_3)_2]^+(aq)$$

The Ag^+ ion is therefore removed from the $AgCl$ equilibrium, causing the dissolution of the $AgCl$. The $[Ag(NH_3)_2]^+$ and Cl^- ions remain in solution.

Adding acid to this system results in an acid-base reaction with the NH_3 $[NH_3(aq) + H^+(aq) \rightarrow NH_4^+(aq)]$. This destroys the $[Ag(NH_3)_2]^+$ complex releasing the Ag^+, which then recombines with the Cl^- to form $AgCl(s)$.

$$[Ag(NH_3)_2]^+(aq) + 2\,H^+(aq) \rightarrow Ag^+(aq) + 2\,NH_4^+(aq)$$
$$Ag^+(aq) + Cl^-(aq) \rightleftarrows AgCl(s)$$

Silver iodide equilibrium. Iodide ion from KI added to the $[Ag(NH_3)_2]^+$ in solution precipitates yellow AgI.

$$[Ag(NH_3)_2]^+(aq) + I^-(aq) \rightleftarrows AgI(s) + 2\,NH_3(aq)$$

This occurs because of the greater *in*solubility of AgI compared to AgCl. The AgI equilibrium is established.

$$AgI(s) \rightleftarrows Ag^+(aq) + I^-(aq)$$

Silver sulfide equilibrium. Silver sulfide, Ag_2S, is even less soluble than AgI. Addition of S^{2-} ion (from Na_2S) precipitates Ag^+ from the AgI equilibrium. The AgI equilibrium shifts right (therefore the AgI dissolves) and the Ag_2S precipitates.

$$2\,Ag^+(aq) + S^{2-}(aq) \rightarrow Ag_2S(s)$$

Common-ion effect. The solubility of a salt is affected by the addition of a common-ion. In the case of ammonium chloride, NH_4Cl, a soluble salt, the addition of concentrated amounts of NH_4^+ ion or Cl^- ion can decrease its solubility.

Common-ion effect: the effect of adding an ion or ions common to those in an existing equilibrium.

$$NH_4Cl(s) \rightleftarrows NH_4^+(aq) + Cl^-(aq)$$

In addition to the ammonium chloride equilibrium, the common-ion effect is also observed for the following equilibria in this experiment.

$$4\,Cl^-(aq) + [Co(H_2O)_6]^{2+}(aq) \rightleftarrows [CoCl_4]^{2-}(aq) + 6\,H_2O(l)$$
$$CH_3COOH(aq) + H_2O(l) \rightleftarrows H_3O^+(aq) + CH_3CO_2^-(aq)$$

The reaction of SO_2 with O_2 producing SO_3 is exothermic by 197 kJ.

Changes in Temperature

$$2\,SO_2(g) + O_2(g) \rightleftarrows 2\,SO_3(g) + 197\ kJ$$

To favor the formation of SO_3, the reaction vessel is kept cool. Removal of heat favors the shift of the equilibrium to the *right* forming additional SO_3.

This experiment examines the effect of temperature on:

$$4\,Cl^-(aq) + \begin{bmatrix} & H_2O & \\ H_2O{,}_{,} & | & \,{,}OH_2 \\ & \diagdown Co \diagup & \\ H_2O & | & OH_2 \\ & H_2O & \end{bmatrix}^{2+}(aq) \rightleftharpoons \begin{bmatrix} & Cl & \\ & | & \\ Cl & Co{,}_{,} & Cl \\ & \diagdown Cl & \end{bmatrix}^{2-}(aq) + 6\,H_2O(aq)$$

$$4\,Cl^-(aq) + [Co(H_2O)_6]^{2+}(aq) \rightleftarrows [CoCl_4]^{2-}(aq) + 6\,H_2O(l)$$

This system involves an equilibrium between the "coordination sphere" about the cobalt(II) ion which is concentration *and* temperature dependent.

PROCEDURE

At each superscript (i.e., [#1]) in the Procedure, **STOP**, and record your observations. Prepare a hot water bath for Part C.2.

A. Silver Ion Equilibria

1. Add 10 drops of 0.1 M Na_2CO_3 to 10 drops of 0.1 M $AgNO_3$ in well A1 of a 24 well plate.[#1] Now add drops of 6 M HNO_3 until change no longer occurs. What did you see?[#2]

2. Now add drops of 0.1 M HCl, again until no further change occurs.[#3] Allow the precipitate to settle and withdraw about one-half of the supernatant with a Beral pipet and properly discard. Add *drops* of conc NH_3 (**Caution:** *avoid inhalation and skin contact with conc* NH_3) until the

precipitate dissolves and the solution clears.[#4] Reacidify the solution with 6 M HNO$_3$ and record.[#5] What happens if an excess of conc NH$_3$ is again added? Try it.[#6]

3. After "trying it", add 10 drops of 0.1 M KI and stir.[#7]

4. To the solution containing the yellow AgI precipitate, add 10 drops of 0.1 M Na$_2$S. Record and explain your observations on the Data Sheet.[#8]

Dispose of the silver salt test solution in the "Waste Silver Salts" container.

B. Saturated NH$_4$Cl Solution

1. Place an excess of solid NH$_4$Cl into well A3, half-filled with deionized H$_2$O and stir to dissolve as much of the salt as possible. Touch underneath the well with your finger.[1] Is the dissolution of NH$_4$Cl an endothermic or exothermic process?[#9]

2. Transfer 10 drops of the concentrated supernatant (none of the solid!) to a well A4. Add 2–3 drops conc HCl (**Caution**: *conc HCl is a severe skin irritant*) to the saturated solution until a "first" change in appearance occurs.[#10] Write a chemical equation to explain this. Warm the well with your finger. What happens to the appearance of the system?[#11]

C. [Co(H$_2$O)$_6$]$^{2+}$ and [CoCl$_4$]$^{2-}$ Equilibrium

1. Place about 10 drops of 1 M CoCl$_2$ in a well B1. Record the color of the solution.[#12] Slowly and carefully add drops of conc HCl (**Caution**: *avoid inhalation or skin contact*) until a color change occurs.[#13] Write an equation that explains your observations. Add drops of water to the system.[#14]

2. Place about 1 mL of 1 M CoCl$_2$ into a 75 mm test tube; place it into a boiling water bath. Compare its color to the original 1 M CoCl$_2$ solution.[#15] Write an equation that accounts for your observation.

D. CH$_3$COOH and CH$_3$CO$_2^-$ Equilibrium (A Buffer System)

Pinch: about the size of a grain of rice.

1. Add two drops of universal indicator to wells C1–C4. Half-fill wells C1 and C2 with 0.1 M CH$_3$COOH; note the color.[#16] Now add 0.05 g (a **pinch**) of solid NaCH$_3$CO$_2$ to wells C1 and C2 and swirl to dissolve the salt. Compare the solution's color with the pH color chart for the universal indicator.[#17]

2. Place 1 mL of deionized water in wells C3 and C4.

3. To wells C1 and C3, add 5–8 drops of 0.1 M NaOH; compare the colors of both solutions.[#18] To wells C2 and C4, add 5–8 drops of 0.1 M HCl; compare the colors of these two solutions.[#19]

Controlling the pH change in a chemical system when an acid or base is added is termed **buffer action**. Explain the action of a buffered system when a strong acid or base is added to it?[#20]

Rinse the 24 well plate with a minimum amount of water over a large beaker. Dispose of the test solution rinse in the "Waste Salts" container.

---

[1]A thermometer may be inserted into the cell to determine the direction of heat flow.

◇ LeChatelier's Principle

Date _____ Name _____ Lab Sec. _____ Desk No. _____

1. Indicate the direction (left, right, or no change) in which the equilibrium shifts in each of the chemical systems when the following stress is placed on each.

   a. Ag^+ is added to $Ag^+(aq) + Cl^-(aq) \rightleftarrows AgCl(s)$ _____

   b. Acid is added to $Ag_2CO_3(s) \rightleftarrows 2\,Ag^+(aq) + CO_3{}^{2-}(aq)$ _____

   c. Acid is added to $Ag^+(aq) + 2\,NH_3(aq) \rightleftarrows [Ag(NH_3)_2]^+(aq)$ _____

   d. Ag^+ is removed from $Ag^+(aq) + I^-(aq) \rightleftarrows AgI(s)$ _____

   e. Heat is added to $2\,NO(g) + Cl_2(g) \rightleftarrows 2\,NOCl(g) + 77.1\ kJ$ _____

   f. Heat is removed from $2\,SO_2(g) + O_2(g) \rightleftarrows 2\,SO_3(g) + 197\ kJ$ _____

   g. Cl_2 is added to $2\,NO(g) + Cl_2(g) \rightleftarrows 2\,NOCl(g) + 77.1\ kJ$ _____

   h. Br^- is added to $4\,Br^-(aq) + [Cu(H_2O)_4]^{2+}(aq) \rightleftarrows [CuBr_4]^{2-}(aq) + 4\,H_2O(l)$ _____

   i. HCO_3^- is added to $CO_3{}^{2-}(aq) + H_2O(l) \rightleftarrows HCO_3{}^-(aq) + OH^-(aq)$ _____

2. Complete the following statements with "increases", "decreases", or "no change".

   a. When Ag^+ is added to $Ag^+(aq) + Cl^-(aq) \rightleftarrows AgCl(s)$, the amount of Cl^- _____.

   b. When Cl_2 is added to $2\,NO(g) + Cl_2(g) \rightleftarrows 2\,NOCl(g) + 77.1\ kJ$, the temperature _____ and the amount of NOCl _____ .

   c. The reaction, $CH_4(g) + 2\,H_2S(g) \rightleftarrows CS_2(g) + 4\,H_2(g)$, is endothermic. If $CH_4(g)$ is added to the system, the amount of H_2S _____, the amount of CS_2 _____, and temperature _____.

   d. When $CO_3{}^{2-}$ is added to the $Mg^{2+}(aq) + CO_3{}^{2-}(aq) \rightleftarrows MgCO_3(s)$, the amount of Mg^{2+} _____ and the amount of insoluble $MgCO_3$ _____.

   e. Adding NH_3 to $4\,NH_3(aq) + [Cu(H_2O)_4]^{2+}(aq) \rightleftarrows [Cu(NH_3)_4]^{2+}(aq) + 4\,H_2O$ causes the amount of $[Cu(H_2O)_4]^{2+}$ to _____ and the amount of $[Cu(NH_3)_4]^{2+}$ to _____ .

   f. Adding H_3O^+ to $HCN(aq) + H_2O(l) \rightleftarrows H_3O^+(aq) + CN^-(aq)$ causes the amount of CN^- to _____. Adding OH^- to the same equilibrium causes the amount of CN^- to _____ and the amount of HCN to _____.

3. Consider the "water tank" in the introduction of the experiment:

a. If water is added to tank #1, what happens initially to the water level in tank #1 and finally in tank #2? Explain in terms of the flow of water molecules through the glass tubing.

b. The stopcock in tank #2 is opened. What happens initially to the water level in tank #2 and subsequently what happens to the water level in tank #1? Again explain in terms of the flow of water molecules through the glass tubing.

◇ LeChatelier's Principle

Date _____ Name _____ Lab Sec. _____ Desk No. _____

A. Silver Ion Equilibria

#1 Equation for Ag_2CO_3 equilibrium system_____

#2 Effect of 6 M HNO_3. Explain._____

Equation_____

#3 Effect of 0.1 M HCl. Explain._____

Equation_____

#4 Effect of conc NH_3. Explain._____

Equation_____

#5 Effect of reacidification with 6 M HNO_3. Explain._____

Equation_____

#6 Effect of additional conc NH_3. Explain._____

#7 Effect of 0.1 M KI. Explain._____

Equation_____

#8 Effect of 0.1 M Na_2S. Explain._____

B. Saturated NH_4Cl Solution

#9 Is the dissolution exothermic or endothermic?_____

Equation for equilibrium system_____

#10 Effect of conc HCl. Explain._____

#11 Effect of heat. Explain._____

C. $[Co(H_2O)_6]^{2+}$ and $[CoCl_4]^{2-}$ Equilibrium

#12 Color of $CoCl_2(aq)$_____

#13 Effect of conc HCl. Explain._____

Equation for equilibrium_____

#14 Effect of added H_2O. Explain._____

#15 Effect of temperature increase. Explain._____

D. CH_3COOH and $CH_3CO_2^-$ Equilibrium (A Buffer System)

Brønsted equation for CH_3COOH in water_____

#16 Color of universal indicator in CH_3COOH _____; pH = _____

#17 Color of universal indicator after $NaCH_3CO_2$ addition _____; pH =_____

Effect of $NaCH_3CO_2$ on the CH_3COOH equilibrium_____

Cell number	$CH_3COOH/CH_3CO_2^-$		H_2O	
	1	2	3	4
#18 Color effect from 0.1 M NaOH		xxx		xxx
Approximate pH		xxx		xxx
Approximate change in pH		xxx		xxx
#19 Color effect from 0.1 M HCl	xxx		xxx	
Approximate pH	xxx		xxx	
Approximate change in pH	xxx		xxx	

#20 How does the magnitude of the pH change when HCl (a strong acid) or NaOH (a strong base) is added to a solution of CH_3COOH and $NaCH_3CO_2$ as compared to the magnitude of the pH change that occurs when HCl or NaOH is added to pure water?

QUESTIONS

1. What is the color of the $[Co(H_2O)_6]^{2+}/[CoCl_4]^{2-}$ equilibrium at 80°C?

2. Explain why the pH change was small when HCl was added to the $CH_3COOH/CH_3CO_2^-$ system, but much larger when added to water.

Experiment 20

◇ pH, Hydration, and Buffers

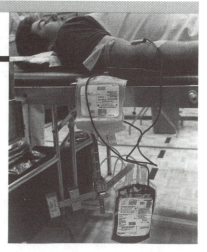

If the solution is aqueous or the substance is damp, then it has a pH. River water, water discharge from a sewage treatment plant, and the fluids in our bodies (see photo, the pH of blood is 7.4 ± 0.1) all have a pH. Most substances dissolved in water affect the pH of the solution; some are obvious—they are called acids and bases—but salts also affect pH. In some aqueous systems it is desirable to change the pH of the solution, in others we purposely add substances so that the pH is virtually unaffected by the addition of acids, bases, or salts. The latter are called buffered solutions. Surface water, such as rivers and lakes, and ground water contain dissolved carbonate and bicarbonate salts that serve as buffering agents.

OBJECTIVES

- To measure the pH of some acids and bases
- To measure the degree of hydrolysis of various ions
- To observe the effectiveness of a buffer system

PRINCIPLES

The acidity or basicity of aqueous solutions is due to small amounts of H_3O^+ or OH^- ions. Low concentrations of H_3O^+ ion are often, and more conveniently, expressed as **pH**, the negative logarithm of the molar concentration of H_3O^+ in an aqueous solution.

pH: the negative logarithm of the molar concentration of the hydronium ion in an aqueous solution.

pH = -log [H_3O^+]

Water dissociates *very* slightly producing equal concentrations of H_3O^+ and OH^-.

$$2 H_2O(l) \rightleftarrows H_3O^+(aq) + OH^-(aq)$$

At 25°C, when the $[H_3O^+] = [OH^-] = 1.0 \times 10^{-7}$ mol/L, the pH is 7.00 and the solution is said to be **neutral**. When there is an imbalance of hydronium and hydroxide ions, $[H_3O^+] \neq [OH^-]$, then the solution is acidic or basic, depending which ion is in higher concentration. Acidic solutions are high in $[H_3O^+]$ (a reduced $[OH^-]$) causing a pH *less than* 7. A base, on the other hand, increases the $[OH^-]$ in solution (decreasing $[H_3O^+]$) producing a pH *greater than* 7.

In this experiment we will measure the pH of some acids and bases using several **acid-base indicators**. An acid-base indicator, a weak organic acid abbreviated HIn, establishes a dynamic equilibrium with its conjugate base, In⁻.

Acid-base indicator: a weak organic acid that has a color different from that of its conjugate base.

$$HIn(aq) + H_2O(l) \rightleftarrows H_3O^+(aq) + In^-(aq)$$

When a solution is acidic, the indicator has the color of the HIn form of the organic acid, but, in a basic solution, it has the color of the In⁻. Since indicators, like all weak acids, are of varying strengths, the dominant form of the indicator (HIn or In⁻) is pH dependent. The pH range over which the color change from HIn to In⁻ for a number of indicators is listed in Table 20.1.

Table 20.1 Acid-Base Indicators: pH Range of Color Changes

Indicator	Acid (HIn) Color	pH Range	Base (In⁻) Color
thymol blue	red	1.2–2.8	yellow
methyl orange	red	3.2–4.4	orange/yellow
bromocresol green	yellow	3.8–5.4	blue
methyl red	red	4.8–6.0	yellow
litmus	red	4.7–8.3	blue
bromocresol purple	yellow	5.2–6.8	purple
bromothymol blue	yellow	6.0–7.6	blue
m-nitrophenol	colorless	6.8–8.6	yellow
thymol blue	yellow	8.0–9.6	blue
phenolphthalein	colorless	8.2–10.0	pink
alizarin yellow R	yellow	10.1–12.0	red

Hydration: water molecules attracted to ions to varying degrees through ion-dipole interactions.

All salts have a solubility in water producing their respective cations and anions in solution. In solution, these ions are **hydrated** to varying degrees depending upon the size and the charge of the ion; ions with a large charge and/or small size tend to be more strongly hydrated. For example, the cations Ba^{2+} and Na^+ are weakly hydrated whereas the Fe^{3+} and Al^{3+} ions, being smaller and having a larger charge, are more strongly hydrated.

Polar molecule: a molecule that has a nonuniform distribution of charge, often designated by δ^+ and δ^-.

A strongly hydrated cation attracts the negative (oxygen) region of the **polar H_2O molecule**; this attraction weakens the O–H bond of the water molecule and increases the probability of free H^+ being released into the solution. This presence of H^+ produces an acidic solution with a pH less than 7. The Fe^{3+} ion produces an acidic solution, H_3O^+, and $FeOH^{2+}$:

A strongly hydrated anion, on the other hand, attracts the positive (hydrogen) region of the polar H_2O molecule; this attraction again weakens the O–H bond and OH^- is released into the solution. This produces a basic solution with a pH greater than 7. The PO_4^{3-} ion produces a basic solution with the release of OH^- and HPO_4^{2-}.

Cations and anions that are strongly hydrated are characterized by their large charge density, i.e., the ion has a large charge and a small volume. Both the Fe^{3+} and the PO_4^{3-} ions have those properties.

Cations and anions with a small charge density have a very weak attraction for polar water molecules and, therefore, do not affect the pH of the solution. These ions are called **spectator ions** with respect to affecting the pH of a solution; examples are:

cations: Group 1A (Na^+, K^+, Rb^+, Cs^+), Group 2A (Mg^{2+}, Ca^{2+}, Sr^{2+}, Ba^{2+}) and all other metal cations in the 1^+ oxidation state.

anions: Cl^-, Br^-, I^-, NO_3^-, ClO_4^-, and ClO_3^-

In this experiment, data are collected to determine the extent of ion hydration for a number of salt solutions. We'll even use the nose as an indicator for the ammonium salts. The ion of the salt most strongly hydrated in solution is identified and an equation is written to account for the observation.

For many aqueous solutions, chemists, biologists, and environmentalists do not want a large pH change when H_3O^+ (from a strong acid) or OH^- (from a strong base) is added or produced in the solution. An aqueous solution that adjusts to the H_3O^+ or OH^- addition without encountering a large change in pH is called a **buffer solution.** (Figure 20.1)

Buffer solution: a solution that resists large changes in pH when small amounts of acid or base are added to the solution. Buffer solutions contain a weak acid and its conjugate base (or weak base and conjugate acid) as solutes.

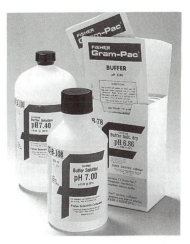

Figure 20.1
Commercial buffer solutions with a desired pH

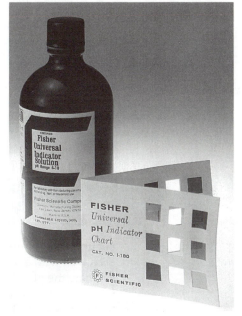

Figure 20.2
A universal indicator solution and color chart for determining pH

A buffer solution must have two components: one is a substance that consumes H_3O^+ (from a strong acid)—this must be a base, a proton acceptor; the other substance must be capable of consuming OH^- (from a strong base)—this must be an acid, a proton donor. In either case, the effect of the addition is minimized. Therefore, the two components of a typical buffer system must be a *weak* acid and its conjugate base (or a weak base along with its conjugate acid). A typical buffer system is the $CH_3COOH/CH_3CO_2^-$ system.

$$CH_3COOH(aq) + H_2O(l) \rightleftarrows H_3O^+(aq) + CH_3CO_2^-(aq) \quad K_a = 1.8 \times 10^{-5}$$

The addition of OH^- to the system, by its reaction with H_3O^+, forms H_2O and more $CH_3CO_2^-$ (a weak base).

$$CH_3COOH(aq) + H_2O(l) \rightleftarrows H_3O^+(aq) + CH_3CO_2^-(aq)$$
$$\underset{\hookrightarrow\ H_2O(l)}{\vert\ OH^-(aq)}$$

The equilibrium for the buffer system shifts *right*, the extent of the shift equals the moles of OH⁻ added.

When hydronium ions from a strong acid are added to the $CH_3COOH/CH_3CO_2^-$ system, the $CH_3CO_2^-$ ion (a proton acceptor) reacts with the H_3O^+ forming CH_3COOH; this causes a shift in the buffer system to the *left*, a shift equal to the moles of H_3O^+ added.

$$CH_3COOH(aq) + H_2O(l) \rightleftarrows H_3O^+(aq) + CH_3CO_2^-(aq)$$
$$\uparrow \quad H_3O^+(aq)$$

The changes in pH caused by the addition of a strong acid and a strong base to buffered and unbuffered solutions are compared in this experiment.

PROCEDURE

A. pH Measurement with Indicators

1. Thoroughly clean a 24 well plate with soap, tap water, and deionized water. Note the labeling on the wells—rows A–D and columns 1–6.

2. Use the wells in rows A–C to test the pH of the 14 solutions and an "other" solution listed on the Data Sheet. Half-fill each well with one of the test solutions (or equivalent substitutes) and add 3–4 drops of universal indicator.[1] Swirl the solutions until the colors are uniform. Compare the color of each solution to the "pH Indicator Chart" (Figure 20.2) or the universal indicator display on the color plate; estimate the pH of each solution.

3. For the first six test solutions write an equation that shows the presence of free H_3O^+ or OH⁻ in solution.

Dispose of the test solutions in the "Waste Salts" container. Rinse the 24 well plate with deionized water and discard the washings as advised by your instructor.

B. Hydration of Ions

1. Set up wells A1–A6 of the 24 well plate. Refer to the Data Sheet; half-fill each well with a test solution. Add 3–4 drops of universal indicator and estimate the pH of the solution. Test with litmus paper and other indicators if universal indicator is unavailable. Identify the hydrated ion that affects the pH of the solution (if any) and write an equation representing the acidity/basicity of the solution.

2. The ammonium ion readily donates a proton to the water vapor of air to release NH_3 gas. Write an equation for this reaction. The extent of this reaction can be detected by the odor of a solid ammonium salt. *Cautiously* smell the $(NH_4)_2CO_3$, NH_4Cl, and $NH_4CH_3CO_2$ salts. The anion having the strongest affinity (the strongest anionic base) for a proton of the NH_4^+ ion causes the greatest release of NH_3. List the anionic bases (CO_3^{2-}, Cl⁻, $CH_3CO_2^-$) in order of increasing strength.

3. A few anions generate H_3O^+ ions in solution. Half-fill wells B1–B3 with 0.1 M $NaHCO_3$, 0.1 M $NaHSO_4$, and 0.1 M NaH_2PO_4. Estimate the pH of the solutions. List the three anions in order of increasing acid strength.

---

[1]If universal indicator is unavailable, use four 1 x 6 arrays of the 24 well plate but test only six solutions at a time. First test each solution with litmus paper. If the red litmus turns blue (indicating that the solution is basic), further identify the pH of the solution by testing additional samples with several indicators, listed in Table 20.1, that change color in the basic range. If the solution tests acidic (blue litmus turns red), similarly test to further identify the exact pH of the solution.

4. Baking powder (Figure 20.3), used for baking bread or making pancakes, consists of a combination of baking soda, $NaHCO_3$, and a dry acid, a proton donor. When the mixture is added to water, CO_2 is produced and the "dough rises."

$$HCO_3^-(aq) + H_3O^+(aq) \rightarrow 2 H_2O(l) + CO_2(g)$$

Some of the acids in baking powders are cream of tartar, $KHC_4H_4O_6$, calcium dihydrogen phosphate, $Ca(H_2PO_4)_2$, and alum, $NaAl(SO_4)_2 \cdot 12H_2O$.

Add from the tip of a spatula a pinch of each solid acid to wells B4–B6 and dissolve with water. Test each acid to determine its relative acidity.

Figure 20.3
Assortment of baking powders

5. In a well C6, mix 1 mL of 0.1 M $NaHCO_3$ from Part B.3 with 1 mL of one of the acid solutions in Part B.4. Observe and record.

C. Buffer Solutions

1. Mix 15 drops of 0.10 M CH_3COOH with 15 drops of 0.10 M $NaCH_3CO_2$ in wells C1 and C2. Half-fill wells D1 and D2 with deionized water. Add 3–4 drops of universal indicator to each well and estimate the pH of each solution.[2]

2. Add 10 drops of 0.10 M HCl to wells C1 and D1, estimate and record the pH, and determine the pH change , ΔpH, of each mixture.[3]

3. Add 10 drops of 0.10 M NaOH to wells C2 and D2 and estimate the pH.[4] What is the pH change in each well?

Dispose of the test solutions in the "Waste Salts" container. Rinse the 24 well plate with deionized water and discard the washings as advised by your instructor.

NOTES AND OBSERVATIONS

---

[2]If universal indicator is unavailable, add 2 drops of methyl orange indicator to wells C1 and D1 and alizarin yellow R to wells C2 and D2.

[3]If methyl orange indicator is used, add, count, and record the drops of 1.0 M HCl needed to reach the methyl orange endpoint (when the color change occurs).

[4]If alizarin yellow R indicator is used, then add, count, and record the drops of 1.0 M NaOH needed to reach its endpoint.

pH, Hydration, and Buffers

Date _____Name _____ Lab Sec. _____Desk No. _____

1. a. What is an acid-base indicator?

    b. How does it function?

2. a. What color is litmus in an acidic solution? _____

    b. What color is litmus in a basic solution? _____

3. Consider the equilibrium for hydrocyanic acid, $HCN(aq) + H_2O(l) \rightleftarrows H_3O^+(aq) + CN^-(aq)$:

    a. State the effect that added H_3O^+ from a strong acid, such as hydrochloric acid, has on the
    equilibrium. Explain why.

    b. State the effect that added OH^- from a strong base, such as sodium hydroxide, has on the
    equilibrium. Explain why.

4. Define and describe the chemical behavior of a buffer solution.

5. Predict the ion that is more strongly hydrated:

    a. Fe^{3+} or Fe^{2+}. _____ Explain.

    b. Li^+ or Na^+._____ Explain.

6. Write an equation that shows how each of the following can generate OH^- in an aqueous solution.

    N_2H_4 _____

    $CH_3CO_2^-$ _____

7. Write an equation that shows how each of the following can generate H_3O^+ in an aqueous solution.

    $(CH_3)_2NH_2^+$ _____

    Al^{3+} _____

8. In preparing a buffer solution at a desired pH, it is advisable to select a weak acid-conjugate base pair in which the pK_a of the acid equals the desired pH $\pm$ 1. Over what pH range is the $CH_3COOH/CH_3CO_2^-$ buffer most effective? The K_a of CH_3COOH is 1.8 x 10^{-5}.

◇ pH, Hydration, and Buffers

Date _____ Name _____ Lab Sec. _____ Desk No. _____

A. pH Measurement with Indicators

Solution	Estimated pH	Equation
0.10 M HCl		
0.00010 M HCl		
0.10 M CH$_3$COOH		
0.10 M NH$_3$		
0.00010 M NaOH		
0.10 M NaOH		
deionized H$_2$O		
tap water		
vinegar		
lemon juice		
household NH$_3$		
detergent		
409™		
7-UP™		
other		

B. Hydration of Ions

1.

Solution	Estimated pH	Ion Affecting pH	Balanced Equation
0.1 M NaCl			
0.1 M NaCH$_3$CO$_2$			
0.1 M Na$_2$CO$_3$			
0.1 M Na$_3$PO$_4$			
0.1 M FeCl$_3$			
0.1 M Al$_2$(SO$_4$)$_3$			

2. Equation for the reaction of NH_4^+ with water _____

List the anionic bases in order of increasing strength according to the increasing strength of NH_3 smell.

_____ < _____ < _____

Write a balanced equation for the reaction of each anion with water (if any).

$CO_3^{2-} (aq)$ _____

$Cl^- (aq)$ _____

$CH_3CO_2^- (aq)$ _____

3.

Acid salt solution	Estimated pH	Ion Affecting pH	Balanced Equation
0.1 M NaHCO$_3$			
0.1 M NaHSO$_4$			
0.1 M NaH$_2$PO$_4$			

List the anions in order of increasing acid strength:

_____ < _____ < _____

4.

Dry acid	Estimated pH	Ion Affecting pH	Balanced Equation
KHC$_4$H$_4$O$_6$			
Ca(H$_2$PO$_4$)$_2$			
NaAl(SO$_4$)$_2 \cdot$12H$_2$O			

5. Observation of reaction of 0.1 M NaHCO$_3$ with an acid from Part B.4.

Balanced equation for the reaction _____

C. Buffer Solutions

	CH$_3$COOH/CH$_3$CO$_2^-$ buffer	Water
initial pH		
pH after 0.10 M HCl		
ΔpH		
(drops of 1.0 M HCl to methyl orange endpoint)		
pH after 0.10 M NaOH		
ΔpH		
(drops of 1.0 M NaOH to alizarin yellow R endpoint)		

Comment on the effectiveness of this buffer solution in resisting large changes in pH.

QUESTIONS

1. Identify those ions in Part B.1 that produce an acidic solution and those that produce a basic solution.

   acidic: _____

   basic: _____

2. a. Which ions in Part B.1 do *not* affect the pH of the solution?

   _____

   b. Which anion has the greatest effect on the pH? _____

   c. Which cation has the greatest effect on the pH? _____

3. At what point does a buffer solution stop resisting a pH change when strong acid is added? Hint: consider the concentration of the components of a buffer solution.

4. Predict whether an aqueous solution of each salt produces a solution with a pH > 7, < 7, or = 7. If the pH >7 or <7, write an equation that justifies your prediction.

Salt	pH	Equation justifying your prediction
$AlCl_3$		
K_3PO_4		
Na_2SO_4		
$Ca(CH_3CO_2)_2$		

5. A student needs 1.00 L of buffer solution with a pH = 7.00. She selects the $H_2PO_4^-$/HPO_4^{2-} buffer system. The $K_a(H_2PO_4^-) = 6.2 \times 10^{-8}$. What must be the $\dfrac{[HPO_4^{2-}]}{[H_2PO_4^-]}$ ratio for the buffer system?

Experiment 21A

◆ A Standard NaOH Solution

Just how effective is an antacid or how acidic is vinegar? How can that be determined in the laboratory? An analysis of any commercial or industrial product is extremely important to the consumer. For example, knowing the amount of impurities or by-product that result from a manufacturing process of an over the counter drug is of value to physicians of internal medicine. The symptoms of any ailment could result from the intake of the impurity of what was thought to be a "safe" drug.

For any analysis it is important to start with a "known" to determine a parameter that is unknown. The known may be a calibrated chemical instrument, such as a pH meter or a spectrophotometer, or a solution having a measured concentration of solute.

• To determine the molar concentration of a sodium hydroxide solution

OBJECTIVE

PRINCIPLES

Standard solution: a solution of known concentration of solute.

The molar concentration of sodium hydroxide in a prepared, aqueous solution is determined in this experiment. The **standard** NaOH **solution** is subsequently used as a titrant to determine the concentration of acids in samples of interest, as suggested in Experiments 21B and 21C.

A titrimetric technique of volumetric analysis can be used to determine the amount of solute in solution. The procedure requires the addition of a solution having a known concentration of solute (a standard solution) to a solution containing the unknown amount of solute until the reaction between the two solutes is complete. The reaction is complete when the mole ratio of the two reacting solutes is the same as that of the balanced equation. This is the **stoichiometric point**[1] in a titration.

Indicator: an acid-base indicator is a weak organic acid that has a color different from that of its conjugate base.

Endpoint: when the volume of titrant dispensed from a buret causes the indicator to change color.

In this experiment, the molar concentration of a prepared sodium hydroxide solution is determined in an acid-base titration. The stoichiometric point is detected using the phenolphthalein **indicator**, which is colorless in an acidic solution but pink (or red) in a basic solution. The point at which the phenolphthalein indicator changes color is the **endpoint** of the titration.[2] Indicators are selected so that the stoichiometric point for the analysis and its endpoint occur at essentially the same point in the titration.

---

[1]The stoichiometric point is also called the **equivalence point**, indicating the point at which stoichiometrically equivalent quantities of the reacting substances are combined.

[2]Refer to the Principles of Experiment 20 for an explanation of the chemistry of indicators.

Nonhygroscopic: a property of not absorbing or retaining water.

Potassium hydrogen phthalate, a **primary standard**[3], is used determine the molar concentration of the NaOH solution. Potassium hydrogen phthalate, $KHC_8H_4O_4$, is a white, crystalline, **nonhygroscopic**, acidic substance with a high degree of purity. For the analysis, a measured mass of dry $KHC_8H_4O_4$ is dissolved in deionized water; to this solution is added a recorded volume (from a buret) of the NaOH solution until the stoichiometric point is reached—when 1 mol NaOH has been added for each measured mole of $KHC_8H_4O_4$ for the analysis, according to the equation:

$$K^+HC_8H_4O_4{}^-(aq) + NaOH(aq) \rightarrow H_2O(l) + Na^+(aq) + K^+(aq) + C_8H_4O_4{}^{2-}(aq)$$

This determination of the stoichiometric point occurs at the endpoint, the point at which the phenolphthalein indicator changes from colorless to pink.

Since the initial measured mass and the molar mass of $KHC_8H_4O_4$ are known, the amount of $KHC_8H_4O_4$ titrated in the analysis is calculated.

$$g\ KHC_8H_4O_4 \times \frac{1\ mol\ KHC_8H_4O_4}{204.2\ g\ KHC_8H_4O_4} = mol\ KHC_8H_4O_4$$

At the stoichiometric point, equal moles of $KHC_8H_4O_4$ and NaOH are present in the reaction system.

$$mol\ NaOH = mol\ KHC_8H_4O_4$$

The molar concentration of the NaOH solution equals the moles of NaOH that react in the analysis divided by the volume of sodium hydroxide solution dispensed from the buret.

$$molar\ concentration\ of\ NaOH = \frac{mol\ NaOH}{liter\ NaOH\ solution}$$

PROCEDURE

You are to complete at least three trials in the standardization of your NaOH solution. To hasten the analysis, clean, dry, and label three 125 mL or 250 mL Erlenmeyer flasks and measure the mass of three $KHC_8H_4O_4$ samples while occupying the balance. If all the balances are occupied, prepare your NaOH solution (Part A.2) and buret (Part B.2).

This is an analytical experiment; you are striving for "good" results. The measured molar concentration of the NaOH for the three trials should be within ±1%. If not, a fourth or fifth trial may be necessary.

Upon completion of the analysis, save your standard NaOH solution for use in the analyses described in Experiments 21B and 21C. Consult with your instructor.

---

[3] A **primary standard** is a substance that has a known purity, is nonhygroscopic, is stable in its pure form and in solution, does not decompose with heat, and has a relatively high molar mass.

1. a. One week before the scheduled laboratory period, dissolve 3–4 g of NaOH (pellets or flakes) in 20 mL of deionized water in a 125 mL rubber-stoppered Erlenmeyer flask.[4] Thoroughly mix and allow the solution to stand for the precipitation of any Na_2CO_3.[5]

   b. Dry about 2 g of $KHC_8H_4O_4$ at 110°C for several hours in a constant temperature drying oven (a dry sample may already be available—ask your instructor). Cool the sample in a desiccator (if available, see Technique T16.d).

2. With a graduated cylinder, carefully transfer 10 mL[6] of the concentrated NaOH solution (**Caution**: *a concentrated NaOH solution causes severe skin burns*) to a 500 mL polyethylene bottle and dilute to 500 mL with previously, boiled, deionized water. The boiled water removes traces of CO_2. Cap the polyethylene bottle to prevent the absorption of CO_2 by the NaOH solution; stir the solution for several minutes and label the bottle. Do *not* shake the bottle (this increases the probability of CO_2 absorption).

A. Preparation of the NaOH Solution

1. Measure 0.3–0.5 g ($\pm$0.001 g) of the dry $KHC_8H_4O_4$ in a clean, dry, 125 mL or 250 mL Erlenmeyer flask. Add 50 mL of deionized water and 2 drops of phenolphthalein.

2. Prepare a clean a buret for titration. Rinse the buret with two 5 mL portions of your NaOH solution, making certain that the NaOH wets its entire inner surface. Have your instructor approve your buret *before* proceeding to Part C. Fill the buret with the NaOH solution and carefully read the meniscus before recording its volume ($\pm$0.02 mL). Place white paper beneath the Erlenmeyer flask.

B. Preparation of the $KHC_8H_4O_4$ for Titration

1. Slowly add the NaOH **titrant** to the $KHC_8H_4O_4$ solution in the flask, swirling the mixture after each addition. Consult the instructor (or Technique 10) for proper techniques of titrating with the left hand (if right-handed), rinsing the wall of the flask with water from the wash bottle, and adding half-drops of NaOH solution from the buret.

2. As the rate of the phenolphthalein color change (pink, where the NaOH has been added, back to colorless, for the acidic solution) decreases (see Figure T.10e), slow the rate of NaOH addition; proceed with drop and half-drop additions of the NaOH until the phenolphthalein endpoint is reached. This occurs when a single half-drop causes the pink color of the phenolphthalein indicator to persist for 30 seconds. Read the meniscus in the buret ($\pm$0.02 mL) and record.

3. Repeat the titration at least two more times with varying but accurately known amounts of $KHC_8H_4O_4$ until $\pm$1% reproducibility in the molar concentration of the NaOH solution is obtained.

C. Analysis of the NaOH Solution

Titrant: solution that is dispensed from the buret.

---

[4]The concentrated NaOH solution in Part A.1 or the more dilute NaOH solution in Part A.2 may have already been prepared. Consult with your instructor.

[5] $2\,NaOH(aq) + CO_2(aq) \rightarrow Na_2CO_3(s) + H_2O(l)$. Na_2CO_3 has a low solubility in a concentrated NaOH solution.

[6]If 1 L of NaOH solution is to be prepared for this experiment, decant all of the concentrated NaOH solution into a 1 L polyethylene bottle and dilute to 1 L. Do *not* transfer any of the Na_2CO_3 precipitate.

Consult with your instructor on the fate of the remaining standard NaOH solution in the 500 mL polyethylene bottle. It should either be saved for an acid analysis (for example, Experiment 21B and/or 21C) or discarded. The standard NaOH in the buret should *not* be returned to the reagent bottle, but discarded.

Dispose of any excess standard NaOH solution in the "Waste Base" container.

NOTES, OBSERVATIONS, AND CALCULATIONS

This is a modern titration apparatus. The titrant is automatically added until a desired (programmed) endpoint is reached. The data can be analyzed by preprogramming the computer and can then be printed. The progress of the titration (i.e., a titration curve) can also be printed out.

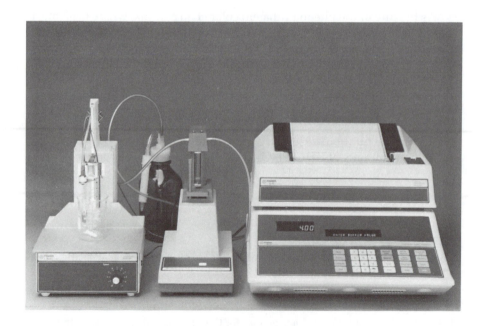

A Standard NaOH Solution

Date _____Name _____ Lab Sec. _____Desk No. _____

1. What are the properties of a primary standard?

2. Explain how CO_2 absorbed from the atmosphere affects the molar concentration of a standard NaOH solution.

3. What is the purpose for placing white paper beneath the Erlenmeyer flask during today's titration? See Procedure, Part B.2.

4. Distinguish between a stoichiometric point and an endpoint.

5. Sulfamic acid can be used as a primary standard for determining the molar concentration of a sodium hydroxide solution. A 0.418 g sample of sulfamic acid, NH_2SO_3H, dissolved in 50.00 mL of water is neutralized by 27.49 mL of NaOH at the phenolphthalein endpoint. What is the molar concentration of the NaOH solution? The molar mass of NH_2SO_3H is 97.1 g/mol.

$$NH_2SO_3H(aq) \ + \ NaOH(aq) \ \rightarrow \ NH_2SO_3^-Na^+(aq) \ + \ H_2O(l)$$

6. A 0.377 g sample of potassium hydrogen phthalate, $KHC_8H_4O_4$, is dissolved in 100 mL of water. If 7.39 mL of a NaOH solution is required to reach the stoichiometric point, what is the molar concentration of the NaOH solution? The molar mass of $KHC_8H_4O_4$ is 204.2 g/mol.

7. a. Where should the meniscus be read when viewing the volume of a solution in a buret (or pipet)?

   b. When rinsing a buret after cleaning it with soap and water, should the rinse be dispensed through the buret tip or at the top of the buret? Explain.

   c. What is the criterion for a clean buret?

   d. In preparing a buret for titration, the final rinse (or two) should be with the solution that is subsequently used in the titration. Why is this solution used for the rinse rather than deionized water?

   e. How is a half-drop dispensed from a buret into the solution?

   f. How long should the color change from an indicator persist to ensure that its endpoint has been reached?

◇ A Standard NaOH Solution

Date _____ Name _____ Lab Sec. _____ Desk No. _____

Maintain at least three significant figures when recording data and performing calculations.

	Trial 1	Trial 2	Trial 3
1. Mass of flask + $KHC_8H_4O_4$ (g)			
2. Mass of flask (g)			
3. Mass of $KHC_8H_4O_4$ (g)			
4. Moles of $KHC_8H_4O_4$ (mol)			
Buret approval by instructor			
5. Buret reading of NaOH, final (mL)			
6. Buret reading of NaOH, initial (mL)			
7. Volume of NaOH used (mL)			
8. Moles of NaOH neutralized (mol)			
9. Molar concentration of NaOH (mol/L)*			

10. Average molar concentration of NaOH (mol/L) _____

*Show calculation for Trial 1

QUESTIONS

1. Is it quantitatively acceptable to titrate all $KHC_8H_4O_4$ samples with the NaOH solution to the same *dark red* endpoint, just as long as the red intensity is consistent from one sample to the next?

2. A student titrates the dissolved $KHC_8H_4O_4$ to a faint pink endpoint which persists for 30 seconds; the buret reading is recorded and the calculations are completed. When he again looks at the receiving flask, the solution is no longer pink. Explain a probable cause of this change. Assume that the $KHC_8H_4O_4$ was pure, completely dissolved, and the stoichiometric point in the titration had been reached.

3. If the endpoint in the titration of the $KHC_8H_4O_4$ with the NaOH solution is mistakenly surpassed (too pink), what effect does this have on the calculated molar concentration of the NaOH solution?

4. Oxalic acid, $H_2C_2O_4$, can also be used as a primary standard in this experiment. What mass of oxalic acid should be measured to neutralize 25.00 mL of your NaOH solution? Oxalic acid is diprotic (Hint: what then is the balanced equation?).

5. If a drop of the NaOH solution adheres to the side of the Erlenmeyer flask during the standardization of the NaOH solution, how does this error affect its reported molar concentration? Explain.

6. If a drop of the NaOH solution adheres to the side of the buret during the standardization of the NaOH solution, how does this error affect its reported molar concentration? Explain.

Experiment 21B

◆ Analysis of Acids

Most of the foods that we eat and the chemicals that we use are acidic. For example, vegetables such as carrots, broccoli, and asparagus have a pH less than 7. Some home cleaning agents, such as tub and shower cleaners, are acidic. Most soils throughout the world are acidic, especially in those areas that have ample rainfall for extensive vegetation. Desert areas tend to be slightly alkaline. Plants most conducive to the pH of the soil tend to thrive. For example, celery (see photo) requires a very acidic soil for best results; hence an attempt to grow celery commercially in arid climates would be disastrous. Therefore, an analysis of various mixtures and formulations for their acidic content is important to the farmer, the rancher, and the consumer. In this experiment we will find out more about the acidic content of various common formulations.

OBJECTIVES

- To determine the molar mass of an acid
- To determine the percent acid in a commercial product
- To measure the percent acetic acid in vinegar

PRINCIPLES

The standard NaOH solution prepared in Experiment 21A is used to analyze for an amount of acid, present in three different forms: (1) as a pure, dry acid, (2) as a dry acid that is part of a commercial product, and (3) as a diluted acid in the solution of a foodstuff.

Molar Mass of an Acid

The molar mass of a pure, dry acid is determined using the standard NaOH solution. The amount of NaOH used for the analysis is determined from the volume of NaOH dispensed from the buret in the titration and its molar concentration.

volume NaOH (L) x molar concentration NaOH (mol/L) = mol NaOH

The moles of dry acid that reacts with the NaOH is calculated from the stoichiometry of the reaction. The acid may be either monoprotic, HA, diprotic, H_2A, or triprotic, H_3A.

$$HA(aq) + NaOH(aq) \rightarrow NaA(aq) + H_2O(l)$$
$$H_2A(aq) + 2\,NaOH(aq) \rightarrow Na_2A(aq) + 2\,H_2O(l)$$
$$H_3A(aq) + 3\,NaOH(aq) \rightarrow Na_3A(aq) + 3\,H_2O(l)$$

From a mass measurement of the acid and the **titrimetric measurement** of the moles of acid present in the sample, the molar mass of the acid can be determined.

Titrimetric measurement: the technique of using a titration procedure for an analysis.

$$\text{molar mass (acid)} = \frac{\text{mass of acid}}{\text{mol acid}}$$

Vanish™, a common household cleaning agent, has a high percentage of sodium bisulfate. Sodium bisulfate, $NaHSO_4$, is a safe, easy-to-use acid that is used to remove rust, calcium deposits from shower stalls, bathtubs, and commodes, and to acidify home swimming pools. The solid $NaHSO_4$ dissolves in water to produce the bisulfate ion, HSO_4^-, which acts as a weak

Percent Acid in Vanish™

acid in aqueous solutions.[1]   A measure of the moles of NaOH (OH⁻) that neutralizes the HSO_4^- in the sample (a 1:1 mole ratio) allows us to determine percent composition of $NaHSO_4$ in the cleaning agent.

$$HSO_4^-(aq) + OH^-(aq) \rightarrow H_2O(l) + SO_4^{2-}(aq)$$
$$\text{mol } OH^- = \text{mol } HSO_4^- = \text{mol } NaHSO_4$$

$$\text{mass of } NaHSO_4 = \text{mol } NaHSO_4 \times \frac{120.1 \text{ g } NaHSO_4}{\text{mol } NaHSO_4}$$

$$\% \ NaHSO_4 = \frac{\text{mass of } NaHSO_4}{\text{mass of sample}} \times 100$$

Percent Acetic Acid in Vinegar

acetic acid

Acetic acid, CH_3COOH, is the acid of vinegar. Its concentration varies slightly in vinegars but must be at least 4% by mass (acetic acid in water) to meet the minimum federal standard. Concentrations of acetic acid may even exceed 5% in some vinegars. Caramel flavoring and coloring may also be added to make the product aesthetically pleasing to the consumer.

The percent by mass of CH_3COOH in vinegar is determined by titrating a measured mass of vinegar to a phenolphthalein endpoint with a measured volume of a standard NaOH solution. The amount of CH_3COOH present in the vinegar is calculated from the balanced equation.

$$CH_3COOH(aq) + OH^-(aq) \rightarrow H_2O(l) + CH_3CO_2^-(aq)$$

At the stoichiometric point, the moles of OH⁻ dispensed from the buret equals the moles of CH_3COOH in the vinegar.

$$\text{mol } OH^- = \text{mol } CH_3COOH$$
$$\text{mass of } CH_3COOH = \text{mol } CH_3COOH \times \frac{60.05 \text{ g } CH_3COOH}{\text{mol } CH_3COOH}$$

$$\% \ CH_3COOH = \frac{\text{mass of } CH_3COOH}{\text{mass of vinegar}} \times 100$$

PROCEDURE

For completing Parts A, B, and C of this experiment you will need approximately 250 mL of the standard NaOH solution prepared in Experiment 21A. Do you need to standardize additional NaOH? Ask your instructor which parts of this experiment you are to complete.

A. Molar Mass of an Acid

1. Three trials are to be completed for Part A; successive results should be within ±1%. To hasten the analysis, clean, dry, and measure the mass (±0.001 g) of three 125 mL or 250 mL Erlenmeyer flasks; make all measurements while occupying the same balance. If all balances are occupied, proceed to Part A.4 and prepare the buret for the titrimetric analysis.

2. Ask the instructor whether your unknown acid is monoprotic, diprotic, or triprotic.

---

[1] The acidic properties of several anions were observed in Experiment 20.

3. In one of your Erlenmeyer flasks, measure 0.3–0.4 g (±0.001 g) of your solid, unknown acid. Add 50 mL of deionized water and 2 drops of phenolphthalein.[2]

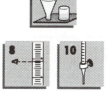

4. Clean and prepare a buret for analysis. Rinse the buret twice with ≈5 mL volumes of the standard NaOH solution; drain each rinse through the buret tip. Have your instructor approve the buret before you fill it.

5. Fill the buret with the standard NaOH solution, remove all air bubbles from the buret tip, and record the initial volume (±0.02 mL). Titrate the sample to the phenolphthalein endpoint. See Experiment 21A, Part C of the Procedure. Read the meniscus in the buret after the endpoint has been reached. Record.

6. Successive trials should have a reproducibility of <1%.

1. Read Part A.1 for increasing the efficiency of your time and technique.

B. Percent Acid in Vanish™ [3]

2. Determine the mass (±0.001 g) of a 125 mL (or 250 mL) Erlenmeyer flask. Measure 0.4–0.5 g (±0.001 g) of sample and dissolve it in 100 mL of deionized H_2O. Add 2 drops of phenolphthalein.

3. Prepare the buret as in Part A.4. Obtain your instructor's approval.

4. Titrate the sample to the phenolphthalein endpoint. Follow the procedure in Part A.5.

1. Read Part A.1 for increasing the efficiency of your time and technique.

C. Percent Acetic Acid in Vinegar

2. Determine the mass (±0.01 g) of a 125 mL (or 250 mL) Erlenmeyer flask. Add 5 mL of a selected brand of vinegar from the reagent shelf to the Erlenmeyer flask and remeasure (±0.01 g). Be sure to use the same balance. Add 2 drops of phenolphthalein to the vinegar and wash down the wall of the flask with 20 mL of deionized water.

3. Prepare the buret as in Part A.4. Obtain your instructor's approval.

4. Titrate the sample to the phenolphthalein endpoint. Follow the procedure in Part A.5.

5. Select another brand of vinegar and complete two analyses to determine its percent acetic acid.

6. Compare the percent acetic acid in the two vinegars and then determine which vinegar has the most acetic acid per unit cost.

Dispose of the test solutions in the "Waste Salts" container.

Consult with your instructor on the fate of the remaining standard NaOH solution in the 500 mL polyethylene bottle. It should either be saved for an another acid analysis (for example, Experiment 21C) or discarded. The

---

[2]The acid may be relatively insoluble, but with the addition of the NaOH titrant, it gradually dissolves. The addition of 10 mL of ethanol may be necessary to hasten the acid's dissolution, especially if the phenolphthalein endpoint is reached before the acid completely dissolves.

[3]Other commercial products containing $NaHSO_4$ may be substituted for the Vanish™.

standard NaOH in the buret should *not* be returned to the reagent bottle, but discarded.

4d Dispose of the excess standard NaOH solution in the "Waste Base" container.

NOTES, OBSERVATIONS, AND CALCULATIONS

Analysis of Acids

Date _____ Name _____ Lab Sec. _____ Desk No. _____

1. Determine the volume (mL) of 0.150 M NaOH that neutralizes 0.188 g of a *di*protic acid having a molar mass of 152.2 g/mol.

2. a. What mass (in grams) of adipic acid, $C_4H_8(COOH)_2$, will neutralize 28.2 mL of 0.188 M NaOH at the phenolphthalein endpoint (adipic acid is diprotic)?

   b. If the mass of the adipic acid sample is 0.485 g, what is the percent purity of the sample?

3. An air bubble initially trapped in the tip of the buret containing the standard NaOH disappears during the titration; how will this affect the reported number of moles in the unknown acid in Part A? (Read the Procedure.)

4. A 0.793 g sample of an unknown monoprotic acid requires 31.90 mL of 0.106 M NaOH to reach the phenolphthalein endpoint. What is the molar mass of this acid?

5. a. A 30.84 mL volume of 0.128 M NaOH is required to reach the phenolphthalein endpoint in titrating 5.441 g of vinegar. Calculate the moles of acetic acid in vinegar.

   b. How many grams of acetic acid are in the vinegar?

   c. What is the percent acetic acid in the vinegar?

Analysis of Acids

Date _____ Name _____ Lab Sec. _____ Desk No. _____

Maintain 3 significant figures when recording data and performing calculations.

Data for Part A, B, or C (circle one) of experiment, entitled _____

	Trial 1	Trial 2	Trial 3	Trial 4
1. Mass of flask + sample (g)				
2. Mass of flask (g)				
3. Mass of sample (g)				
4. Instructor's approval of buret				
5. Buret reading of NaOH, final (mL)				
6. Buret reading of NaOH, initial (mL)				
7. Volume of NaOH used (mL)				
8. Molar concentration of NaOH (mol/L)				
9. Moles of NaOH used (mol)				

A. Molar Mass of an Acid, Calculations

1. Unknown number or name of acid _____; molecular form of acid: HA, H_2A, or H_3A _____

2. Balanced equation for the reaction of the acid with NaOH:

	Trial 1	Trial 2	Trial 3
3. Moles of unknown acid (mol)			
4. Molar mass of acid (g/mol)			
5. Average molar mass (g/mol)			

B. Percent Acid in Vanish™, Calculations

Sample number _____

	Trial 1	Trial 2	Trial 3
1. Moles of $NaHSO_4$ in Vanish™ (mol)			
2. Mass of $NaHSO_4$ in Vanish™ (g)			
3. Percent $NaHSO_4$ in Vanish™ (%)			
4. Average percent $NaHSO_4$ in Vanish™ (%)			

C. Percent Acetic Acid in Vinegar, Calculations

Brand of Vinegar: _____

	Trial 1	Trial 2	Trial 1	Trial 2
1. Mass of flask and vinegar sample (g)				
2. Mass of flask (g)				
3. Mass of vinegar sample (g)				
4. Moles of CH_3COOH in vinegar sample				
5. Mass of CH_3COOH in vinegar sample (g)				
6. Percent CH_3COOH in vinegar by mass (%)				
7. Average percent CH_3COOH in vinegar by mass (%)			

Comment on the availability of CH_3COOH per unit cost in the two vinegars.

QUESTIONS

1. If the endpoint is surpassed in analyzing for the percent $NaHSO_4$ in Vanish™, will its reported percent concentration be high or low? Explain.

2. A drop of standardized NaOH solution adheres to the side of the Erlenmeyer flask and is not washed down into the vinegar with the wash bottle; how does this error affect the reported percent CH_3COOH in vinegar?

3. In determining the percent CH_3COOH in vinegar, the mass of each sample was determined rather than its volume. Explain.

Experiment 21C

◇ Antacid Analysis

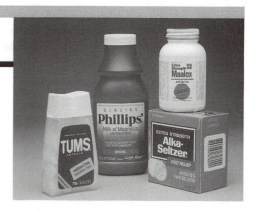

Acid indigestion! What a terrible feeling! Many of the foods we eat are naturally acidic but, in addition, many also stimulate acidic secretions from the lining of the stomach causing an excess of acid to build, leading to the upset stomach. Various commercial antacids (see photo) claim to give the "best relief" for acid indigestion, but often they do not substantiate their claims with quantitative data. All antacids, which as the name implies must be bases, neutralize acid; in this experiment we shall quantitatively determine the effectiveness of various antacids.

• To determine the amount of antacid in commercial antacids

OBJECTIVE

Stomach secretions generally have a pH ranging from 1.0 to 2.0; acid indigestion occurs at a lower pH. Antacids neutralize (or buffer) the excess hydronium ion, H_3O^+, in the stomach to relieve this discomfort. The amount of antacid needed for relief is dependent upon its "strength," although there is little serious danger of consuming too much antacid.

PRINCIPLES

Some of the "old" antacids are still among the most effective, the safest, and the cheapest. Milk of magnesia, an aqueous, milky suspension of magnesium hydroxide, $Mg(OH)_2$, while not very tasty, is an antacid that provides hydroxide ions to neutralize hydronium ions (Figure 21C.1).

$$Mg(OH)_2(s) + 2 H_3O^+(aq) \rightarrow Mg^{2+}(aq) + 2 H_2O(l)$$

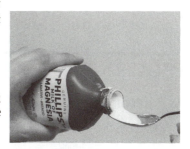

Maalox™, a "double strength" antacid, contains equal masses of $Mg(OH)_2$ and $Al(OH)_3$. The formulations of other common antacids are listed in Table 21C.1

Table 21C.1 Common Antacids

Principal Active Ingredient(s)	Representative Antacid
$NaHCO_3$/ $KHCO_3$/citric acid	Alka-Seltzer
$Mg(OH)_2$	Phillips' Milk of Magnesia
$Mg(OH)_2$/$Al(OH)_3$	Extra Strength Maalox
$NaAl(OH)_2CO_3$	Rolaids
$CaCO_3$	Titralac and Tums
$Al(OH)_3$/$MgCO_3$	Gaviscon
$Al(OH)_3$/$MgCO_3$/$Mg(OH)_2$	Di-Gel

Figure 21C.1
Milk of magnesia is a saturated solution of magnesium hydroxide

Buffer system: a solution that resists large changes in acidity.

The advertised antacids that *buffer* excess acid in the stomach are those containing calcium carbonate, $CaCO_3$, or sodium bicarbonate, $NaHCO_3$. These antacids establish the HCO_3^-/CO_3^{2-} **buffer system** in the stomach.

$$CO_3^{2-}(aq) + 2 H_3O^+(aq) \rightleftarrows HCO_3^-(aq) + 2 H_2O(l)$$

Baking soda, which is pure $NaHCO_3$, is an inexpensive antacid. The acidity of the stomach acid converts the bicarbonate to $CO_2(aq)$ and the warmth of the stomach converts the $CO_2(aq)$ to $CO_2(g)$ creating "gas on the stomach" with belching as a natural result.

$$HCO_3^-(aq) + H_3O^+(aq) \rightleftarrows 2 H_2O(l) + CO_2(aq)$$

Rolaids™, containing dihydroxyaluminum sodium carbonate, $NaAl(OH)_2CO_3$, is a combination antacid that also reacts with stomach acid.

$$NaAl(OH)_2CO_3(aq) + 3 H_3O^+(aq) \rightarrow$$
$$Na^+(aq) + Al^{3+}(aq) + 5 H_2O(l) + HCO_3^-(aq)$$

This experiment determines the effectiveness of several antacids using a strong acid-strong base titration. To avoid a buffer system[1] from being established and thus affecting the analysis, an excess of $HCl(aq)$ is added to the dissolved antacid, driving the $HCO_3^-(aq) + H_3O^+(aq) \rightleftarrows 2 H_2O(l) + CO_2(aq)$ equilibrium far to the right. The solution is then heated to expel the $CO_2(aq) \rightarrow CO_2(g)$. The unreacted $HCl(aq)$ is titrated with a standardized NaOH solution.

The procedure by which an excess of a standard solution (in this case, $HCl(aq)$) is used to drive a reaction to completion and then analyzing for the amount of unreacted (excess) standard solution is known as a **back titration analysis**.

Since an antacid has the same neutralizing effect on stomach acid as does NaOH, the amount of antacid in a sample is called its $NaOH_{equivalent}$.[2] To determine the $NaOH_{equivalent}$ for an antacid, we subtract the excess moles of $HCl(aq)$ from the total moles of $HCl(aq)$ added to the antacid.

$$NaOH_{equivalent} = HCl(aq)_{total} - HCl(aq)_{excess}$$

PROCEDURE

A. Dissolving the Antacid

1. a. Determine the mass (±0.001 g) of a 250 mL Erlenmeyer flask. Accurately measure approximately 0.7 g (±0.001 g) of a pulverized commercial antacid tablet in the flask.

    b. Pipet 25.0 mL of standardized 0.1 M HCl into the flask and swirl to dissolve the antacid.[3] Record the actual HCl concentration on the Data Sheet.

2. Slowly heat the solution to boiling and continue to heat at a *gentle boil* for at least 1 minute to expel the dissolved CO_2. Add 4–8 drops of bromophenol blue indicator.[4] If the solution is blue, add an additional 15.0 mL of 0.1 M HCl and boil again.

---

[1]In our analysis, we "swamp" the system with excess $HCl(aq)$ to remove the buffering effect of the commercial antacid and then analyze for the $HCl(aq)$ that is *not* neutralized by the antacid.

[2]The $NaOH_{equivalent}$ can also be referred to as the number of "equivalents" of antacid in the sample. An **equivalent** of any substance is merely an expression of its amount in the system, just as a **mole** of a substance indicates an amount of that substance in the system.

[3]The inert ingredients, such as the binder used in manufacturing the tablet, may not dissolve.

[4]Bromophenol blue is yellow in an acidic solution and blue in a basic solution.

1. Prepare a clean 50 mL buret. Rinse the clean buret with two 5 mL portions of the standard NaOH solution prepared in Experiment 21A. Fill the buret with the NaOH solution; read and record its initial volume (±0.02 mL). Remember to use the correct technique for reading the meniscus. Place a white sheet of paper beneath the Erlenmeyer flask.

2. Titrate the excess standard HCl(*aq*) to the blue endpoint of the bromophenol indicator. Read and record the final volume of NaOH in the buret.

3. Repeat the experiment for a second trial.

4. Select a second antacid for analysis and repeat the procedure. Compare the strengths (amount of antacid per gram of tablet) of the two antacids.

Dispose of the test solutions in the "Waste Salts" container. Dispose of the excess standard NaOH in the "Waste Base" container

---

NOTES, OBSERVATIONS, AND CALCULATIONS

Antacid Analysis

Date _____ Name _____ Lab Sec. _____ Desk No. _____

1. a. Write the balanced equation for the reaction of one mole of the active ingredient in Rolaids® with *excess* H_3O^+ ion.

    b. Write a balanced equation that represents the antacid effect of sodium citrate, $Na_3C_6H_5O_7$, on an excess of stomach acid. In an aqueous solution sodium citrate dissociates into Na^+ ions and citrate, $C_6H_5O_7^{3-}$, ions.

2. Assuming the acidity of the stomach is due to hydrochloric acid, what molar concentration of hydrochloric acid gives a pH of 1.5, the average for that of stomach acid?

3. What acid-base indicator is used in this experiment? _____

    Its color in an acidic solution is _____ ; in a basic solution it is _____ .

4. a. How much time should be allowed for the titrant to drain from the wall of a buret before a reading is made?

    b. What color should be the background of the receiving flask in today's titration?

5. A volume of 50.0 mL of 0.104 M HCl is added to an unknown base. The HCl *not* neutralized by the base (the excess HCl) is titrated to a bromophenol blue endpoint with 26.7 mL of 0.0841 M NaOH. Calculate the NaOH$_{equivalent}$ for the unknown base.

◇ Antacid Analysis

Date _____ Name _____ Lab Sec. _____ Desk No. _____

A. Dissolving the Antacid

Commercial Antacid

	Trial 1	Trial 2	Trial 1	Trial 2
1. Mass of flask + crushed tablet (g)				
2. Mass of flask (g)				
3. Mass of crushed tablet (g)				
4. Volume of HCl added (mL)				
5. Molar conc. of HCl solution (mol/L)				
6. Moles of HCl added (mol)				

B. Analysis of the Antacid Sample

	Trial 1	Trial 2	Trial 1	Trial 2
7. Buret reading, final (mL)				
8. Buret reading, initial (mL)				
9. Volume of NaOH added (mL)				
10. Molar conc. of stnd. NaOH (mol/L)				
11. Moles of NaOH added (mol)				
12. Moles of excess HCl (mol)				
13. NaOH$_{equivalent}$ of antacid in tablet				
14. NaOH$_{equivalent}$/g tablet				
15. Cost of antacid/g tablet (¢/g)				
16. Cost of antacid/NaOH$_{equivalent}$ (¢/$NaOH_{equivalent}$)				

Which antacid is the better buy (¢/$NaOH_{equivalent}$)?   _____

QUESTIONS

1. If the CO_2 is *not* removed by boiling after the 0.1 M HCl is added to the antacid, how will this affect the amount of NaOH needed to reach the bromophenol blue endpoint? Explain.

2. If the results from Trials 1 and 2 differ by a substantial amount (>2%), what should you do before presenting your results to your laboratory instructor?

3. If the endpoint in the titration is surpassed, will the reported amount of antacid in the sample be too high or too low? Explain.

Experiment 22

◇ An Equilibrium Constant

In a manner of speaking, all natural processes tend to reach a "steady state"—to arrive at a position of equilibrium. For example, the traffic flow in both directions across a bridge appears to be constant; a ball rolls and comes to rest; the corrosion of metal eventually takes the metal back to its natural state, to some equilibrium state. Chemical reactions also begin with substances that eventually react and when left alone reach some final steady state condition of reactants and products. Industrial chemists and engineers oftentimes wage a war on chemical reactions, trying to prevent them from coming to an equilibrium position so that product can be continuously produced (see photo).

Knowing this position and the conditions of equilibrium is valuable information for a chemist in analyzing and controlling chemical reactions, and for obtaining desirable results, i.e., more product formation. This knowledge of equilibrium position conditions is available as the equilibrium constant for the reaction system.

OBJECTIVES

- To develop the techniques for the care and operation of a spectrophotometer
- To determine the equilibrium constant for a soluble ionic system

PRINCIPLES

The magnitude of an equilibrium constant, K_c, expresses the equilibrium position for a chemical system. For example, a small K_c indicates large amounts of reactants at equilibrium conditions, whereas a large K_c indicates large amounts of product. The value of K_c is constant for a chemical system at a constant temperature.

Complex ion: a specie donates a pair of electrons for bonding to a metal ion forming a larger, more stable ion, called a complex ion.

This experiment determines K_c for a system in which all species are ionic and soluble. The equilibrium involves the hydrated iron(III) ion, the thiocyanate ion, and the iron(III)-thiocyanate **complex ion**:

$$Fe(H_2O)_6^{3+}(aq) + SCN^-(aq) \rightleftarrows Fe(H_2O)_5NCS^{2+}(aq) + H_2O(l)$$

Because the concentration of H_2O is essentially constant in dilute aqueous solutions, we can ignore the waters of hydration and write the equilibrium.

$$Fe^{3+}(aq) + SCN^-(aq) \rightleftarrows FeNCS^{2+}(aq)$$

Its mass action expression for the system at equilibrium, equal to K_c, is

$$K_c = \frac{[FeNCS^{2+}]}{[Fe^{3+}][SCN^-]}$$

To determine K_c for the system, known amounts of Fe^{3+} and SCN^- are mixed and react to form an equilibrium condition with $FeNCS^{2+}$, a deep, blood-red ion with an absorption maximum at about 447 nm. Because of its intense color the concentration of $FeNCS^{2+}$ is determined spectrophotometrically.[1] By measuring out an *initial* number of moles of Fe^{3+} and SCN^- and measuring the *equilibrium* number of moles of $FeNCS^{2+}$ spectrophotometrically, the equilibrium moles of Fe^{3+} and SCN^- can be calculated.

$$\text{moles } Fe^{3+}_{\text{at equilibrium}} = \text{moles } Fe^{3+}_{\text{initial}} - \text{moles } FeNCS^{2+}_{\text{at equilibrium}}$$
$$\text{moles } SCN^-_{\text{at equilibrium}} = \text{moles } SCN^-_{\text{initia}} - \text{moles } FeNCS^{2+}_{\text{at equilibrium}}$$

The equilibrium molar concentrations of Fe^{3+}, SCN^-, and $FeNCS^{2+}$ can be calculated with the known volume of the test solution.

Using the equilibrium molar concentrations of each ion in the mass action expression, the K_c for the chemical equilibrium system is calculated.

Before beginning the Procedure for this experiment, it is strongly recommended that the Principles of Experiment 14 be read and understood.

Absorbance: the amount of light that is absorbed when it passes through a sample; its value is directly proportional to the concentration of the absorbing substance in the sample.

In Part A, a set of standard solutions for the $FeNCS^{2+}$ ion (known concentrations of $FeNCS^{2+}$) is prepared and the percent transmittance, $\%T$, of each solution is measured. The **absorbance**, A, for each solution is then calculated[2] and plotted against the known molar concentration of $FeNCS^{2+}$ to establish a **calibration curve**. Because absorbance is directly proportional to molar concentration, a linear calibration curve is expected. This plot of data is then used to determine the molar concentration of $FeNCS^{2+}$ at equilibrium conditions for the systems in Part B.

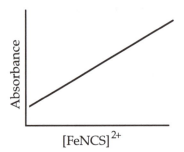

[FeNCS]$^{2+}$

To prepare the standard solutions for Part A, the Fe^{3+} concentration is added in a *large excess* relative to the SCN^- concentration; this large amount of Fe^{3+} drives the equilibrium ($Fe^{3+}(aq) + SCN^-(aq) \rightleftarrows FeNCS^{2+}(aq)$) far to the *right*. Because of this procedure, we assume that the moles of $FeNCS^{2+}$ that form at equilibrium conditions approximates the number of moles of SCN^- initially placed in the system, i.e., we assume that all of the SCN^- is in the form of the $FeNCS^{2+}$ ion at equilibrium (Figure 22.1)

Data for the determination of K_c is collected in Part B. In the chemical systems that are analyzed, the initial molar concentrations of Fe^{3+} and SCN^- are nearly the same (Figure 22.2). In reaching equilibrium conditions, the molar concentrations of Fe^{3+} and SCN^- decrease and $FeNCS^{2+}$ increases, all of which are present in the system and measurable.

PROCEDURE

Listen carefully to your instructor's advice on the care and operation of the spectrophotometer used for collecting data. Ask your instructor about working with a partner(s) for Parts A and/or B.

Two sets of solutions using different concentrations of reactants are prepared for this experiment.

- For Part A, the set of standard solutions, five 25 mL volumetric flasks,[3] one 10 mL graduated (1.0 mL) pipet, and one 10 mL pipet are needed. The calibration curve is created with the data.

---

[1]The principles of a spectrophotometric analysis are presented in Experiment 14.

[2]$A = \log \dfrac{I_o}{I_t} = a \bullet b \bullet c = \log \dfrac{100}{\%T}$

[3]150 mm test tubes may be substituted for the 25 mL volumetric flasks, provided each test solution has a final volume of 25 mL (see Table 22.1). The volume of 0.1 M HNO_3 must be adjusted and delivered with a calibrated pipet accordingly.

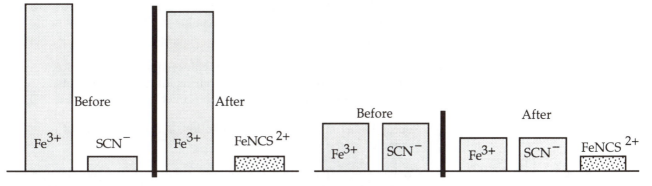

Figure 22.1
Assumption: a large excess of Fe^{3+} consumes the SCN^- forming $FeNCS^{2+}$ with a minimal change in the amount of Fe^{3+}

Figure 22.2
To reach equilibrium, the concentrations of Fe^{3+} and SCN^- decrease and $FeNCS^{2+}$ increase

- For Part B, a set of equilibrium solutions of unknown concentrations, one 5 mL pipet and two 10 mL graduated pipets are required. The calibration curve is used to determine the molar concentration of $FeNCS^{2+}$ in an array of prepared samples.

A. A Set of Standard $FeNCS^{2+}$ Solutions

1. Prepare the solutions in Table 22.1: use the 10 mL graduated pipet to pipet 0, 1, 2, 3, and 4 mL of 0.001 M NaSCN[4] into separate, (clean) labeled 25 mL volumetric flasks (see footnote 3). Pipet 10.0 mL of 0.2 M $Fe(NO_3)_3$ into each flask[5] and dilute to the "mark" with 0.1 M HNO_3. These solutions are used to establish a plot of absorbance vs. $[FeNCS^{2+}]$, the calibration curve. An additional test solution of 5 mL of 0.001 M NaSCN is suggested.

Table 22.1. A Set of Standard $FeNCS^{2+}$ Solutions

Solution	0.2 M $Fe(NO_3)_3$ (in 0.1 M HNO_3)	0.001 M NaSCN (in 0.1 M HNO_3)	0.1 M HNO_3
Blank	10.0 mL	0 mL	dilute to 25 mL
1	10.0 mL	1 mL	dilute to 25 mL
2	10.0 mL	2 mL	dilute to 25 mL
3	10.0 mL	3 mL	dilute to 25 mL
4	10.0 mL	4 mL	dilute to 25 mL

2. Set the wavelength on the spectrophotometer at 447 nm. Fill a cuvet approximately three-fourths full with the **blank solution**[6] and dry the outside with a clean Kimwipe to remove water droplets and fingerprints. Handle only the lip of the cuvet since any foreign material on the cuvet affects the amount of the transmitted light.

3. Calibrate the spectrophotometer with the blank solution as follows: place the cuvet containing the blank solution in the cuvet holder and set the spectrophotometer to read 100%T. Remove the cuvet and set the

Figure 22.3
The higher the concentration of the light-absorbing specie, the more intense is the color of the solution.

---

[4]Record the *exact* molar concentration of the NaSCN on the Data Sheet.

[5]Record the *exact* molar concentration of the $Fe(NO_3)_3$ on the Data Sheet.

[6]The **blank solution** is used to calibrate the spectrophotometer; the solution contains all light absorbing species in the test solutions *except* the one of interest, in this case $[FeNCS^{2+}]$. When the spectrophotometer is calibrated with the blank solution all "stray" absorbances are "blanked out."

spectrophotometer to read 0%T. Repeat until no further adjustments are necessary. Be sure to position the cuvet in the cuvet holder the same way each time.

4. Empty the cuvet and rinse the cuvet with several portions of Solution 1 and then fill it approximately three-fourths full and dry the outside with a clean Kimwipe. Place the cuvet in the cuvet holder and read its %T.

5. Repeat the %T measurements for Solutions 2, 3, and 4 (and 5, if prepared).

6. Convert your %T readings to absorbance values, A, and plot them *vs.* [FeNCS^{2+}]. Draw the best straight line through the four (or five if a fifth solution was prepared) points *and* the origin. Ask the instructor to approve your graph.

B. Set of Equilibrium Solutions

This set of equilibrium solutions is independent of the solutions outlined in Table 22.1. The concentrations of the reactants are different—be sure to use the reactant solutions with the suggested concentrations.

1. Prepare the test solutions in Table 22.2 in a clean, *dry* 150 mm test tubes.[7] Use three pipets, one for each solution, for the volumetric measurements. Be careful not to mix the pipets. Thoroughly stir each solution with a clean stirring rod.

Table 22.2. A Set of Equilibrium Solutions

Solution	0.002 M Fe(NO$_3$)$_3$* (in 0.1 M HNO$_3$)	0.002 M NaSCN (in 0.1 M HNO$_3$)	0.1 M HNO$_3$
5	5 mL	1 mL	4 mL
6	5 mL	2 mL	3 mL
7	5 mL	3 mL	2 mL
8	5 mL	4 mL	1 mL
9	5 mL	5 mL	• • • •

* If 0.002 M Fe(NO$_3$)$_3$ is not available, dilute 1.0 mL (measure with a 1.0 mL pipet) of the 0.2 M Fe(NO$_3$)$_3$ used in Part A with 0.1 M HNO$_3$ in a 100 mL volumetric flask and then share the leftover solution with other students.

2. Again use the blank solution (from Table 22.1) to check the calibration of the spectrophotometer (see Part A.3).

3. Rinse the cuvet with 1–2 portions of Solution 5 and fill three-fourths full; wipe dry the cuvet and insert it (properly aligned) into the cuvet holder. Read and record its %T. Be careful in handling the cuvet—handle only the lip of the cuvet and don't drop it!

4. Repeat the %T measurements for Solutions 6–9.

5. Use the calibration curve from Part A to determine the equilibrium molar concentration of FeNCS^{2+} for solutions 5–9.

Dispose of the waste iron(III)-thiocyanate solutions in the "Waste Iron Salts" container.

---

[7]10 mL volumetric flasks can be substituted for the 150 mm test tubes and the solutions can then be diluted to the mark with the 0.1 M HNO$_3$.

An Equilibrium Constant

Date _____ Name _____ Lab Sec. _____ Desk No. _____

1. a. What is white light?

    b. What is monochromatic light?

2. a. What is a blank solution?

    b. Describe its use in this experiment.

3. Describe the procedure for setting 0%T and 100%T on the spectrophotometer for its calibration.

4. A maximum absorption of visible light occurs at 447 nm for the $FeNCS^{2+}$ ion. What is the dominant

    color of light absorbed?_____ and the dominant color of light transmitted?_____

5. What effect do fingerprints on a cuvet containing the test sample have on the transmittance (%T) of visible light through the solution?

6. A 5.0 mL volume of 0.0200 M $Fe(NO_3)_3$ is mixed with 5.0 mL of 0.00200 M NaSCN; the blood-red $FeNCS^{2+}$ ion forms and an equilibrium is established between reactants and products.

$$Fe^{3+}(aq) + SCN^-(aq) \rightleftarrows FeNCS^{2+}(aq)$$

The molar concentration of $FeNCS^{2+}$, measured spectrophotometrically, is determined to be 7.0×10^{-4} mol/L at equilibrium. To calculate the K_c for the equilibrium system, proceed through the following steps.

a. moles of Fe^{3+}, initial

b. moles of SCN^-, initial

c. moles of $FeNCS^{2+}$ at equilibrium

d. moles of Fe^{3+} reacted

e. moles of SCN^- reacted

f. moles of Fe^{3+} remaining unreacted, at equilibrium (a-d)

g. moles of SCN^- remaining unreacted, at equilibrium (b-c)

h. $[Fe^{3+}]$ at equilibrium

i. $[SCN^-]$ at equilibrium

j. $[FeNCS^{2+}]$ at equilibrium 7.0×10^{-4} mol/L

k. K_c of $FeNCS^{2+}$

7. Using the value of K_c in Question 6, determine the equilibrium $[SCN^-]$ after 90.0 mL of 0.100 M Fe^{3+} are added to 10.0 mL of a SCN^- solution; the equilibrium $[FeNCS^{2+}]$ is 1.0×10^{-6} mol/L.

◇ An Equilibrium Constant

Date _____ Name _____ Lab Sec. _____ Desk No. _____

A. A Set of Standard FeNCS^{2+} Solutions

Exact molar concentration of NaSCN _____

Molar concentration of Fe(NO$_3$)$_3$ _____

Standard solutions	Blank	1	2	3	4
Volume of NaSCN (*mL*)					
Moles of SCN$^-$ (*mol*)					
[SCN$^-$] (25 mL solution)					
[FeNCS^{2+}]					
%T					
Absorbance, A					

Calibration curve, (A vs [FeNCS^{2+}]); instructor's approval _____

B. Set of Equilibrium Solutions

Exact molar concentration of Fe(NO$_3$)$_3$ _____

Exact molar concentration of NaSCN _____

Solutions	5	6	7	8	9
Volume of Fe(NO$_3$)$_3$ (*mL*)					
Moles of Fe^{3+}, initial (*mol*)					
Volume of NaSCN (*mL*)					
Moles of SCN$^-$, initial (*mol*)					
%T					
Absorbance, A					

C. Determination of K_c

Solutions	5	6	7	8	9
[FeNCS^{2+}], from calibration curve					
moles of FeNCS^{2+} in solution at equilibrium					

Calculation for [Fe^{3+}] at equilibrium

	5	6	7	8	9
moles of Fe^{3+}, reacted					
moles of Fe^{3+}, unreacted					
[Fe^{3+}], equilibrium (unreacted)	*				

Calculation of [SCN$^-$] at equilibrium

	5	6	7	8	9
moles of SCN$^-$, reacted					
moles of SCN$^-$, unreacted					
[SCN$^-$], equilibrium (unreacted)	**				
$K_c = \dfrac{[\text{FeNCS}^{2+}]}{[\text{Fe}^{3+}][\text{SCN}^-]}$					
Average K_c					

*Show calculation for [Fe^{3+}] at equilibrium for Solution 5.

**Show calculation for [SCN$^-$] at equilibrium for Solution 5.

Calculate the standard deviation, σ, of K_c using the expression:

$$\sigma = \sqrt{\frac{d_1{}^2 + d_2{}^2 + d_3{}^2 + \ldots + d_n{}^2}{(n-1)}} = \underline{\hspace{2cm}}$$

d = difference between a single K_c value and the average K_c

n = number of K_c values determined

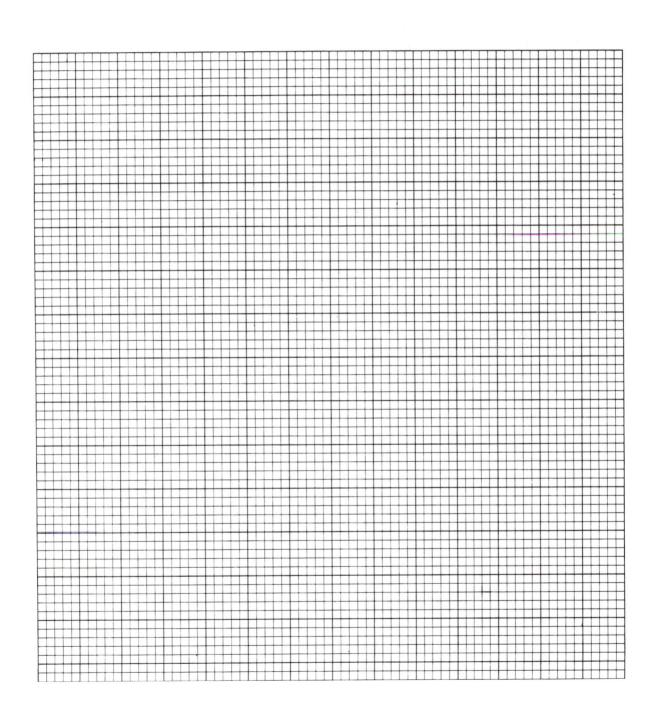

QUESTIONS

1. What effect does a dirty cuvet (caused by fingerprints, water spots, or lint) have on the value of K_c in this experiment?

2. In our calculations the thickness of the test solution and the probability of light absorption by the $FeNCS^{2+}$ ion were not considered. Explain.
    See the discussion of Beer's Law in Experiment 14.

3. Why is 0.2 M $Fe(NO_3)_3$ in 0.1 M HNO_3 used as the blank solution for calibrating the spectrophotometer instead of deionized water?

4. Considering Solution 8, suppose the 1.00 mL of 0.1 M HNO_3 is not added.
    a. How will this affect the $\%T$ of the solution?

    b. How will this affect the reported the equilibrium molar concentration of $FeNCS^{2+}$ for the solution?

    c. Will the reported K_c be greater or less than it should have been for this sample? Explain.

Experiment 23

◇ Solubility Constant and Common-Ion Effect

Rocks never seem to change in appearance..."once a rock, always a rock." Many national parks, such as Bryce Canyon, Utah (see photo) or Garden of the Gods, Colorado, have a large number of unusual rock formations that have maintained their appearance for centuries. Rocks are composed of various minerals, or complex salts, that have a very low solubility in water. Even though the solubility of each of these minerals is low, it is not zero. The mineral slowly dissolves with each bit of rainfall because of its very slight solubility; slowly the edges of the rock formations become rounded and the height of the rock formation diminishes. Geologists refer to the Appalachian Mountains as "old" formations, but the Rocky Mountains as "young," because of their time and extent of exposure to rainfall (and wind) and because of the slight solubility of the minerals in water.

OBJECTIVES

- To determine the molar solubility and solubility constant of $Ca(OH)_2$
- To determine the molar solubility of $Ca(OH)_2$ in the presence of Ca^{2+}

PRINCIPLES

Salts which have a low solubility in water are called **slightly soluble salts**. A saturated solution of a slightly soluble salt is a dynamic equilibrium between the solid salt and the very low concentrations of its ions in solution. Limestone rock appears to be insoluble in its natural environment; however, with time, the limestone slowly dissolves due to its slight solubility. The major component of limestone is calcium carbonate, $CaCO_3$. Its equilibrium with the Ca^{2+} and CO_3^{2-} favors the solid $CaCO_3$, or the equilibrium lies far to the *left*.

Slightly soluble salts: also referred to as "insoluble" salts.

$$CaCO_3(s) \rightleftarrows Ca^{2+}(aq) + CO_3^{2-}(aq)$$

The mass action expression for this system at equilibrium is a constant, called the solubility constant, K_s, for the salt. Since the concentration of a solid remains a constant, the mass action expression, equal to the solubility constant, is

$$K_s = [Ca^{2+}][CO_3^{2-}]$$

K_s: also called the solubility product constant and designated as K_{sp}.

[ ]: used to indicate the molar concentration of the indicated species.

This equation states that a product of the molar concentrations of the Ca^{2+} ion and the CO_3^{2-} ion in aqueous solution is a constant at equilibrium; it does *not* state that the molar concentrations of the Ca^{2+} ion and the CO_3^{2-} ion must be equal in solution. If a solution has a high $[Ca^{2+}]$, then its $[CO_3^{2-}]$ must be necessarily low—it is their product that is experimentally found to be a constant.

According to the CRC's *Handbook of Chemistry and Physics*, the K_s of $CaCO_3$ is 4.95×10^{-9} at 25°C. In a saturated solution, where the $[Ca^{2+}] = [CO_3^{2-}]$, the molar concentration of each ion is

Handbook of Chemistry and Physics: a reference book that contains chemical data, published by the Chemical Rubber Company. See Experiment 11.

$$K_s = [Ca^{2+}][CO_3^{2-}] = 4.95 \times 10^{-9}$$

$$[Ca^{2+}] = [CO_3^{2-}] = \sqrt{4.95 \times 10^{-9}} = 7.04 \times 10^{-5} \text{ mol/L}$$

Molar solubility: the solubility of a salt, expressed mol/L.

The **molar solubility** of $CaCO_3$ is also 7.04×10^{-5} mol/L, because for each mole of $CaCO_3$ that dissolves 1 mol of Ca^{2+} and 1 mol of CO_3^{2-} are in solution.

Common-Ion Effect

What happens to the solubility of a slightly soluble salt when an ion, cation or anion that is common to the salt, is added to a saturated solution? According to LeChatelier's Principle, its equilibrium shifts to compensate for the added (common) ion to favor the formation of more solid salt, reducing the salt's molar solubility.

> **Example.** Suppose 0.0010 mol CO_3^{2-} is added to a saturated $CaCO_3$ solution. What happens?
> The molar solubility decreases because the added CO_3^{2-} shifts the equilibrium,
> $CaCO_3(s) \rightleftarrows Ca^{2+}(aq) + CO_3^{2-}(aq)$, to the *left*, meaning that less moles of $CaCO_3$ dissolve. The new molar solubility of $CaCO_3$ equals the $[Ca^{2+}]$ in solution or
>
> $$\text{molar solubility of } CaCO_3 = [Ca^{2+}] = \frac{K_s}{[CO_3^{2-}]} = \frac{4.95 \times 10^{-9}}{0.0010} = 4.95 \times 10^{-6} \text{ mol/L}$$
>
> The added CO_3^{2-} decreases the molar solubility of $CaCO_3$ from 7.04×10^{-5} mol/L to 4.95×10^{-6} mol/L.

In this experiment you will determine the solubility constant, K_s, of $Ca(OH)_2$, and the molar solubilities of $Ca(OH)_2$ in a saturated solution and in a saturated solution with added Ca^{2+}.

Stoichiometric point: the point at which stoichiometric amounts of reactants are present in the reaction mixture.

The hydroxide ion, OH^-, of the saturated $Ca(OH)_2$ solution is titrated with a standard hydrochloric acid solution to the **stoichiometric point**, using bromocresol green as the indictor. The stoichiometric point occurs, for this titration, when equal moles of H_3O^+ and OH^- are present in the receiving flask. Bromocresol green is an acid-base indicator that is blue in its basic form and yellow in its acidic form.

According to the equation

$$Ca(OH)_2(s) \rightleftarrows Ca^{2+}(aq) + 2OH^-(aq)$$

for each mole of $Ca(OH)_2$ that dissolves, 1 mol Ca^{2+} and 2 mol OH^- are present in solution. Thus by determining the $[OH^-]$, the $[Ca^{2+}]$, the molar solubility of $Ca(OH)_2$, and the K_s for $Ca(OH)_2$ can be calculated.

$$[Ca^{2+}] = \tfrac{1}{2}[OH^-]$$

$$\textbf{molar solubility of } Ca(OH)_2 = [Ca^{2+}] = \tfrac{1}{2}[OH^-]$$

$$K_s = [Ca^{2+}][OH^-]^2 = \left(\tfrac{1}{2}[OH^-]\right)[OH^-]^2$$

Likewise the same procedure is used to determine the molar solubility of $Ca(OH)_2$ with the common-ion Ca^{2+} ion added to the saturated $(Ca(OH)_2$ solution.

Three trials for Part A and three trials for Part B are to be completed. To hasten the analysis, clean and label three 125 mL or 250 mL Erlenmeyer flasks. In Part A.3b, pipet 25 mL of the saturated $Ca(OH)_2$ solution into each flask.

PROCEDURE

A. K_s and Molar Solubility of $Ca(OH)_2$

1. Prepare a saturated $Ca(OH)_2$ solution one week[1] before the experiment by adding about 3 g of solid $Ca(OH)_2$ to 120 mL of boiled, deionized water in a 125 mL Erlenmeyer flask. Stir the solution and stopper.

2. Prepare a 50 mL buret for titration. Rinse the clean buret and tip with two 5 mL portions of a standard 0.05 M HCl solution. Fill, read (±0.02 mL), and record the volume of 0.05 M HCl in the buret. Record the exact molar concentration of the HCl solution.

3. a. Allow the undissolved $Ca(OH)_2$ to remain settled on the bottom of the flask. Without disturbing the insoluble $Ca(OH)_2$, *carefully* decant about 90 mL of the saturated solution into a second 125 mL Erlenmeyer flask or 150 mL beaker.

Decantate: the solution that is transferred from above the settled precipitate.

   b. Rinse a 25 mL pipet with 1–2 mL of the saturated $Ca(OH)_2$ solution and discard. Pipet 25 mL of the decantate into a clean 125 mL Erlenmeyer flask and add 2 drops of bromocresol green indicator. Record the temperature of the solution.

4. Titrate the saturated $Ca(OH)_2$ solution with the standard HCl solution. Record the volume (±0.02 mL) needed to just turn the blue color of the bromocresol green indicator yellow.[2] Follow the techniques of titration as described in Technique 10.

5. Repeat the titration with two new samples of the saturated $Ca(OH)_2$ solution.

B. Solubility of $Ca(OH)_2$ in the Presence of Ca^{2+}

1. Mix 3 g of solid $Ca(OH)_2$ and 1 g of solid $CaCl_2 \cdot 2H_2O$ with 120 mL of boiled, deionized water in a 125 mL Erlenmeyer flask one week[3] before the experiment, stir, and stopper the flask. Carefully decant about 90 mL of the solution into a second flask or beaker.

2. To complete the analysis, titrate a 25 mL sample of the decantate from Part B.1 to the bromocresol green endpoint with the standard 0.05 M HCl solution as was described in Parts A.3–5.

Dispose of the test solutions in the "Waste Limewater" container.

---

[1]This solution may have been prepared for you. Ask your instructor.

[2]The color change at the endpoint for bromocresol green (blue at pH = 4.6 to yellow at pH = 3.0) is subtle. Keep the titrated solutions from successive trials to compare the colors at the endpoint. The methyl orange indicator, turning from yellow to red in the analysis, may also be used in the titration.

[3]This solution may have also been prepared for you. Ask your instructor.

NOTES,
OBSERVATIONS,
AND
CALCULATIONS

The formation of stalagmites and stalactites results from the limited solubility of $CaCO_3$. As the water evaporates, the $CaCO_3$ precipitates. . . but ever so slowly!

Solubility Constant and Common-Ion Effect

Date _____Name _____ Lab Sec. _____Desk No. _____

1. How is the molar concentration of OH⁻ ion in a saturated $Ca(OH)_2$ solution measured in this experiment?

2. a. What indicator is used in today's titration?

    b. What color change is observed at the endpoint in the titration?

    c. Does this color change occur as a result of the indicator changing from an acidic to basic form or basic to acidic form?

    d. How should the rate of addition of titrant be changed as the color fade of the indicator slows during the addition of titrant? Explain.

3. Write the mass action expression for these slightly soluble salt equilibria:

    a. $CuS(s) \rightarrow Cu^{2+}(aq) + S^{2-}(aq)$ $K_s =$

    b. $BaSO_4(s) \rightarrow Ba^{2+}(aq) + SO_4^{2-}(aq)$ $K_s =$

    c. $Ca_3(PO_4)_2(s) \rightarrow 3\,Ca^{2+}(aq) + 2\,PO_4^{3-}(aq)$ $K_s -$

4. a. Calculate the molar solubility of AgI. $K_s = 1.5 \times 10^{-16}$

   b. Calculate the molar solubility of AgI in the presence of 0.020 M KI.

5. The pH of a saturated $Ni(OH)_2$ solution at equilibrium is 8.92.
   a. Calculate the $[OH^-]$ in a saturated $Ni(OH)_2$ solution.

   b. Write the equation that represents the $Ni(OH)_2$ equilibrium. Calculate the $[Ni^{2+}]$ in a saturated $Ni(OH)_2$ solution.

   c. Calculate the K_s of $Ni(OH)_2$.

◇ Solubility Constant and Common-Ion Effect

Date _____ Name _____ Lab Sec. _____ Desk No. _____

A. K_s and Molar Solubility of $Ca(OH)_2$

Concentration of standard HCl solution (*mol/L*) _____

	Trial 1	Trial 2	Trial 3
1. Buret reading, final (*mL*)			
2. Buret reading, initial (*mL*)			
3. Volume of standard HCl used (*mL*)			
4. Moles of HCl added (*mol*)			
5. Moles of OH^- in sat'd $Ca(OH)_2$ solution (*mol*)			
6. Volume of sat'd $Ca(OH)_2$ solution (*mL*)	25.0	25.0	25.0
7. Temperature of $Ca(OH)_2$ solution (°C)			
8. [OH^-] at equilibrium (*mol/L*)			
9. [Ca^{2+}] at equilibrium (*mol/L*)			
10. Molar solubility of $Ca(OH)_2$ at ___°C			

11. Average molar solubility of $Ca(OH)_2$ at ___°C　　_____

12. K_s of $Ca(OH)_2$　　_____

B. Solubility of $Ca(OH)_2$ in the Presence of Ca^{2+}

	Trial 1	Trial 2	Trial 3
1. Buret reading, final (*mL*)			
2. Buret reading, initial (*mL*)			
3. Volume of standard HCl used (*mL*)			
4. Moles of HCl added (*mol*)			
5. Moles of OH^- in $Ca(OH)_2/Ca^{2+}$ solution (*mol*)			
6. Volume of $Ca(OH)_2/Ca^{2+}$ solution titrated (*mL*)	25.0	25.0	25.0
7. Temperature of $Ca(OH)_2/Ca^{2+}$ solution (°C)			
8. [OH^-] at equilibrium (*mol/L*)			
9. [Ca^{2+}] at equilibrium (*mol/L*)	*		

10. Molar solubility of $Ca(OH)_2$ in $Ca(OH)_2/Ca^{2+}$ solution at ___°C _____

*Show calculations for Trial 1

QUESTIONS

1. How did the addition of $CaCl_2$ affect the molar solubility of $Ca(OH)_2$? Explain.

2. a. Suppose in Part A.3 that some solid $Ca(OH)_2$ was inadvertently transferred into the 125 mL Erlenmeyer flask. How would this error affect the volume of standard HCl used to reach an endpoint?

   b. How will this affect the reported K_s value of $Ca(OH)_2$? Explain.

3. If the endpoint of the titration is surpassed in Part A.4, will the K_s value for $Ca(OH)_2$ be higher or lower than the accepted value? Explain.

4. Does adding boiled, deionized water to the Erlenmeyer receiving flask, in order to wash the side of the flask and the buret tip, affect the value of K_s for $Ca(OH)_2$? Explain.

5. How would the use of tap water, instead of boiled, deionized water, for preparing the saturated solution in Part A affect the value of K_s for $Ca(OH)_2$?

Experiment 24

◇ Hard Water Analysis

What does it mean when we say that water is "hard?" Hard water is obviously not hard in the same sense that wood or metal is hard, nor difficult to understand as is the creation of the universe. Hard water contains the dissolved salts of calcium, magnesium, and iron ions which are called hardening ions. In low concentrations these ions are not considered harmful for domestic use, but at higher concentrations these ions interfere with the cleansing action of soaps and accelerate the corrosion of steel pipes, especially those carrying hot water (see photo).

The source of hardening ions in natural waters is the result of slightly acidic rainwater flowing over mineral deposits of varying compositions; the acidic rainwater reacts with the *very* slightly soluble carbonate salts of calcium and magnesium and with various iron-containing rocks. A partial dissolution of these salts releases the ions into the water supply, which may be surface water or ground water.

- To learn the cause and effects of hard water
- To determine the hardness of a water sample

OBJECTIVES

PRINCIPLES

Hardening ions, such as Ca^{2+}, Mg^{2+}, and Fe^{3+}, form insoluble compounds with soaps. Soaps, which are sodium salts of fatty acids such as sodium stearate, $C_{17}H_{35}CO_2^- Na^+$, are very effective cleansing agents so long as they remain soluble; the presence of the hardening ions however causes the formation of a gray, insoluble soap scum such as $(C_{17}H_{35}CO_2)_2Ca$.

$$2\,C_{17}H_{35}CO_2^-Na^+(aq)\ +\ Ca^{2+}(aq)\ \rightarrow (C_{17}H_{35}CO_2)_2Ca(s)\ +\ 2\,Na^+(aq)$$

This gray precipitate appears as a "bathtub ring" and it also clings to clothes, causing white clothes to appear gray.

Hard water is also responsible for the appearance and undesirable formation of "boiler scale" on tea kettles and pots used for heating water. The boiler scale is a poor conductor of heat and thus reduces the efficiency of transferring heat. Boiler scale also builds on the inside of hot water pipes to decrease the flow of water; in extreme cases, this buildup causes the pipe to break. The problems associated with boiler scale, while a problem in residential plumbing, is especially costly to the chemical industry where cooling water cycles are used in processing.

Boiler scale consists primarily of the carbonate salts of the hardening ions and is formed according to

$$Ca^{2+}(aq) + 2\,HCO_3^-(aq) \xrightarrow{\Delta} CaCO_3(s) + CO_2(g) + H_2O(l)$$

Ground water becomes hard as it flows through underground limestone ($CaCO_3$) deposits; generally, the water from deep wells has a higher hardness than that from shallow wells because of a longer time of contact with the limestone. **Surface water** similarly accumulates hardening ions as a

Surface water: water that is collected from a watershed, e.g., lakes, rivers, and streams.

result of it flowing over limestone deposits. In either case the CO_2 dissolved in rainwater[1] solubilizes limestone deposits.

$$CO_2(aq) + H_2O(l) + CaCO_3(s) \rightarrow Ca^{2+}(aq) + 2\,HCO_3^-(aq)$$

Notice that this reaction is just the reverse of the reaction for the formation of boiler scale. The same two reactions are key to the formation of stalactites and stalagmites for caves located in regions with large limestone deposits.

Because of the relatively large natural abundance of limestone deposits and other calcium minerals, such as gypsum, $CaSO_4 \bullet 2H_2O$, it is not surprising that Ca^{2+} ion, in conjunction with Mg^{2+}, is a major component of the dissolved solids in water.

ppm: parts per million by mass, e.g., 15 ppm $CaCO_3$ is 15 g $CaCO_3$ in 10^6 g solution.

The concentration of the hardening ions in a water sample is commonly expressed as though the hardness is due exclusively to $CaCO_3$. The units for hardness is mg $CaCO_3$/L, which is also **ppm** $CaCO_3$.[2] A general classification of hard waters is listed in Table 24.1.

Table 24.1 Hardness Classification of Water

Hardness (*ppm* $CaCO_3$)	Classification
< 15 ppm	very soft water
15 ppm - 50 ppm	soft water
50 ppm - 100 ppm	medium hard water
100 ppm - 200 ppm	hard water
>200 ppm	very hard water

Theory of Analysis

Na_2H_2Y

Complex ion: a specie that donates a pair of electrons for bonding to a metal ion to form a larger, more stable ion, called a complex ion.

In this experiment a titration technique is used to measure the combined Ca^{2+} and Mg^{2+} concentrations in a water sample. The titrant is the disodium salt of ethylenediaminetetraacetic acid (abbreviated Na_2H_2Y).[3]

In aqueous solution Na_2H_2Y dissociates into Na^+ and H_2Y^{2-} ions. The H_2Y^{2-} ion reacts with the hardening ions, Ca^{2+} and Mg^{2+}, to form very stable **complex ions**, especially in a solution buffered at a pH of about 10. An ammonia–ammonium ion buffer is often used for this pH adjustment in the analysis.

As H_2Y^{2-} **titrant** is added to the **analyte**, it complexes with the "free" Ca^{2+} and Mg^{2+} of the water sample to form the respective complex ions:

$$Ca^{2+}(aq) + H_2Y^{2-}(aq) \rightarrow [CaY]^{2-}(aq) + 2\,H^+(aq)$$
$$Mg^{2+}(aq) + H_2Y^{2-}(aq) \rightarrow [MgY]^{2-}(aq) + 2\,H^+(aq)$$

From the balanced equations, it is apparent that once the molar concentration of the Na_2H_2Y solution is known, the moles of hardening ions in a water sample can be calculated, a 1:1 stoichiometric ratio.

---

[1]CO_2 dissolved in rainwater makes rainwater slightly acidic.

$$CO_2(g) + 2\,H_2O(l) \rightarrow H_3O^+(aq) + HCO_3^-(aq)$$

[2]ppm means "parts per million"—1 mg of $CaCO_3$ in 1 000 000 mg (or 1 kg) solution is 1 ppm $CaCO_3$. Assuming the density of the solution is 1 g/mL (or 1 kg/L), then 1 000 000 mg solution = 1 L solution.

[3]**E**thylene**d**iamine**t**etr**aa**cetate is often simply referred to as EDTA.

Titrant: the solution placed in the buret in a titrimetric analysis.

Analyte: the solution containing the substance being analyzed, generally in the receiving flask in a titration setup.

volume H_2Y^{2-} x molar concentration of H_2Y^{2-} = moles H_2Y^{2-}
= moles hardening ions

The hardening ions, for reporting purposes, are assumed to be exclusively Ca^{2+} from the dissolving of $CaCO_3$. Since one mole of Ca^{2+} forms from one mole of $CaCO_3$, the hardness of the water sample expressed as mg $CaCO_3$ per liter of sample is

$$\text{moles hardening ions} = \text{moles } Ca^{2+} = \text{moles } CaCO_3$$

$$\text{ppm } CaCO_3 \left(\frac{\text{mg } CaCO_3}{\text{L sample}} \right) = \frac{\text{mol } CaCO_3}{\text{L sample}} \times \frac{100.1 \text{ g } CaCO_3}{\text{mol}} \times \frac{\text{mg}}{10^{-3}\text{g}}$$

The Indicator for the Analysis

A special indicator is used to detect the endpoint in the titration. Called Eriochrome Black T (EBT), it also forms complex ions with the Ca^{2+} and Mg^{2+} ions, but binds more strongly to Mg^{2+} ions. Because only a small amount of EBT is added, only a small quantity of Mg^{2+} is complexed; no Ca^{2+} ion complexes to EBT—therefore, most of the hardening ions remain "free" in solution. The EBT indicator is sky-blue in solution but forms a wine-red complex with $[Mg\text{-}EBT]^{2+}$.

$$Mg^{2+}(aq) + EBT(aq) \rightarrow [Mg\text{–}EBT]^{2+}(aq)$$
$$\text{sky-blue} \quad\quad \text{wine-red}$$

Therefore even before any H_2Y^{2-} titrant is added for the analysis, the analyte is wine-red because of the $[Mg\text{–}EBT]^{2+}$ complex ion.

Once the H_2Y^{2-} complexes all of the "free" Ca^{2+} and Mg^{2+} from the water sample, it then removes the trace amount of Mg^{2+} from the wine-red $[Mg\text{-}EBT]^{2+}$ complex; the solution changes from wine-red back to the sky-blue color of the EBT indicator, and the endpoint is reached—all of the hardening ions have been complexed with H_2Y^{2-}.

$$[Mg^{2+}\text{–}EBT]^{2+}(aq) + H_2Y^{2-}(aq) \rightarrow MgY^{2-}(aq) + 2 H^+(aq) + EBT(aq)$$
$$\text{wine-red} \quad\quad\quad\quad\quad\quad\quad\quad\quad\quad\quad\quad\quad\quad \text{sky-blue}$$

Eriochrome Black T

Therefore, Mg^{2+} *must* be present in the water sample for the EBT indicator to "work." Oftentimes, a small amount Mg^{2+} as MgY^{2-} is initially added to the analyte to ensure/enhance the formation of the wine-red color with EBT by the equation,

$$MgY^{2-}(aq) + 2 H^+(aq) + EBT(aq) \rightarrow [Mg\text{–}EBT]^{2+}(aq) + H_2Y^{2-}(aq)$$

This reaction is then reversed at the stoichiometric point.

$$[Mg\text{–}EBT]^{2+}(aq) + H_2Y^{2-}(aq) \rightarrow MgY^{2-}(aq) + 2 H^+(aq) + EBT(aq)$$

Because H_2Y^{2-} is freed initially but then consumed at the endpoint (previous two equations), no additional H_2Y^{2-} titrant is required for the analysis of hardness in the water sample because of this added Mg^{2+}.

PROCEDURE

A. A Standard 0.01 M Disodium Ethylenediaminetetraacetate, Na_2H_2Y, Solution

Three trials should be completed for the standardization of the Na_2H_2Y solution. To save time, initially prepare three Erlenmeyer flasks for Part A.3.

1. Measure about 0.5 g (±0.01 g) of $Na_2H_2Y\bullet 2H_2O$ (molar mass = 372.24 g/mol) on weighing paper; transfer it to a 250 mL volumetric flask containing about 100 mL of deionized water and stir to dissolve. Dilute to the "mark" of the volumetric flask with deionized water.

2. Prepare a buret for titration. Rinse the buret with the Na_2H_2Y solution and then fill. Record the volume (±0.02 mL) of the titrant.

3. Obtain about 80 mL of the standard Ca^{2+} solution from the reagent shelf and record its molar concentration. Pipet 25.0 mL into a 125 mL Erlenmeyer flask, add 1 mL of buffer (pH = 10) solution, and 2 drops of EBT indicator.[4]

4. Titrate the standard Ca^{2+} solution with the Na_2H_2Y titrant; swirl continuously. Near the endpoint, slow the rate of addition to drops; the last few drops should be added at 3–5 s intervals. The solution changes from wine-red to purple to blue—no tinge of the wine-red color should remain; the solution is *blue* at the endpoint.

5. a. Repeat the titrations on the other two samples.
    b. Calculate the molar concentration of the Na_2H_2Y solution.

B. Analysis of Water Sample

Complete three trials for your analysis. The first trial is an indication of the hardness of your water sample—the hardness values may vary tremendously, depending upon your water sample. You may want to adjust the volume of water for the analysis of the second and third trials.

1. a. Obtain about 100 mL of a water sample from your instructor. You may use your own water sample or simply the tap water in the laboratory.

    b. If the water sample is from a lake, stream, or ocean, you will need to gravity filter the sample before the analysis.

    c. If your sample is acidic, add 1 M NH_3 until it is basic to litmus.

2. Pipet 25.0 mL of your (filtered, if necessary) water sample[5] into a 125 mL Erlenmeyer flask, add 1 mL of the buffer (pH = 10) solution, and 2 drops of EBT indicator.

3. Repeat Parts A.4 and A.5 to determine the hardness of your water sample.

Dispose of the analyzed solutions in the "Waste EDTA" container.

---

[4]The indicator, calgamite, can be substituted for the Eriochrome Black T indicator.

[5]If your water is known to have a high hardness, decrease the volume of the water proportionally until it takes about 15 mL of Na_2H_2Y titrant for your second and third trials. Similarly if your water sample is known to have a low hardness, increase the volume of the water proportionally.

Hard Water Analysis

Date _____ Name _____ Lab Sec. _____ Desk No. _____

1. What ions are responsible for water hardness?

2. a. How do hardening ions cause soap to be less effective?

    b. What is and what causes boiler scale?

3. The analysis procedure in this experiment requires that the titration be conducted at a pH = 10. How is the solution adjusted to this pH?

4. a. Which hardening ion, Ca^{2+} or Mg^{2+}, binds more tightly to (forms a stronger complex ion with) the Eriochrome Black T indicator used for today's analysis?

    b. What is the color change at the endpoint?

5. A 50.0 mL water sample required 5.14 mL of 0.0100 M Na_2H_2Y to reach the endpoint for the titration procedure described in this experiment.

    a. Calculate the moles of hardening ions in the water sample.

    b. Assuming the hardness is due exclusively to $CaCO_3$, express the hardness concentration in mg $CaCO_3$/L sample.

    c. What is this hardness concentration expressed in ppm $CaCO_3$?

    d. Classify the hardness of this water according to Table 24.1.

6. The hardness of a water sample is known to be 500 ppm $CaCO_3$. In an analysis of a 50 mL water sample, what volume of 0.0100 M Na_2H_2Y is needed to reach the endpoint?

◇ Hard Water Analysis

Date _____ Name _____ Lab Sec. _____ Desk No. _____

A. A Standard 0.01 M Disodium Ethylenediaminetetraacetate, Na_2H_2Y, Solution

		Trial 1	Trial 2	Trial 3
1	Mass of weighing paper + Na_2H_2Y (g)			
2.	Mass of weighing paper (g)			
3.	Mass of Na_2H_2Y (g)			
4.	Volume of standard Ca^{2+} solution (mL)	25.0	25.0	25.0
5.	Concentration of standard Ca^{2+} solution			
6.	Mol Ca^{2+} = mol Na_2H_2Y (mol)			
7.	Buret reading, final (mL)			
8.	Buret reading, initial (mL)			
9.	Volume of Na_2H_2Y titrant (mL)			
10.	Molar concentration of Na_2H_2Y solution (mol/L)			
11.	Average molar concentration of Na_2H_2Y solution (mol/L)			

B. Analysis of Water Sample

		Trial 1	Trial 2	Trial 3
1.	Volume of water sample (mL)			
2.	Buret reading, final (mL)			
3.	Buret reading, initial (mL)			
4.	Volume of Na_2H_2Y titrant (mL)			
5.	Mol Na_2H_2Y = mol hardening ions, Ca^{2+} and Mg^{2+} (mol)			
6.	Mass of equivalent $CaCO_3$ (g)			
7.	ppm $CaCO_3$ $(mg\ CaCO_3/L$ sample$)$			
8.	Average ppm $CaCO_3$			

QUESTIONS

1. Because the chemist couldn't accurately detect the endpoint, the addition of Na_2H_2Y titrant was discontinued before the endpoint was reached. How will this affect the reported hardness of the water sample? Explain.

2. Explain what might happen in the analysis if the indicator had been omitted from the procedure.

*3. Washing soda, $Na_2CO_3 \cdot 10H_2O$, is often used to "soften" hard water, i.e., to remove hardening ions. Assuming hardness is due to Ca^{2+}, the CO_3^{2-} ion precipitates the Ca^{2+}.

$$Ca^{2+}(aq) + CO_3^{2-}(aq) \rightarrow CaCO_3(s)$$

How many grams and pounds of washing soda are needed to remove the hardness from 500 gallons of water having a hardness of 150 ppm $CaCO_3$ (1 gal = 4 qt; see Appendix B for additional conversion factors)?

Experiment 25

◇ Redox; Activity Series

Man! Look at that old car! What a piece of junk! One sign of an "old" car is its external appearance. The door panels are beginning to rust (and perhaps disappear!) and the paint is losing its luster. Oftentimes only professional care and repair will return the exterior to its original beauty. All one needs to do is to visit a junk yard of old cars to see what happens when the finish is exposed to the harsh conditions of rain, salt, various air contaminants, and sunlight; the surface and underlying steel alloys become "oxidized" and the coating looks dull and ugly.

All metals have a tendency to **oxidize**; some are easily oxidized by water, others require harsher chemicals such as nitric acid. In this experiment we will use oxidizing agents of various strengths to differentiate the ease of oxidation of a series of metals. For example, should we use aluminum instead of iron to guard against the corrosion of our automobile? Or would zinc be better?

OBJECTIVES

- To learn rules for determining the oxidation numbers of the elements in compounds
- To develop a general understanding of redox reactions
- To determine the relative chemical reactivity of several metals

PRINCIPLES

Most elements exist in chemical combination with other elements, forming any number of compounds. An element chemically bound to another element in the formation of a compound no longer has its elemental physical or chemical properties because of changes in its electronic structure—it has either lost or gained valence electrons to form an ionic compound or has shared its valence electrons to form a molecular compound.

Oxidize: to combine with oxygen or to increase in oxidation number.

An atom in its use of **valence electrons** to form a molecule (or ion) assumes an "apparent" (or assigned) charge. This apparent charge, the charge an atom would have if all the electrons in each bond were *assigned* to the more electronegative element, may be either positive (+) or negative (–). This apparent charge is called the **oxidation number** (or **oxidation state**) of the atom in the molecule (or ion). When the atom appears to have *lost* valence electrons in bond formation, its oxidation number increases and we say that the atom is *oxidized*; but when the atom appears to have *gained* valence electrons in bond formation, its oxidation number decreases, it is *reduced*.

Valence electrons: electrons in the highest principal energy level of the atom.

Reduce: to remove oxygen from a compound or to decrease the oxidation number of an element.

Oxidation numbers are very handy "bookkeeping" devices for keeping track of valence electrons when elements combine to form compounds. From a few rules about the assignments of oxidation numbers to many common elements, we can not only determine which substance is oxidized and which substance is reduced in a chemical reaction, but also write the correct chemical formulas for a large number of compounds. These rules are:

1. Any element in the elemental or "free" state (not combined with any other element) has an oxidation number of zero, regardless of the complexity of the molecule in which it occurs. Each atom in Ne, O_2, P_4, and S_8 has an oxidation number of zero.

2. The oxidation number of any ion (monatomic or polyatomic) equals the charge on the ion. The Ca^{2+}, NH_4^+, S^{2-}, and PO_4^{3-} ions have oxidation numbers of 2^+, 1^+, 2^-, and 3^- respectively.

3. Oxygen in compounds has an oxidation number of 2^- (except 1^- in peroxides, e.g., H_2O_2, and 2^+ in OF_2). The oxidation number of oxygen is 2^- in $KMnO_4$, Fe_2O_3, CaO, and N_2O.

4. Hydrogen in compounds has an oxidation number of 1^+ (except 1^- in metal hydrides, e.g., NaH or CaH_2). Its oxidation number is 1^+ in HCl, $NaHCO_3$, and NH_3.

Group 1A: *called the alkali metals*
Group 2A: *called the alkaline-earth metals*

5. Some elements exhibit only one oxidation number in certain compounds.
   a. **Group 1A** elements (Li, Na, K, Rb, and Cs) always have a 1^+ oxidation number in compounds.
   b. **Group 2A** elements (Be, Mg, Ca, Sr, and Ba) always have a 2^+ oxidation number in compounds.
   c. Boron and aluminum always possess a 3^+ oxidation number in compounds.

Group 6A: *called the chalcogens*
Group 7A: *called the halogens*

   d. **Group 6A** elements (O, S, Se, and Te) exhibit a 2^- oxidation number in all *binary* compounds with metal and polyatomic cations.
   e. **Group 7A** elements (F, Cl, Br, and I) exhibit a 1^- oxidation number in all *binary* compounds with metal and polyatomic cations.

6. The oxidation number of any other element does not follow set rules, it may vary depending upon the elements with which it combines to form the compound. Generally one or more of the elements in a compound follow Rules 1–5; since the compound must be neutral, the oxidation number of most *other* elements can be calculated.

   Example 1. What is the oxidation number of Fe in Fe_2O_3?
   Since each oxygen is known to have a 2^- oxidation number (for a total of 6^-) and the sum of the oxidation numbers of Fe_2O_3 is zero, the combined oxidation numbers of the two iron atoms must be 6^+, or *each* Fe atom must be 3^+.

7. The oxidation number of polyatomic ions is the charge of the ion (see Rule 2). The oxidation numbers of the constituent atoms in the ion are determined as in Rule 6.

   Example 2. What is the oxidation number of Cr in the dichromate ion, $Cr_2O_7^{2-}$?
   Each oxygen has a 2^- oxidation number (for a total of 14^-) and the oxidation number of $Cr_2O_7^{2-}$ is 2^-. Since $2\ Cr + 14^- = 2^-$, the combined oxidation number of the two chromium atoms must be 12^+, or *each* Cr atom is 6^+.

Additional practice questions are available in the Lab Preview.

The relative tendency for a metal to acquire a positive oxidation number by losing valence electrons (referred to as its **chemical activity**) is measured in this experiment. The selected metals are tested with various oxidizing agents, starting with water, a *very* weak oxidizing agent, and finishing with nitric acid, a strong oxidizing agent, and other metal ions. Observations, such as the rate of evolution of a gas and the rate of disappearance of a metal, are used to determine the relative activity of the metals.

The metals listed in order of decreasing activity comprise an abbreviated **activity series**.

Water at room temperature readily oxidizes a very active metal (M), **Metal and Water** resulting in the evolution of H_2 gas and heat.

oxidation:	$M(s) \rightarrow M^{n+}(aq) + n\ e^-$
reduction:	$2\,H_2O(l) + 2\ e^- \rightarrow 2\,OH^-(aq) + H_2(g)$

The rate of evolution of H_2 gas is the criterion used to measure the reactivity of the metal; however it is not the only criterion of the chemical activity of the metal because the reaction involves several variables, e.g., particle size of a metal. It is, however, a qualitative method for observing chemical activity.

Less active metals in hot water cause a similar reaction to occur. The rate of **Metal and Hot Water** evolution of H_2 gas is often so slow that the rate of disappearance of the metal is used as the measure of activity, even though the particle size of the metal does influence the reaction rate.

By comparison, some metals evolve H_2 gas only with the addition of a **Metal and Nonoxidizing** nonoxidizing acid, such as $HCl(aq)$ or $H_2SO_4(aq)$. The metal (M) is oxidized **Acid** and the H^+ from the acid is reduced. For this, the rate of hydrogen gas evolution is used as a criterion of chemical reactivity.

oxidation:	$M(s) \rightarrow M^{n+}(aq) + n\ e^-$
reduction:	$2\,H^+(aq) + 2\ e^- \rightarrow H_2(g)$

Metals that are oxidized only by an oxidizing acid have a very low chemical **Metal and Oxidizing** activity. Common oxidizing acids are nitric acid, $HNO_3(aq)$, and perchloric **Acid** acid, $HClO_4(aq)$. For conc nitric acid, the reactions are:

oxidation:	$M(s) \rightarrow M^{n+}(aq) + n\ e^-$
reduction:	$2\,H^+(aq) + NO_3^-(aq) + e^- \rightarrow NO_2(g) + H_2O(l)$

The rate of NO_2 evolution is a measure of the relative reactivity of a metal. Some metals, such as gold and mercury, are virtually unaffected by common oxidizing agents. This property of gold is evident in the preserved appearance of objects, such as jewelry, made of gold or covered with gold foil.

The activity of one metal relative to another can be determined by placing a **Metal and More** metal into a solution containing the cation of the other. If metal **M** is more **Reactive Metal Cation** active than metal **R**, then **M** displaces R^{n+} from the aqueous solution. **M** goes into solution as M^{n+} and R^{n+} forms the metal, **R**.

$$M(s) + R^{n+}(aq) \rightarrow M^{n+}(aq) + R(s)$$

For example, when iron wire is placed in a lead(II) solution, a reaction occurs because the iron is more active than lead, causing the formation of Fe^{2+} and the reduction of Pb^{2+} to Pb metal.

oxidation:	$Fe(s) \rightarrow Fe^{2+}(aq) + 2\ e^-$
reduction:	$Pb^{2+}(aq) + 2\ e^- \rightarrow Pb(s)$

Conversely, when lead wire is placed in an iron(II) solution, no reaction occurs; lead metal, the less active of the two, remains as the metal and is incapable of displacing the iron(II) ion from solution:

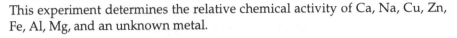

$$Pb(s) + Fe^{2+}(aq) \rightarrow \text{no reaction}$$

Since iron metal displaces lead(II) ion from solution, it is said to have a greater chemical activity than lead.

PROCEDURE

This experiment determines the relative chemical activity of Ca, Na, Cu, Zn, Fe, Al, Mg, and an unknown metal.

You and your partner will need a number of metals (wire or strips) and solutions for this experiment. Have them readily available so the testing can proceed quickly. Set up a hot water bath for use in Part B.

A. Metal and Water

1. **Demonstration Only**. Ca, Na. (**Caution**: *never touch* Na or Ca; *each causes a severe skin burn*.)[1]

   a. Wrap a pea-sized (*no larger!*) piece of the freshly cut calcium in aluminum foil. Fill a 200 mm Pyrex test tube with water and invert it in an 800 mL beaker $^3/_4$-filled with water. Add 2 drops of phenolphthalein.[2] Set the beaker and test tube behind a safety shield.

   b. Punch 5 pin-sized holes in the aluminum foil. With a pair of tongs or tweezers, place the wrapped calcium in the mouth of the test tube, keeping it under water (Figure 25.1). Observe the rate of $H_2(g)$ evolution.

2. Repeat the test for sodium and compare its rate of $H_2(g)$ evolution. Record these observations.

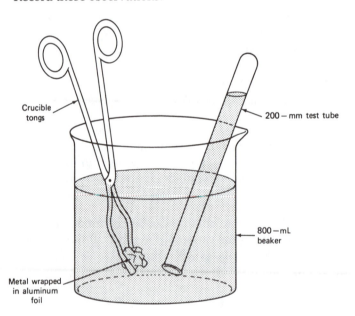

Figure 25.1
Collection of $H_2(g)$ from the reaction of an active metal with water

---

[1]It is strongly recommended that the reactions of calcium and sodium with water be completed as a laboratory demonstration by the laboratory instructor, using a safety shield.

[2]Phenolphthalein is an acid-base indicator that has a pink color in a basic solution but colorless in acidic and neutral solutions. A color change from colorless to pink indicates the generation of the hydroxide ion, OH⁻.

B. Metal and Hot Water

Oxide coating: the dull finish on a metal caused by its reaction with the oxygen in air.

1. Al, Mg, Cu, Zn, Fe, and unknown. Thoroughly clean a 2 cm wire/strip of each metal with steel wool to remove any **oxide coating**. This is especially crucial for Al and Mg since they quickly form tough, protective oxide coatings. Rinse a 24 well plate with boiling, deionized water. Add 1 drop of phenolphthalein to each well and then (with a Beral pipet) half-fill wells A1–A6 with boiling water.

2. *Quickly* place each metal into the well, Al and Mg first (Figure 25.2). Look for $H_2(g)$ evolution, discoloration of the metal surface, a color change of the solution (due to the phenolphthalein), or a disappearance of the metal. Evidence of any reactions may not be immediately obvious. Allow the reaction systems to set for several minutes and then record your observations.

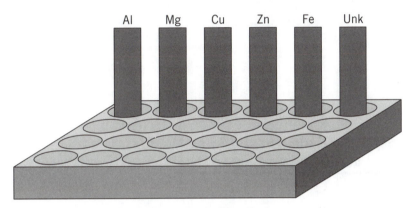

Al   Mg   Cu   Zn   Fe   Unk

Figure 25.2

Arrangement of metals for testing reactivity

C. Metal and Nonoxidizing Acid

1. Add 5 drops of conc HCl (**Caution**: *avoid skin contact*) to wells A1–A6 from Part B in which no reaction was observed. Swirl the solutions; allow 10–15 minutes for evidence of a reaction (evolution of H_2 gas) to appear. Record.

D. Metal and Oxidizing Acid

1. If a metal shows no reaction in Parts B or C, draw off the water/acid solution with a Beral pipet and discard it. Add 1 mL of 6 M HNO_3 (**Caution**: *avoid skin contact*). If evidence of a reaction (evolution of NO_2 gas) is now observed, record your observation.

E. Metal and More Reactive Metal Cation

1. Cu, Zn, Fe, and unknown. Place a small amount of a freshly cleaned and polished metal wire/strip in wells C1–C4 of the 24 well plate—one metal in each well. Half-fill each well with 0.1 M $Cu(NO_3)_2$. Any tarnishing or dulling of the metal or changing of the color of the solution indicates a reaction. Allow 5–10 minutes for a reaction to be observed.

2. Repeat the procedure, using 0.1 M test solutions of $Zn(NO_3)_2$ and $Fe(NH_4)_2(SO_4)_2$ on each metal. In each case the same metal wire/strip may be reused if it remains unreacted from the previous test, rinsed with deionized water, and cleaned with steel wool. Record.

F. Establishment of Activity Series

1. Using the observations from Parts A through E, list the eight metals in order of decreasing activity.

Dispose of the metal wire/strips in the "Waste Solids" container and the solutions in the "Waste Salt Solution" container.

Activity Series for Hydrogen and Some Metals

Element	Reduced State	Oxidized State
Cesium	Cs	Cs^+
Rubidium	Rb	Rb^+
Potassium	K	K^+
Barium	Ba	Ba^{2+}
Strontium	Sr	Sr^{2+}
Calcium	Ca	Ca^{2+}
Sodium	Na	Na^+
Magnesium	Mg	Mg^{2+}
Aluminum	Al	Al^{3+}
Manganese	Mn	Mn^{2+}
Zinc	Zn	Zn^{2+}
Chromium	Cr	Cr^{3+}
Iron	Fe	Fe^{2+}
Cadmium	Cd	Cd^{2+}
Cobalt	Co	Co^{2+}
Tin	Sn	Sn^{2+}
Lead	Pb	Pb^{2+}
Hydrogen	H_2	H^+
Copper	Cu	Cu^{2+}
Silver	Ag	Ag^+
Mercury	Hg	Hg^{2+}
Gold	Au	Au^{3+}

Redox; Activity Series

Date _____Name _____ Lab Sec. _____Desk No. _____

1. Indicate the oxidation number of the bold faced element.

   a. **C**O _____   d. **N**$_2$O _____   g. **P**Cl$_3$ _____   j. **P**F$_5$ _____   m. **Cu**$_2$SO$_4$ _____

   b. **C**O$_2$ _____   e. O**F**$_2$ _____   h. **S**O$_3^{2-}$ _____   k. H**P**O$_3^-$ ____   n. K$_2$**Mn**O$_4$ _____

   c. **I**F$_7$ _____   f. **N**I$_3$ _____   i. **I**O$_3^-$ _____   l. **Au**Cl$_3$ ____   o. **Cr**O$_3$ _____

2. From the following displacement reactions, arrange, at right, these hypothetical metals in order of increasing activity.

   $A + B^+ \rightarrow A^+ + B$ _____ least reactive

   $C + B^+ \rightarrow$ no reaction   _____

   $D + C^+ \rightarrow D^+ + C$ _____

   $F + G^+ \rightarrow$ no reaction   _____

   $A + F^+ \rightarrow$ no reaction   _____

   $F + D^+ \rightarrow F^+ + D$ _____ most reactive

   $D + B^+ \rightarrow$ no reaction

   $G + A^+ \rightarrow G^+ + A$

3. a. List four methods by which the chemical activity of a metal can be observed.

   b. Which observation is characteristic of the most reactive metals?

4. Cesium is a very reactive metal, reacting with cold water to produce H$_2$ (g).
   a. Write a balanced equation for the reaction.

   b. The oxidizing agent is _____; the substance oxidized is _____.

5. Considering the trend in ionization energies of the Group 1A elements (see Experiment 13), predict their relative reactivity in cold water?

6. Based upon your experiences, would you predict a reaction to occur when
   a. Zn metal is placed in a solution containing Ag^+? (Consider the relative activity of Zn *versus* Ag.)

   b. Au metal is placed in a solution containing Zn^{2+}?

7. Distinguish the properties of an oxidizing acid from a nonoxidizing acid. Give an example of each.

◇ Redox; Activity Series

Date _____ Name _____ Lab Sec. _____ Desk No. _____

A. Metal and Water

Ca and Na

1. From the rate of $H_2(g)$ evolution, arrange Ca and Na in order of decreasing activity.

2. What does the phenolphthalein indicate about a product produced in the reaction?

3. Write a balanced equation for each reaction of the metal with water.

B. Metal and Hot Water

Al, Mg, Cu, Zn, Fe, and unknown

1. Which metals show a definite reaction with hot water?

2. What does the phenolphthalein indicate about a product produced in the reaction?

3. If possible, arrange the metals that do react in order of decreasing activity according to the rate of disappearance of the metal.

4. Write a balanced equation for each reaction.

C. Metal and Nonoxidizing Acid

Al, Mg, Cu, Zn, Fe, and unknown

1. Which remaining metals show a definite reaction with HCl?

2. If possible, arrange these metals in order of decreasing activity according to the rate of evolution of H_2 gas.

3. Write a balanced equation for each reaction.

D. Metal and Oxidizing Acid

1. Which of the remaining metals react with HNO_3?

E. Metal and More Reactive Metal Cation

Cu, Zn, Fe, and unknown

Complete the table with NR (no reaction) or R (reaction) where appropriate.

For all reactions observed, write a balanced net ionic equation. Use additional paper if necessary.

Test Reagents	Cu	Zn	Fe	unknown
HCl				
$Cu(NO_3)_2$	none			
$Zn(NO_3)_2$		none		
$Fe(NH_4)_2(SO_4)_2$			none	

Balanced Equations

F. Establishment of Activity Series

List the eight metals studied in this experiment in order of *decreasing* activity.

QUESTIONS

1. Tons of aluminum and magnesium are used annually where light-weight construction is required (e.g., for airplanes) even though each reacts rapidly with oxygen. Explain why this use is possible.

2. Zinc is used as the protective covering of steel on "galvanized iron." Explain its function in terms of chemical activity. Is zinc or iron more reactive?

3. a. Does hydrochloric acid added to copper metal generate $H_2(g)$? Explain.

    b. Which metals studied in this experiment would generate $H_2(g)$ in a reaction with HCl?

4. Magnesium metal is used as a sacrificial metal to reduce the corrosion of underground storage tanks (made of iron, or more accurately, steel which is an iron/carbon alloy). As a result the sacrificial metal corrodes instead of the iron. Describe the chemical reaction that occurs.

5. Circle the metals that react with

    a. Na^+:   Fe, Cu, Mg

    b. Fe^{2+}:   Na, Mg, Cu

    c. Cu^{2+}:   Ca, Zn, Al

6. Write a balanced equation for only one (if any) of the reactions that occurs in each part of Question 5.

    a. Na^+ + _____ →

    b. Fe^{2+} + _____ →

    c. Cu^{2+} + _____ →

◇ A Standard Na₂S₂O₃ Solution

All packaged foods are now required by law to list amounts of fat, sodium, carbohydrate, and protein in the food. All manufactured products must meet certain quality standards, whether the standards are required by law or set internally by the manufacturing company. Where a specific chemical or a general class of chemicals is the focus of the quality control program for the manufacturing process, a chemical analysis is required. The method of analysis may require a spectrophotometer, a mass spectrometer, chromatographic principles, or acid-base or oxidation-reduction titrimetric techniques. Several of these techniques and procedures appear in this manual.

OBJECTIVE

- To prepare and standardize a sodium thiosulfate solution

PRINCIPLES

A **standard solution** is used in the analysis of a sample containing an undetermined amount of substance in a sample, whether it be the amount of illicit drug or an amount of phosphate. Sodium thiosulfate is an excellent, safe reducing agent, readily undergoing oxidation by the equation:

Standard solution: a solution having a best known concentration of a solute in solution.

$$2\,S_2O_3^{2-}(aq) \rightarrow S_4O_6^{2-}(aq) + 2\,e^-$$

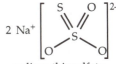

In this experiment a standard solution of sodium thiosulfate, $Na_2S_2O_3$, is prepared. Potassium iodate, KIO_3, is used as the **primary standard** for the analysis. The standard sodium thiosulfate solution is subsequently used for an oxidation-reduction titrimetric analysis of ascorbic acid (also, vitamin C) and/or bleach in a sample (see Experiments 26B and 26C).

sodium thiosulfate

*Primary standard: a solid chemical that has a high (and known) degree of purity, a relatively large molar mass, and is **nonhygroscopic** (Figure 26A.1).*

A measured mass of KIO_3 is dissolved in an acidic solution containing an excess of KI. Potassium iodate, a strong oxidizing agent, oxidizes I^- to I_2. With an excess amount of iodide ion present in solution, the molecular I_2 binds to the iodide ion to form a red-brown,[1] soluble triiodide ion, I_3^-. A quantitative amount of I_3^- is generated in the reaction:

Standardization of a Na₂S₂O₃ Solution

Nonhygroscopic: a property of not adsorbing or retaining water.

$$IO_3^-(aq) + 8\,I^-(aq) + 6\,H^+(aq) \rightarrow 3\,I_3^-(aq) + 3\,H_2O(l)$$

The I_3^- is titrated to the stoichiometric point with a prepared solution of $Na_2S_2O_3$. Sodium thiosulfate reduces the molecular I_2 of the triiodide ion to I^-. Starch is used as an indicator to detect the stoichiometric point. Just prior to the disappearance of the red-brown I_3^- in the titration, starch is added; this forms the deep-blue ion, $[I_3 \bullet starch]^-$.

Figure 26A.1

Potassium iodate has the necessary properties of a primary standard

$$3\,I_3^-(aq) + 3\,starch(aq) \rightarrow 3\,[I_3 \bullet starch]^-(aq, deep\text{-}blue)$$

---

[1] The red-brown color is due to the presence of I_2; the excess I^- combines with I_2 to form a water soluble I_3^- ion. It is common to refer to $I_2(aq)$ and $I_3^-(aq)$ as the same substance in aqueous systems.

The addition of the $S_2O_3^{2-}$ titrant is continued until the $[I_3 \bullet \text{starch}]^-$ is reduced to I^-; the solution appears colorless at the endpoint.

$$3\,[I_3\bullet\text{starch}]^-(aq) + 6\,S_2O_3^{2-}(aq) \rightarrow 9\,I^-(aq) + 3\,S_4O_6^{2-}(aq) + 3\,\text{starch}(aq)$$

The chemical analysis for the standardization of the sodium thiosulfate solution is summarized in the net ionic equation:

Net ionic equation: $IO_3^-(aq) + 6\,S_2O_3^{2-}(aq) + 6\,H^+(aq) \rightarrow$
$$I^-(aq) + 3\,S_4O_6^{2-}(aq) + 3\,H_2O(l)$$

PROCEDURE

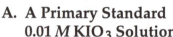

A. A Primary Standard 0.01 M KIO$_3$ Solution

B. A Standard 0.1 M Na$_2$S$_2$O$_3$ Solution

Three trials are necessary for the standardization of the $Na_2S_2O_3$ solution, for subsequent use in Experiments 26B or 26C. Quantitative, reproducible data are objectives of this experiment; practice good laboratory techniques.

1. Measure about 0.5 g ($\pm$0.001 g) of KIO_3 (dried at 110°C for one hour) on weighing paper, transfer the solid to a 250 mL volumetric flask,[2] dissolve and dilute to the mark. Calculate and record the molar concentration of the KIO_3 solution.

1. The $Na_2S_2O_3$ solution[3] should be prepared about one week in advance because of the unavoidable decomposition of the sodium thiosulfate. Dissolve about 6 g ($\pm$0.01 g) of $Na_2S_2O_3\bullet 5H_2O$ with freshly boiled, deionized water and dilute to 250 mL. Agitate until the salt dissolves.

2. Properly prepare a clean, 50 mL buret for titration. Fill it with your $Na_2S_2O_3$ solution, drain the tip of air bubbles, and, after 30 s, read and record the volume ($\pm$0.02 mL).

3. Pipet 25 mL of the standard KIO_3 solution into a 125 mL Erlenmeyer flask and add about 1 g ($\pm$0.01 g) of solid KI. Add about 5 mL of 0.5 M H_2SO_4 and 0.1 g of $NaHCO_3$.[4]

4. Immediately begin titrating with the $Na_2S_2O_3$ solution. When the red-brown solution (due to I_3^-) changes to a pale yellow color, add 2 mL of starch solution. Stirring constantly, continue titrating slowly until the blue color disappears.

5. Repeat the titrimetric procedure twice by rapidly adding the $Na_2S_2O_3$ titrant until 1 mL before the endpoint. Add the starch solution and continue titrating until the solution is colorless. Repeated analyses should be within $\pm$1%.

6. Save the standard $Na_2S_2O_3$ solution for Experiments 26B and/or 26C.

Save the standard KIO$_3$ solution for Experiment 26B. Dispose of any waste KIO$_3$ solution from this experiment into the "Waste Oxidants" container.

---

[2]Experiment 26C does *not* require the use of a KIO_3 solution. If Experiment 26B is to be completed, but not Experiment 26C, then reduce the preparation of the standard KIO_3 solution to 0.2 g KIO_3 dissolved in water in a 100 mL volumetric flask.

[3]Ask your instructor about this solution; it may have already been prepared.

[4]The $NaHCO_3$ reacts in the acidic solution to produce $CO_2(g)$, providing an inert atmosphere above the solution and minimizing the possibility of the air oxidation of I^- ions.

A Standard $Na_2S_2O_3$ Solution

Date _____Name _____ Lab Sec. _____Desk No. _____

1. a. What is the color change at the stoichiometric point in today's analysis?

    b. What is the cause of the color change?

2. A 0.428 g sample of KIO_3 is transferred to a 250 mL volumetric flask, dissolved and diluted to the mark with deionized water. What is the molar concentration of the KIO_3 solution?

3. A 25 mL aliquot of a 0.0104 M KIO_3 solution is titrated to the stoichiometric point with 17.72 mL of a sodium thiosulfate, $Na_2S_2O_3$, solution. What is the molar concentration of the $Na_2S_2O_3$ solution?

4. Describe "how" the starch solution is used to detect the stoichiometric point in today's experiment.

5. a. Which finger should be used to control the delivery of a solution from a pipet?

    b. How much time should elapse between the time that the stopcock on a buret is closed and the time that a volume reading should be made and recorded?

    c. What criterion is used to determine if a buret is clean?

    d. What does the phrase "Properly prepare a clean, 50 mL buret for titration" mean? (see Part B.2)

◇ A Standard $Na_2S_2O_3$ Solution

Date _____ Name _____ Lab Sec. _____ Desk No. _____

A. A Primary Standard 0.01 M KIO$_3$ Solution

1. Mass of KIO_3 + weighing paper (g)   _____

2. Mass of weighing paper (g)   _____

3. Mass of KIO_3 (g)   _____

4. Moles of KIO_3 (mol)   _____

5. Molar concentration of standard KIO_3 solution (mol/L) _____

B. A Standard 0.1 M Na$_2$S$_2$O$_3$ Solution

	Trial 1	Trial 2	Trial 3
1. Volume of KIO_3 solution (mL)	25.0	25.0	25.0
2. Moles of KIO_3 titrated (mol)			
3. Moles of I_3^- generated (mol)			
4. Buret reading, final (mL)			
5. Buret reading, initial (mL)			
6. Volume of $Na_2S_2O_3$ added (mL)			
7. Moles of $Na_2S_2O_3$ added (mol)			
8. Molar concentration of $Na_2S_2O_3$ solution (mol/L)	*		

9. Average molar concentration of $Na_2S_2O_3$ solution (mol/L) _____

*Show calculations for Trial 1.

QUESTIONS

1. How will the addition of 2 g of KI to the sample in Part B.3, instead of the 1 g, affect the molar concentration of the $Na_2S_2O_3$ solution? Explain.

2. The triiodide ion, I_3^-, is a soluble ion that has the color (violet) of molecular iodine; the iodide ion, I^-, is colorless. Could the starch solution have been omitted from the experimental analysis and still have the stoichiometric point detected? Explain why or why not.

Experiment 26B

◇ Vitamin C Analysis

The human body does not synthesize vitamins; therefore the vitamins that we need for catalyzing specific biochemical reactions are gained only from the food that we eat. We are generally aware that vitamin C can be obtained from citrus fruits, but it can also be obtained from a variety of fresh fruit and vegetables. However, storage and processing causes vegetables to lose a part of their vitamin C content. Cooking (boiling or steaming) leaches the water soluble vitamin C from the vegetables; in addition, high temperatures accelerates its degradation by air oxidation. Therefore to maximize the intake of vitamin C, only freshly harvested fruits or vegetables should be consumed; their natural protective coverings (e.g., orange peel) should only be removed just before its consumption.

- To determine the amount of vitamin C in a vitamin tablet, a fresh fruit, **OBJECTIVE**
  or a fresh vegetable sample

Vitamin C, also called **ascorbic acid**, is one of the more abundant and easily **PRINCIPLES** obtained vitamins in nature. It is a colorless, water-soluble acid that, in addition to its acidic properties, is a powerful biochemical reducing agent, meaning it readily undergoes oxidation, even from the oxygen of the air.

Table 26B.1 lists the concentration ranges of ascorbic acid for various vegetables (Figure 26B.1).

Table 26B.1 Ascorbic Acid in Foods

<10 mg/100 g	beets, carrots, eggs, milk
10-25 mg/100 g	asparagus, cranberries, cucumbers, green peas, lettuce, pineapple
25-100 mg/100 g	Brussels sprouts, citrus fruits, tomatoes, spinach
100-350 mg/100 g	chili peppers, sweet peppers, turnip, greens

Figure 26B.1
Vegetables are a good supply of vitamin C

Even though ascorbic acid is an acid, its *reducing* properties are used in this experiment to analyze its concentration in various samples. There are many other acids present in foods (e.g., citric acid) that would interfere with an acid analysis and not permit us to selectively determine the ascorbic acid content. The equation for the oxidation of ascorbic acid is

or $C_6H_8O_6(aq) + H_2O(l) \rightarrow C_6H_8O_7(aq) + 2H^+(aq) + 2e^-$

Vitamin C Analysis

The sample containing ascorbic acid is dissolved in water and treated with a measured amount of iodate ion, IO_3^-, in an acidic solution containing an excess of I^-; the red-brown triiodide ion, I_3^-, a milder oxidizing agent than IO_3^-, forms in solution (Step 1):

$$IO_3^-(aq) + 8\,I^-(aq) + 6\,H^+(aq) \rightarrow 3\,I_3^-(aq) + 3\,H_2O(l) \quad \text{(Step 1)}$$

For the analysis, ascorbic acid from the sample reduces a portion of the known amount of I_3^- generated in solution (Step 2).

$$C_6H_8O_6(aq) + I_3^-(aq) + H_2O(l) \rightarrow$$
$$C_6H_8O_7(aq) + 3\,I^-(aq) + 2\,H^+(aq) \quad \text{(Step 2)}$$

The remainder of the I_3^- (or the excess, "xs") is titrated with a standard thiosulfate, $S_2O_3^{2-}$, solution, producing the colorless I^- and $S_4O_6^{2-}$ ions (Step 3).

$$2\,S_2O_3^{2-}(aq) + (xs)\,I_3^-(aq) \rightarrow 3\,I^-(aq) + S_4O_6^{2-}(aq) \quad \text{(Step 3)}$$

Therefore the difference between the I_3^- generated (from the IO_3^- in Step 1) and that which is titrated as an excess (Step 3) is a measure of the ascorbic acid content of the sample.

The stoichiometric point is detected using starch as an indicator; the detection is described in Experiment 26A.

PROCEDURE

Three trials are necessary to analyze for ascorbic acid in a sample assigned by your instructor. Quantitative, reproducible data are objectives of this experiment; practice good laboratory techniques.

A. Sample Preparation

1. **Vitamin C tablet.** Read the label on the bottle to determine the approximate mass of vitamin C in each tablet. Measure (± 0.001 g) the fraction of the total mass of a tablet that corresponds to 100 mg of ascorbic acid. Dissolve the sample in a 250 mL Erlenmeyer flask with 40 mL of 0.5 M H_2SO_4[1] and then add about 0.5 g $NaHCO_3$.[2] Kool-Aid™, Tang™, or Gatorade™ may be substituted as dry samples, even though their ascorbic acid concentrations are much lower. Fresh Fruit™ has a very high concentration of ascorbic acid. Proceed immediately to Part B.

2. **Fresh fruit sample.** Filter 125–130 mL of freshly squeezed juice through several layers of cheesecloth (or vacuum-filter). Measure the mass (± 0.01 g) of a clean, dry 250 mL Erlenmeyer flask. Add about 100 mL of filtered juice and again determine the mass. Add 40 mL of 0.5 M H_2SO_4 and 0.5 g $NaHCO_3$. Concentrated fruit juices may also be used as samples. Proceed immediately to Part B.

---

[1]Remember that vitamin tablets contain binders and other material that may be insoluble in water—do not heat in an attempt to dissolve the tablet!

[2]The $NaHCO_3$ reacts in the acidic solution to produce $CO_2(g)$, providing an inert atmosphere above the solution, minimizing the possibility of the air oxidation of the ascorbic acid.

3. **Fresh vegetable sample.** Measure about 100 g (±0.01 g) of a fresh vegetable. Transfer the sample to a mortar[3] (Figure 26B.2) and grind. Add 5 mL of 0.5 M H_2SO_4 and continue to pulverize the sample. Add another 15 mL of 0.5 M H_2SO_4, stir, and filter through several layers of cheesecloth (or vacuum-filter). Add 20 mL of 0.5 M H_2SO_4 to the mortar, stir, and pour through the same filter. Repeat the washing of the mortar with 20 mL of freshly boiled, deionized water. Combine all of the washings in a 250 mL Erlenmeyer flask and add 0.5 g $NaHCO_3$. Proceed immediately to Part B.

Figure 26B.2
A mortar

B. Vitamin C Analysis

1. Pipet 25.0 mL of the standard KIO_3 solution (from Experiment 26A) into the sample solution from Part A and add 1 g of KI. Add about 5 mL of 0.5 M H_2SO_4 and 0.1 g of $NaHCO_3$. Titrate the excess I_3^- in the sample with the standard $Na_2S_2O_3$ solution as described in Experiment 26A, Part B.4. Read and record the final buret reading (±0.02 mL).

2. Repeat the analysis twice in order to complete the three trials. Repeated analyses of the ascorbic acid content of the sample should be ±1%.

Dispose of the remaining KIO_3 solution in the "Waste Oxidants" container.
Dispose of the excess $Na_2S_2O_3$ solution in the "Waste Reducing Agent" container.

NOTES, OBSERVATIONS, AND CALCULATIONS

---

[3] A blender may be substituted for the mortar and pestle.

Vitamin C Analysis

Date _____Name _____ Lab Sec. _____Desk No. _____

1. Explain why cooked fruits and vegetables have a lower vitamin C content than fresh fruits and vegetables.

2. a. What is the oxidizing agent in Part B of the procedure?

    b. What will be the color change of the starch indicator that signals the end of the titration in today's experiment?

    c. Vitamin C is an acid (ascorbic acid) and a reducing agent. Which property is utilized for its analysis in the experiment?

3. Six ounces (1 fl. oz. = 29.57 mL) of a well-known vegetable juice contains 35% of the recommended daily allowance of vitamin C (equal to 60 mg). How many milliliters of the vegetable juice will provide 100% of the recommended daily allowance?

4. A 25.0 mL volume of 0.010 M KIO_3, containing an excess of KI, is added to a 0.246 g sample of a Real Lemon™ solution containing vitamin C. The red-brown solution, caused by the presence of excess I_3^-, is titrated to a colorless starch endpoint with 10.7 mL of 0.100 M $Na_2S_2O_3$.

a. How many moles of KIO_3 were added to the Real Lemon™ solution?

b. Calculate the moles of I_3^- that are generated from the KIO_3?

c. How many moles of I_3^- reacted with the 0.100 M $Na_2S_2O_3$ in the titration?

d. How many moles (of the total moles) of I_3^- had reacted with the vitamin C in the Real Lemon™ sample?

e. Calculate the moles and grams of vitamin C, $C_6H_8O_6$, in the sample.

f. Calculate the percent (by mass) of vitamin C in the Real Lemon™ sample.

Vitamin C Analysis

Date _____Name _____ Lab Sec. _____Desk No. _____

A. Sample Preparation

Sample Name: _____

1. Mass of sample (g)   _____

B. Vitamin C Analysis

	Trial 1	Trial 2	Trial 3
1. Volume of KIO_3 added (mL)	25.0	25.0	25.0
2. Moles of IO_3^- added (mol)			
3. Moles of I_3^- generated, total (mol)			
4. Buret reading, final (mL)			
5. Buret reading, initial (mL)			
6. Volume of $Na_2S_2O_3$ added (mL)			
7. Molar concentration of $Na_2S_2O_3$ (mol/L)			
8. Moles of $S_2O_3^{2-}$ added (mol)			
9. Moles of I_3^- reduced by $S_2O_3^{2-}$ (mol)			
10. Moles of I_3^- reduced by $C_6H_8O_6$ (mol)			
11. Moles of $C_6H_8O_6$ in sample (mol)			
12. Mass of $C_6H_8O_6$ in sample (g)			
13. Percent of $C_6H_8O_6$ in sample (%)			

14. Average percent of $C_6H_8O_6$ in sample (%)   _____

QUESTIONS

1. If the blue color does not appear when the starch solution is added during the titration, should you continue titrating or discard the sample (or leave for a cup of coffee)? Explain (but with a serious response).

2. After adding the KIO_3 solution and KI to the sample (in Part B) and stirring, the sample solution remains colorless; what modification of the procedure can be made to correct for this unexpected observation in order to complete the analysis.

Experiment 26C

◇ Bleach Analysis

Whiter than white! How can that be? "To remove that stubborn strawberry stain from baby's new white shirt, just apply some of our new and improved liquid detergent, now with bleach." Chemical companies do extensive advertising in an attempt to convince the consumer that their detergent is stronger, gets clothes whiter, and is safe to the environment. Of all the formulations that have been placed on the market that claim to remove stains from most clothing, perhaps the most effective is still simply bleach, such as Clorox®, Purex®, or some similar generic brand. In addition many laundry detergents also add "whiteners" which actually fluoresce in the presence of ultraviolet radiation, giving the appearance of a whiter wash.

• To determine the percent "available Cl_2" in a bleach	**OBJECTIVE**

PRINCIPLES

Commercial bleaching agents contain the hypochlorite ion, ClO^-, as the "active ingredient" for making clothes whiter or removing stains. This ion is generally in the form of the sodium salt, $NaClO$, or the calcium salt, $Ca(ClO)_2$.

Normally the oxidizing *strength* of a bleach is rated relative to chlorine, Cl_2. This rating, called the **available Cl_2**, is the oxidizing strength of the bleach that is equivalent to the oxidizing strength of the same mass of Cl_2 per unit volume of bleach solution (or mass of the powder). Occasionally, the strength of a bleach is expressed as percent chlorine by mass.

Liquid laundry bleach is a $NaClO$ solution, approximately 5.25% (by mass); bleaching powder is a mixture of $CaCl_2$, $Ca(ClO)Cl$, and $Ca(ClO)_2$.

Read Experiments 12 and 14 for an explanation of visible light and colors.

Substances (fabrics and stains) have a color because energy from the visible light excites electrons to a higher energy level in the molecule, causing the unabsorbed visible radiation to be transmitted and detected with the eye. In both dry and liquid bleaches, the hypochlorite ion removes these "excitable" electrons from the substance.

$$ClO^-(aq) + H_2O(l) + 2\,e^- \rightarrow Cl^-(aq) + 2\,OH^-(aq)$$

Since the electrons are no longer present, visible radiation is not absorbed and, because the eye detects only visible light, the clothes appear whiter. The hypochlorite ion is an oxidizing agent to the stains.

Analysis of Bleach

In this experiment the oxidation-reduction analysis of a bleach involves the reaction of the hypochlorite ion, an oxidizing agent, with iodide ion. The I^- ion is oxidized to I_2, which in the presence of excess I^-, forms I_3^-.

$$ClO^-(aq) + 3\,I^-(aq) + H_2O(l) \rightarrow I_3^-(aq) + Cl^-(aq) + 2\,OH^-(aq)$$

The solution is acidified and the triiodide ion, I_3^-, generated in the reaction, is titrated with a standard sodium thiosulfate, $Na_2S_2O_3$, solution until the red-brown color of I_3^- (or $I_2 \bullet I^-$)[1] nearly disappears just prior to the stoichiometric point.

$$I_3^-(aq) + 2\,S_2O_3^{2-}(aq) \rightarrow 3\,I^-(aq) + S_4O_6^{2-}(aq)$$

The chemical analysis for the amount of hypochlorite ion in a bleach solution is summarized in the net ionic equation:

Net ionic equation: $ClO^-(aq) + 2\,S_2O_3^{2-}(aq) + H_2O(l) \rightarrow$
$$Cl^-(aq) + S_4O_6^{2-}(aq) + 2\,OH^-(aq)$$

Starch is added as the indicator to detect the stoichiometric point (see Experiment 26A for details).

The available Cl_2 in the bleach sample is calculated as if free Cl_2 per unit volume (or mass, if the bleach is a powder) reacts with the I^- ion. The chlorine is assumed to be analyzed with $S_2O_3^{2-}$, just as the hypochlorite ion:

$$Cl_2(aq) + 3\,I^-(aq) \rightarrow 2\,Cl^-(aq) + I_3^-(aq)$$
$$I_3^-(aq) + 2\,S_2O_3^{2-}(aq) \rightarrow 3\,I^-(aq) + S_4O_6^{2-}(aq)$$

Net ionic equation: $Cl_2(aq) + 2\,S_2O_3^{2-}(aq) \rightarrow 2\,Cl^-(aq) + S_4O_6^{2-}(aq)$

This equation serves as the basis for determining the strengths of bleaching agents in this experiment. Notice that for both reactions, the net ionic equation indicates 1 mol ClO^- and 1 mol Cl_2 react with 2 mol $S_2O_3^{2-}$.

PROCEDURE

A. Preparation of a Liquid Bleach

B. Preparation of a Powdered Bleach

Glacial acetic acid: also called concentrated acetic acid.

The concentrations of the oxidizing agent in two bleaching agents are determined in this experiment; two trials are required for each determination. Therefore, obtain about 12 mL of each liquid bleach (or 5 g of each powdered bleach) from your instructor.

1. Pipet 10.0 mL of bleach[2] into a 100 mL volumetric flask and dilute to the mark with boiled, deionized water. Mix thoroughly. Pipet 25.0 mL of this diluted bleach solution into a 250 mL Erlenmeyer flask and add 20 mL of deionized water, 2 g of KI, and 10 mL of 3 M H_2SO_4. A yellow color indicates the presence of I_2 as I_3^-. Proceed immediately to Part C.

1. Place about 5 g of powdered bleach into a mortar (Figure 26C.1) and grind. Measure the exact mass (± 0.01 g) of the pulverized sample on weighing paper and transfer it to a 100 mL volumetric flask fitted with a funnel. Dilute to the mark with deionized water. Mix the solution thoroughly.[3] Pipet 25.0 mL of this solution into a 250 mL Erlenmeyer flask and add 20 mL of deionized water, 2 g of KI, and 15 mL **glacial acetic acid**. A yellow color indicates the presence of I_2 as I_3^-. Proceed immediately to Part C.

---

[1] The red-brown color is due to the presence of I_2; the excess I^- combines with I_2 to form a soluble I_3^- ion. It is common to refer to $I_2(aq)$ and $I_3^-(aq)$ as the same substance.

[2] The density of liquid bleach is 1.084 g/mL.

[3] All of the powder may not dissolve because of the insoluble abrasive in the bleach.

1. *Immediately* titrate the liberated I_2 with your standardized $Na_2S_2O_3$ solution until the red-brown color *nearly* disappears, as described in Experiment 26A, Part B.4. Consult your instructor at this point if you are uncertain. Add 2 mL of starch solution to form the deep-blue $[I_3 \bullet starch]^-$ ion.[4] While swirling the flask, continue titrating *slowly* until the deep-blue color disappears. Record the final volume (±0.02 mL) of the titrant in the buret.

2. Repeat the experiment with the same bleaching agent. Calculate the amount of available chlorine in the sample.

C. Titration of Bleach

Dispose of the remaining $Na_2S_2O_3$ solution in the "Waste Reducing Agent" container.

1. Analyze a similar bleaching agent to determine the amount of available Cl_2. Compare the percent Cl_2 per unit price to determine the best buy.

D. A Comparative Determination

Dispose of the excess bleach solution in the "Waste Bleach" container.

NOTES, OBSERVATIONS, AND CALCULATIONS

---

[4]If you are analyzing a powdered bleach, the 2 mL of starch may be added after the glacial acetic acid solution in Part B.

Bleach Analysis

Date _____ Name _____ Lab Sec. _____ Desk No. _____

1. a. Write the formula for the hypochlorite ion.

   b. In today's chemical analysis of a bleach solution, what substance is oxidized by the hypochlorite ion?

   c. Write the equation for the reaction in **b**.

2. Define available Cl_2.

3. a. The number of moles of ClO^- that react with 1 mol of I^- is _____.

   b. The number of moles of Cl_2 that react with 1 mol of I^- is _____.

   c. The number of moles of ClO^- equivalent to 1 mol of available Cl_2 is _____.

4. Sodium thiosulfate is the standardized titrant for the bleach analysis.

   a. Write the formula for sodium thiosulfate. _____

   b. Does sodium thiosulfate serve as an oxidizing agent or a reducing agent? _____

   c. What does sodium thiosulfate oxidize (or reduce)? _____

   d. One mole of sodium thiosulfate titrant is equivalent to _____ moles of ClO^- and _____ moles of Cl_2 in the bleach.

5. A 0.684 g sample of a powdered bleach is analyzed according to the procedure in this experiment. A volume of 31.7 mL of 0.100 M $Na_2S_2O_3$ is required to reach the stoichiometric point.
   a. How many moles of I_2 (or I_3^-) did the powdered bleach produce?

b. Calculate the moles and grams of available Cl_2 in the bleach.

c. What is the percent (by mass) of available Cl_2 in the bleach?

6. A 10.0 mL bleach sample is diluted to 100 mL in a volumetric flask. A 25 mL aliquot of this solution is analyzed according to the procedure in this experiment. If 11.3 mL of 0.30 M $Na_2S_2O_3$ is needed to reach the stoichiometric point, what is the percent (by mass) of available Cl_2 in the *original* sample? Assume that the density of the bleach solution is 1.084 g/mL.

◈ Bleach Analysis

Date _____ Name _____ Lab Sec. _____ Desk No. _____

A. Preparation of Liquid Bleach

Sample and unit price (¢/g)

	Trial 1	Trial 2	Trial 1	Trial 2
1. Volume of original liquid bleach titrated (mL)	2.5	2.5	2.5	2.5

B. Preparation of a Powdered Bleach

Sample and unit price (¢/g)

1. Mass of weighing paper and bleach (g)			
2. Mass of weighing paper (g)			
3. Mass of bleach (g)			

C. Titration of Bleach

1. Buret reading, final (mL)			
2. Buret reading, initial (mL)			
3. Volume of $Na_2S_2O_3$ added (mL)			
4. Molar concentration of $Na_2S_2O_3$ (mol/L)			
5. Moles of $Na_2S_2O_3$ added (mol)			
6. Moles of available Cl_2 (mol)			
7. Mass of available Cl_2 (g)			
8. Percent (by mass) available Cl_2 (assume the density of liquid bleach to be 1.084 g/mL)			
9. Average % available Cl_2			

10. Best buy (%Cl_2/unit price) _____

QUESTIONS

1. The starch solution can be added at the same time as the KI solution in Part A. However, the $[I_3 \bullet starch]^-$ ion does not readily dissociate as the $Na_2S_2O_3$ titrant is added. What difficulties might you have encountered if the starch solution had been added at the same time as the KI solution in Part A?

2. If an air bubble initially trapped in the tip of the buret is released during the titration, will the reported percent available Cl_2 be too high or too low? Explain.

Experiment 27

◇ The Electrolytic Cell

The next time you purchase a soft drink in a can and before you take that first inviting drink, stop for just a moment and think of all of the chemistry that went into producing that aluminum can. The ore containing the aluminum, called bauxite ore, is mined from the Earth's crust and the ore is shipped to where it is refined to alumina, a white powder with a formula of Al_2O_3. The alumina is then melted in a stainless steel container ("the cell") in the presence of cryolite, Na_3AlF_6, at about 900°C. Electric current (the movement of electrons) is forced through the molten alumina and aluminum metal is produced at the cathode of the cell. Thereafter, the aluminum metal is rolled and shaped into the can. It is subsequently lined with a thin plastic liner before the carbonated soft drink is added and the can is sealed.

The passage of an electric current through molten salts or aqueous solutions is very important to the economy of any industrialized nation; for example, the large quantities of sodium, chlorine, and sodium hydroxide are produced by electrolysis. Now you can take a drink . . . maybe you can appreciate it a little more!

OBJECTIVES

- To identify the reactions occurring at the anode and cathode in an electrolytic cell
- To determine the amount of charge passed through an electrolytic cell

PRINCIPLES

An oxidation-reduction reaction that requires energy occurs in an **electrolytic cell**. Electrical energy is used to direct a flow of electrons through a chemical system causing an otherwise nonspontaneous reaction to occur. For example, most active metals, such as sodium and aluminum, are prepared by the **electrolysis** of their molten salts—the directed current causes the reduction of the Na^+ ion or Al^{3+} ion at the **cathode** and the oxidation of its anion at the **anode**. The products are isolated at the electrodes as they form.

Electrolysis: the process of using energy to force a nonspontaneous oxidation-reduction reaction to occur.

Cathode: electrode at which reduction occurs in a cell.

Anode: electrode at which oxidation occurs in a cell.

The polarities of the electrodes in an electrolytic cell are determined by their connection to the direct current (dc) energy source. The cathode is the negative electrode—cations flow to the cathode and are reduced by the rich supply of electrons. The anode, the positive electrode, is deficient of electrons—anions flow to and release electrons to the anode and thus become oxidized.

When molten NaCl is subjected to a dc source in an electrolytic cell, Na^+ ions migrate toward and are reduced to sodium metal at the cathode and Cl^- ions migrate toward and are oxidized to chlorine gas at the anode .

reduction, cathode:	$Na^+(l) + e^- \rightarrow Na(l)$
oxidation, anode:	$2\,Cl^-(l) \rightarrow Cl_2(g) + 2\,e^-$

In the electrolysis of an aqueous $CuBr_2$ solution (Figure 27.1), the Cu^{2+} ions are reduced to copper metal at the cathode and Br^- ions are oxidized to bromine at the anode.

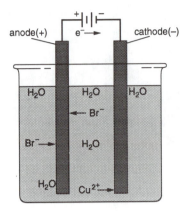

Figure 27.1
Electrolysis of CuBr$_2$(aq)

reduction, cathode: $Cu^{2+}(aq) + 2\ e^- \rightarrow Cu(s)$
oxidation, anode: $2\ Br^-(aq) \rightarrow Br_2(aq) + 2\ e^-$

But for electrolysis reactions in aqueous solutions, we must also consider the reduction of H$_2$O at the cathode and the oxidation of H$_2$O at the anode.

reduction, cathode: $2\ H_2O(l) + 2\ e^- \rightarrow H_2(g) + 2\ OH^-(aq)$
oxidation, anode: $2\ H_2O(l) \rightarrow O_2(g) + 4\ H^+(aq) + 4\ e^-$

If reduction of H$_2$O occurs at the cathode, H$_2$ gas is evolved and the solution near the cathode becomes basic due to the formation of OH$^-$. When oxidation of H$_2$O occurs at the anode, O$_2$ gas is evolved and the solution near the anode becomes acidic because of the formation of H$^+$. The production of H$^+$ and/or OH$^-$ can be detected using litmus paper or any acid-base indicator with an endpoint near seven.

With two competing reactions at the cathode, the reaction having the greatest natural tendency to undergo reduction[1] is the one that occurs. Conversely, for two competing reactions at the *anode*, the reaction having the greatest natural tendency to undergo oxidation[2] (or the lowest probability to undergo reduction) is the one that occurs. An analysis of these competing reactions is undertaken in Parts A and B of this experiment.

Movement of Charge Through an Electrolytic Cell

Electric current moves through two different paths in the electrical circuit of an electrolytic cell—the internal path is the movement of ions carrying charge in the electrolyte (or aqueous solution) and the external path is the movement of electrons carrying charge through the external wire that connects the electrodes and the dc power source. The quantity of charge passing through the cell can be measured through either pathway.

Coloumb: the SI unit for electric charge.

Ampere: unit of electrical current equal to one coulomb per second.

For the external circuit, the quantity of charge (measured in **coulombs, C**) is determined by measuring the passage of electrons (the current, measured in **amperes, C/s**) over a time period (measured in seconds, s) The charge is then calculated from the equation,

$$\textbf{coulombs, C = amperes (C/s) x time (s)}$$

Faraday: one mole of electrons, a total charge of 96 500 C.

For the internal circuit, the quantity of charge is determined by measuring a mass loss at the anode (an oxidation of the metal anode) or a mass gain at the cathode (a deposition of a metal ion from solution onto the cathode). For example, when Zn as the anode is oxidized, its mass loss is proportional to the charge that passes through the cell—one mole of Zn is oxidized when 2 moles of electrons (or 2 **faradays**) pass, causing a mass loss of 65.38 g Zn from the anode.

$$Zn(anode) \rightarrow Zn^{2+}(aq) + 2\ e^-$$

Since one mole of electrons has a charge of 96 500 C, the charge passing through the cell for the oxidation of one mole of Zn is 2(96 500 C) or 1.93×10^5 C.

---

[1] A reaction that has a high probability to undergo reduction is said to have a high **reduction potential**. Reduction potentials are often used to compare the probability for half-reactions to occur in an electrolysis process.

[2] A reaction with a high probability for oxidation has a high oxidation potential but a low reduction potential.

In this experiment you will electrolyze a number of solutions with a dc energy source and determine which of the competing **half-reactions** occurs at each electrode from a test or observation of the products formed during the electrolysis.

Half-reaction: the oxidation or reduction reaction that occurs in a redox reaction.

In addition, the quantity of charge that passes through a cell over an established time period is determined by measuring both the electron movement and the mass change of an electrode.

PROCEDURE

A. Electrolysis of Salt Solutions, Carbon Electrodes

1. Dissolve about 3 g of NaCl in 150 mL of deionized water and transfer it to the U-tube (Figure 27.2; this apparatus can be constructed with a U-shaped bent piece of glass tubing and graphite "lead" from a pencil). Place two carbon (graphite) electrodes in the solution and connect them to a dc energy source.[3] Identify cathode (the negative terminal) and the anode (the positive terminal) of the electrolytic cell.

2. Electrolyze the solution for 5 minutes. During the electrolysis, observe any evidence of a reaction occurring in the anode and cathode chambers:

   • Does the pH of the solution change at the electrodes? Test the solution at each electrode with litmus paper. Compare the color with a similar test on the original salt solution.
   • Is a gas evolved at one of the electrodes? Which one?
   • Does a metal deposit on the cathode? Is there any discoloration of the cathode and/or anode?

3. On the basis of your observations, write a balanced equation for the half-reaction at the cathode reaction and for that at the anode.

4. Substitute $CuSO_4$, $ZnCl_2$, KI, NaBr, and other salts suggested by your laboratory instructor for NaCl and repeat Parts A.1–3. Thoroughly clean the U-tube and electrodes after each electrolysis.

5. Rinse the graphite electrodes with deionized water and return to an appropriately labeled container.

Dispose of all of the test solutions in the "Waste Salts" container.

B. Electrolysis of a $CuSO_4$ Solution, Copper Electrodes

1. Fill the U-tube three-fourths full with 0.1 M $CuSO_4$. Clean, with steel wool, two copper electrodes (copper wire is satisfactory), connect them to the dc energy source, insert them into the two arms of the U-tube and electrolyze the solution for 5 minutes. Observe closely the cathode and anode chambers to detect any reactions and test with litmus; write balanced half-reactions that are consistent with your observations.[4]

Dispose of the test solution in the "Waste Salts" container.

---

[3]The most convenient dc power source is a 9V transistor battery.

[4]This electrolysis procedure is similar to that used for the industrial electrolytic refining of copper.

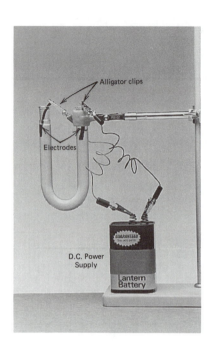

Figure 27.2
An electrolysis apparatus

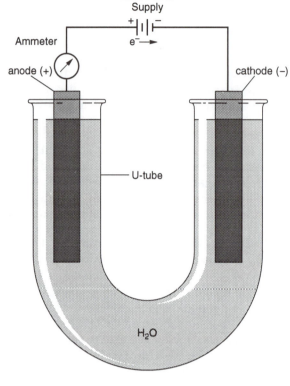

Figure 27.3
Apparatus for the electrolysis of
$CuSO_4$

C. Determination of the Quantity of Charge Passed Through the Cell

1. Set up the apparatus in Figure 27.3. The dc energy source must provide 3–5V[5] and the ammeter range should be from 0.2–1.0 amperes.

2. a. Polish a strip of Cu metal with steel wool, dip it into a 6 M HNO_3 (**Caution:** *avoid skin contact!*) solution (for ≈30 s) to remove any oxide coating, rinse with deionized water and acetone, air-dry, and measure its mass (±0.0001 g).[6] Use this Cu metal for the anode (the positive terminal). Similarly prepare a Cu metal strip as the cathode. Three-fourths fill the U-tube with 0.1 M $CuSO_4$ to a level just below the top of the Cu electrodes.

   b. Electrolyze the solution for *exactly* 15 minutes. Record the average current from the ammeter during the 15 minute period.

   c. Calculate the coulombs of charge that passed through the cell.

3. *Carefully* remove the Cu anode from the solution, rinse with acetone, air-dry, and again determine its mass. Determine the mass loss and calculate the coulombs of charge that passed through the solution.

4. Rinse the copper electrodes with deionized water and return to an appropriately labeled container.

Dispose of the test solution in the "Waste Salts" container.

---

[5] A 3-5V direct current power source can be two flashlight batteries, a lantern battery, or a transistor battery.

[6] A balance with 0.1 mg precision is best; however, if only ±0.001 g sensitivity is available, extend the time for electrolysis to 30 min.

◇ The Electrolytic Cell

Date _____ Name _____ Lab Sec. _____ Desk No. _____

1. In an electrolytic cell

    a. What is the sign of the anode? _____

    b. What is the sign of the cathode? _____

    c. What type of reaction occurs at an anode? _____

    d. What type of reaction occurs at a cathode? _____

    e. Toward which electrode do cations migrate? _____

    f. Toward which electrode do anions migrate? _____

    g. Electrons migrate from the _____ to the _____ .

2. In the electrolysis of an aqueous solution, water can be oxidized and/or reduced.

    a. Write the equation for the electrolysis of water at the anode.

    b. Write the equation for the electrolysis of water at the cathode.

    c. If water is electrolyzed at the cathode, what will be the color of the litmus paper in a test?

3. A current of 0.5 A passes through an electrolytic cell for 15 min.

a. How many coulombs of charge pass through the cell?

b. If this charge deposits zinc metal at the cathode, how many grams of zinc will electroplate (1 mole of electrons = 96 500 C)?

◇ The Electrolytic Cell

Date _____ Name _____ Lab Sec. _____ Desk No. _____

A. Electrolysis of Salt Solutions, Carbon Electrodes

Observations, Anode Chamber (+)

Salt	Litmus Test	Evidence of Reaction	Equation for the Reaction
NaCl			
$CuSO_4$			
$ZnCl_2$			
KI			
NaBr			

Observations, Cathode Chamber (-)

Salt	Litmus Test	Evidence of Reaction	Equation for the Reaction
NaCl			
$CuSO_4$			
$ZnCl_2$			
KI			
NaBr			

B. Electrolysis of a $CuSO_4$ Solution, Copper Electrodes

Experimental observation at	Equation
Anode	
Cathode	
Cell reaction	

C. Determination of the Quantity of Charge Passed Through the Cell

Time (*min*)	Current (*A*)	Time (*min*)	Current (*A*)	Time (*min*)	Current (*A*)
1		6		11	
2		7		12	
3		8		13	
4		9		14	
5		10		15	

Electrical Measurement

1. Average current (*amperes* = *C/s*) _____

2. Time for electrolysis (*s*) _____

3. Coulombs passed (*C*) _____

Chemical Measurement

1. Mass of polished Cu anode before electrolysis (*g*) _____

2. Mass of Cu anode after electrolysis (*g*) _____

3. Moles of Cu oxidized (*mol*) _____

4. Moles of e^- passed (*faradays*) _____

5. Coulombs passed (*C*) _____

Compare the chemical and electrical measurements of charge passed through the electrolytic cell. Comment on any differences.

QUESTIONS

1. a. From the chemical measurement in Part C, how many (total) electrons were generated at the anode in 15 minutes?

   b. From the electrical measurement in Part C, and the answer in **a**, calculate the charge of an electron.

2. a. If nickel electrodes are used instead of carbon electrodes in Part A, could the reaction at the anode have changed? Explain.

   b. If nickel electrodes are used instead of carbon electrodes in Part A, could the reaction at the cathode have changed? Explain.

Experiment 28

◆ The Galvanic Cell

Over one billion batteries are manufactured in the United States each year, from the "cheap" throw aways used in toys, flashlights, calculators, and automatic-focusing cameras to the rechargeable batteries used in power tools and tooth brushes. The flashlight battery revolutionized the toy industry in the mid 1960s; its technology promises to make even greater changes in the future, especially in the development of rechargeable batteries for personal transportation. All batteries are constructed for the same purpose, that is to provide a convenient and portable source of energy. The energy is provided by a movement of electrons generated from a spontaneous oxidation-reduction reaction.

OBJECTIVES

- To measure the reduction potentials for several redox couples
- To follow the movement of electrons, cations, and anions in a galvanic cell

PRINCIPLES

Redox: abbreviation for oxidation-reduction.

The transfer of electrons from one substance to another occurs in a **redox** reaction. In a redox reaction, oxidation cannot occur unless reduction also occurs—a substance cannot lose electrons unless another substance accepts the electrons. The substance that loses electrons is oxidized (its oxidation number increases); the substance that gains electrons is reduced (its oxidation number decreases). In the reaction

$$Zn(s) + 2\,Ag^+(aq) \rightarrow 2\,Ag(s) + Zn^{2+}(aq)$$

The oxidation and reduction parts of the overall reaction can be divided into two half-reactions, where each half-reaction is called a **redox couple**.

oxidation half-reaction	$Zn \rightarrow Zn^{2+} + 2\,e^-$	Zn^{2+}/Zn **redox couple**
reduction half-reaction	$2\,Ag^+ + 2\,e^- \rightarrow 2\,Ag$	Ag^+/Ag **redox couple**

Zn metal is oxidized and Ag^+ ions are reduced. When Zn loses electrons, the Ag^+ is reduced; hence, Zn is called the **reducing agent**. Conversely, when Ag^+ gains electrons, the Zn metal is oxidized and Ag^+ is the **oxidizing agent**.

In a galvanic cell, aqueous solutions of the reducing agent (the oxidation half-reaction) and the oxidizing agent (the reduction half-reaction) are placed in separate compartments called **half-cells**. The two half-cells are connected by a salt bridge which allows for the movement of ions from one half-cell to another.[1] Each half-cell has an electrode; the two electrodes are connected with a wire through a voltmeter. An electrode, which must be a part of each half-cell, serves to conduct electrons away from the reducing agent (substance being oxidized) through the wire to the oxidizing agent (substance being reduced). The electrode at which oxidation occurs is called

---

[1]A porous barrier which also prevents the free mixing of the two redox couples can be substituted for a salt bridge. A commonly used porous barrier is an unglazed porcelain cup.

the **anode**, the (−) electrode in a galvanic cell; reduction occurs at the **cathode**, the (+) electrode.

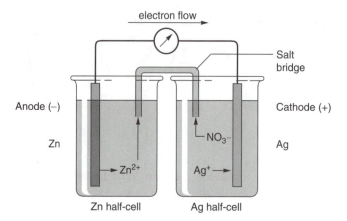

Figure 28.1
Schematic diagram of a galvanic cell

Let's see what happens in a galvanic cell consisting of the Zn^{2+}/Zn and Ag^+/Ag redox couples (Figure 28.1): Zn oxidizes to Zn^{2+}; electrons are deposited on the Zn anode and migrate through the wire to the Ag cathode; the Ag^+ in solution accepts the electrons at the Ag cathode to form Ag metal. In the meantime, the Zn^{2+} migrates away from the Zn anode and through the salt bridge—this occurs because as the positive charge of Ag^+ is being removed from the Ag half-cell, the Zn^{2+} diffuses in and replaces the positive charge. Conversely the anion of the Ag^+, such as NO_3^-, can move into the Zn half-cell and out of the Ag half-cell. Both ion movements serve to maintain electrical neutrality in each half-cell of the galvanic cell.

Different metals, such as Zn and Ag, have different tendencies (or potentials) to oxidize; their cations have different tendencies (or potentials) to reduce. We can consider the Zn^{2+} as having a potential (call it $E_{(Zn^{2+}/Zn)}$) to exist in its reduced state, Zn, and also Ag^+ as having a potential (call it $E_{(Ag^+/Ag)}$) to exist in its reduced state, Ag.

The ion having the greatest potential to be in the reduced state is the one that undergoes reduction when the two redox couples are connected to form a galvanic cell; oxidation must then occur for the other redox couple. In this case, Ag^+ has a higher potential for existing in the reduced state than does Zn^{2+}. The difference in these two reduction potentials is measured as the **cell potential**, E_{cell}, for the galvanic cell. The positive cell potential is written as

$$E_{cell} = E_{(Ag^+/Ag)} - E_{(Zn^{2+}/Zn)}$$

*Standard reduction potential: reduction potential measurements at molar concentrations of 1 mol/L for all ions and 1 atm pressure for all gases; also called **standard conditions**.*

Cell potentials are measured with an instrument called a **potentiometer** (or voltmeter). The standard reduction potential, $E°$,[2] for the Zn^{2+}(1 mol/L)/Zn redox couple is -0.76V and that for the Ag^+(1 mol/L)/Ag redox couple is +0.80V. A voltmeter connected to the electrodes would show the difference between these potentials, a cell potential of

$$E°_{cell} = 0.80V - (-0.76V) = 1.56V$$

---

[2]The superscript "o" in $E°$ refers to standard conditions.

Deviations from the predicted (or theoretical) cell potential may be due to surface activity at the electrodes, activity of the ions in solution, and/or current drawn by a voltmeter.

The cell potential, E_{cell}, for a galvanic cell measured at ionic concentrations other than 1 mol/L is related to the standard cell potential, $E°_{cell}$, by the **Nernst** equation.

Walther Nernst (1864-1941), a German chemist, derived his equation from thermodynamics in 1889. He received the Nobel Prize in chemistry in 1920 for his discovery of the third law or thermodynamics. He also contributed to our understanding of buffer solutions and the solubility of slightly soluble salts.

$$E_{cell} = E°_{cell} - \frac{0.0592}{n} \log Q$$

Q is the reaction quotient for the cell reaction and n identifies the moles of electrons transferred in the cell reaction. For the Zn-Ag galvanic cell (where $n = 2$) the Nernst equation becomes

$$E_{cell} = 1.56V - \frac{0.0592}{2} \log \frac{[Zn^{2+}]}{[Ag^+]^2}$$

Notice that the concentrations of Zn and Ag are omitted since they, as solids, maintain a constant concentration.

The E_{cell} of a selected number of redox couples are measured in the experiment; from the data the redox couples will be listed in order of their relative reduction potentials.

In addition the Nernst equation is used to determine an unknown Cu^{2+} concentration using the Zn^{2+}/Zn redox couple as a reference potential. The cell reaction is

$$Cu^{2+}(aq) + Zn(s) \rightarrow Cu(s) + Zn^{2+}(aq)$$

The corresponding Nernst equation is

$$E_{cell} = 1.10V - \frac{0.0592}{2} \log \frac{[Zn^{2+}]}{[Cu^{2+}]}$$

If $0.1\,M\,Zn^{2+}$ is used for the measurement, the Nernst equation becomes

$$E_{cell} = 1.10V - \frac{0.0592}{2} \log [0.1] + \frac{0.0592}{2} \log [Cu^{2+}] \text{ or}$$
$$E_{cell} = 1.13V + 0.0296 \log [Cu^{2+}]$$

Therefore, E_{cell} is directly proportional to $\log [Cu^{2+}]$; cell potentials can therefore be used to determine the concentrations of ions in a solution. This is the basic principle used for pH determinations using a pH meter.

The latter equation represents that of a straight line, $y = b + mx$. Plotted data of E_{cell} vs. $\log [Cu^{2+}]$ produces a straight line with slope of 0.0296 and a y-intercept of 1.13V.

PROCEDURE

A. Reduction Potentials of Selected Half-Reactions

1. a. Transfer about 2 mL of 0.1 M $CuSO_4$, 0.1 M $SnCl_2$ 0.1 M $FeSO_4$, and 0.1 M $Zn(NO_3)_2$ respectively to wells A1, A2, B1, and B2 of a 24 well plate (see Figure 28.2). Place a polished[3] strip of each metal, Cu, Sn, Fe, and Zn, in the respective wells.

    b. Prepare several strips of filter paper, cut to about 1" x 1", roll and fold, and wet the strips with a sodium sulfate solution from a dropper bottle. The wetted strip of filter paper serves as a salt bridge. *Reproduce, as best possible, the construction of the salt bridges and the positioning of the electrodes in the wells.*

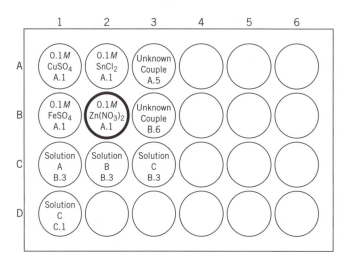

Figure 28.2
Arrangement of the half-cells in the 24 well plate

2. Measure the potential of the Cu–Fe cell.
    a. Connect wells A1 and B1 with a salt bridge; connect one electrode to the positive terminal of the voltmeter and the other electrode to the negative terminal.[4] If the needle shows a positive deflection, the connections are correct; if not, reverse the connections of the electrodes.

    b. Once the needle of the voltmeter has stabilized, record the E_{cell} .[5] Which electrode is connected to the positive (+) terminal (cathode)? Does oxidation occur at the Fe or Cu electrode? Write an equation for the reaction at each electrode.

3. Repeat the voltage measurements for all possible combinations of cells from the remaining half-cells, using a freshly prepared salt bridge each time in connecting the half-cells.

4. Assuming the reduction potential of the Cu^{2+} (0.1 M)/Cu redox couple is +0.31V, calculate the reduction potentials of all other redox couples.[6]

---

[3]Polish each metal strip with steel wool to remove any oxide coating.

[4]The voltmeter should have an expanded range from 0 to about 2.5 volts.

[5]A positive E_{cell} indicates that the redox couple at the cathode, the one connected to the (+) terminal of the voltmeter, has the higher reduction potential by the voltage read on the voltmeter.

[6]These are not standard reduction potentials because 1 M solutions at 25°C were not used for the measurements.

5. Other redox couples: Determine the reduction potential for the following redox couples designated by your instructor. Polish each electrode.

Ag$^+$/Ag   (0.1 M AgNO$_3$ solution)
Ni^{2+}/Ni   (0.1 M Ni(NO$_3$)$_2$ solution)
Mg^{2+}/Mg   (0.1 M MgSO$_4$ solution)
M^{n+}/M (unknown)   (0.1 M solution of M^{n+})

Rinse the wells with deionized water and discard in the "Waste Salts" container.

1. Obtain a 1 mL pipet and three 100 mL volumetric flasks. To prepare solution **A**, pipet 1 mL of 1 M CuSO$_4$ solution into a 100 mL volumetric flask and dilute to **the mark** with deionized water. Prepare solution **B** by pipetting 1 mL of **A** into a second 100 mL volumetric flask and diluting to the mark. Similarly, prepare solution **C**. Calculate [Cu^{2+}] and log [Cu^{2+}] for solutions **A**, **B**, and **C**.

B. Measuring an Unknown Concentration

"The mark": the etched line that calibrates the volume of a volumetric flask.

Share this set of solutions with other students in the laboratory.

1 M CuSO$_4$	–1 mL→	dilute to 100 mL A	–1 mL→	dilute to 100 mL B	–1 mL→	dilute to 100 mL C

2. Use the equation, E_{cell} = 1.13V + 0.0296 log [Cu^{2+}], to calculate the theoretical E_{cell} for the 1 M CuSO$_4$ and solutions **A**, **B**, and **C**. Plot the data of E_{cell} (calculated) vs. log [Cu^{2+}]. Have your instructor approve your graph.

3. a. Place about 2 mL of solution **A** in well C1; 2 mL of solution **B** in well C2, and 2 mL of solution **C** in well C3.

   b. Connect the polished Zn electrode in well B2 to the negative (–) terminal of the voltmeter; connect a polished Cu electrode in well C1 to the positive (+) terminal. Join well B2 and C1 with a freshly prepared salt bridge (see Part A.1). Record E_{cell}.

4. Repeat Part B.3b to record the E_{cell} for solutions **B** and **C** relative to the Zn^{2+}(0.1 M)/Zn redox couple. Be sure to use a freshly prepared salt bridge for each measurement. Record E_{cell} for each cell and the Cu–Zn cell in Part A.

5. Plot the data of E_{cell} (exp'tl) vs. log [Cu^{2+}] for the four cells on the same graph as in Part B.2. Draw the *best* straight line through the plotted data.

6. Obtain a solution with an unknown [Cu^{2+}] from your instructor. Place about 2 mL of the solution in well B3. As before, determine the E_{cell}. From an analysis of your plotted data in Part B.5, estimate the [Cu^{2+}] in your unknown.

C. Concentration Cell, Cu^{2+}(0.01 mol/L)/Cu and Cu^{2+} (1 x 10⁻⁶ mol/L)/Cu Redox Couples

1. Place about 2 mL of solution C in well D1; connect well C1 to D1 with a freshly prepared salt bridge. Connect the electrodes to the voltmeter. Determine the E_{cell}, the anode reaction, and the cathode reaction. Explain *why* a potential is recorded. Write an equation for the reaction at each electrode.

2. Add about 20 drops of 6 M NH₃ (**Caution**: *avoid inhalation*) to solution C. Observe any changes in E_{cell}. Explain.

Rinse the 24 well tray with deionized water and discard in the "Waste Salts" container.

NOTES, OBSERVATIONS, AND CALCULATIONS

The Galvanic Cell

Date _____ Name _____ Lab Sec. _____ Desk No. _____

1. a. State the purpose of the salt bridge in today's experiment.

    b. How is the salt bridge constructed?

2. What is the sign of the anode in a galvanic cell? _____

    What electrochemical process occurs at the anode? _____

3. Consider a galvanic cell consisting of the Au^{3+} (1 mol/L)/Au and Sn^{4+} (1 mol/L)/Sn^{2+} (1 mol/L) redox couples.

    $Au^{3+} + 3\,e^- \rightleftarrows Au$ $E° = +1.42$ V
    $Sn^{4+} + 2\,e^- \rightleftarrows Sn^{2+}$ $E° = +0.15$ V

    a. Which redox couple undergoes reduction? _____

    b. Write an equation for the reaction that occurs at the

        anode _____

        cathode _____

    c. Write the equation for the cell reaction.

    d. What is the cell potential, $E°_{cell}$, for the cell?

4. Determine the cell potential, E°_{cell}, of a galvanic cell consisting of the Sn^{4+} (1 mol/L)/Sn^{2+} (1 mol/L) and Cr^{3+} (1 mol/L)/Cr redox couples.

$$Sn^{4+} + 2\,e^- \rightleftarrows Sn^{2+} \qquad E^\circ = +0.15\text{ V}$$
$$Cr^{3+} + 3\,e^- \rightleftarrows Cr \qquad E^\circ = -0.74\text{ V}$$

5. Use the Nernst equation to determine the cell potential, E_{cell}, of the galvanic cell consisting of the following redox couples:

$$Sn^{4+}(10^{-3}\text{ mol/L}) + 2\,e^- \rightleftarrows Sn^{2+}(10^{-5}\text{ mol/L})$$
$$Fe^{3+}(1\text{ mol/L}) + e^- \rightleftarrows Fe^{2+}(10^{-5}\text{ mol/L})$$

$$Fe^{3+} + e^- \rightleftarrows Fe^{2+} \qquad E^\circ = +0.77\text{ V}$$
$$Sn^{4+} + 2\,e^- \rightleftarrows Sn^{2+} \qquad E^\circ = +0.15\text{ V}$$

The Galvanic Cell

Date _____ Name _____ Lab Sec. _____ Desk No. _____

A. Reduction Potentials of Selected Half-Reactions

Cell	E_{cell}, exp'tl	anode	anode reaction	cathode	cathode reaction
Cu–Fe					
Cu–Sn					
Cu–Zn					
Fe–Sn					
Fe–Zn					
Zn–Sn					
other					
unknown					

1. Compare the sum of the Cu–Fe + Fe–Zn cell potentials with the Cu–Zn cell potential. Explain.

2. Compare the sum of the Fe–Zn + Fe–Sn cell potentials with the Zn–Sn cell potential. Explain.

3. Arrange the redox couples in order of *decreasing* reduction potentials, the most positive first. List the reduction potential for each redox couple relative to that of the Cu^{2+} (0.1 mol/L)/Cu couple, assuming its reduction potential is +0.31V.

Redox Couple	Reduction Potential

B. Measuring an Unknown Concentration

Solution	Molar Concentration	log [Cu^{2+}]	E_{cell} (calc)	E_{cell} (exp'tl)
1 M CuSO$_4$	1.0	0		
A				
B				
C				

2. *Two* plots (see Parts B.2 and B.5) of E_{cell} *vs.* log [Cu^{2+}], instructor's approval _____

3. Compare E_{cell} (calc) to E_{cell} (exp'tl). Are differences consistent? Is the trend of cell potentials as predicted? Explain.

4. E_{cell} for the unknown galvanic cell: Zn^{2+}(0.1 mol/L)/Zn–Cu^{2+}(? mol/L)/Cu. _____

    From the plotted data, what is the [Cu^{2+}] in solution in unknown? _____

C. Concentration Cell, Cu^{2+}(0.01 mol/L)/Cu and Cu^{2+} (1 x 10^{-6} mol/L)/Cu Redox Couples

1. E_{cell} _____

    anode reaction (oxidation)   _____

    cathode reaction (reduction)   _____

    Why is a potential recorded?

2. E_{cell} _____
    Account for the ΔE_{cell} from Part C.1

QUESTIONS

1. Identify the oxidizing and reducing agent of each cell in Part A.

Cell	Oxidizing Agent	Reducing Agent
Cu–Fe		
Cu–Sn		
Cu–Zn		
Fe–Sn		
Fe–Zn		
Zn–Sn		

2. List two factors accounting for the experimental cell potential *not* being equal to the calculated cell potential.

3. *If* in Part B, the $Zn(NO_3)_2$ solutions were successively diluted instead of the $CuSO_4$ solutions, how would the relative cell potentials have differed?

4. The newly discovered metal squeedeldinium (Sq) has the redox couple, $Sq^{2+} + 2\,e^- \rightarrow Sq$. When the Sq^{2+} (1 mol/L)/Sq redox couple is connected to the Cu^{2+} (0.1 mol/L)/Cu redox couple the cell potential is +0.56 V with Sq serving as the anode. What is the $E°$ of the Sq^{2+}/Sq redox couple?

5. a. What is n, the number of moles of electrons exchanged, for a galvanic cell that utilizes these half-reactions?

$$Cr_2O_7^{2-} + 14\,H^+ + 6\,e^- \rightarrow 2\,Cr^{3+} + 7\,H_2O \qquad E^\circ = +1.33\ V$$
$$Cr^{3+} + 3\,e^- \rightarrow Cr \qquad\qquad\qquad\qquad E^\circ = -0.74\ V$$

b. Which half-reaction occurs in the cathodic chamber?

c. Write the equation for the cell reaction.

d. How does lowering the pH of the dichromate solution affect the E_{cell}?

e. What is the oxidizing agent in the galvanic cell reaction?

Experiment 29

◆ Reaction Rates

What is the secret to quickly building a roaring fire in a fireplace on a cold, romantic evening? Lets see, we need some wood, preferably some small pieces (kindling) and a few larger pieces, a match, some starter fluid (nice but not really necessary), and some soft music (you decide if that is really necessary!). The match head has a few chemicals that quickly become thermally activated with friction causing the ignition of the match (paper or wood). The kindling (old newspaper also works) has a large surface area exposed to the oxygen in air, thereby making it easy to reach its ignition temperature from the heat of the match. Fanning the burning kindling causes it to burn faster and therefore generate more heat; a larger piece of wood, now at a higher temperature and with a plentiful supply of air also ignites. Opening the draft now causes the larger and larger pieces to catch fire and before you know it, BINGO, you have a nice roaring fire! Now, the rest of the evening is not a part of this course or manual!

OBJECTIVE

- To identify factors that affect the rate of a chemical reaction

PRINCIPLES

A number of conditions can alter the rate of a chemical reaction. Experimental parameters that we, as chemists, can quickly change include temperature, pressure (if gases), and reactant concentrations. In addition the substitution of a more reactive reactant, the subdividing of a reactant (if a solid or liquid), or the inclusion of a catalyst can increase a reaction rate. Each of these factors are observed and studied in this experiment.

Temperature Effects

A temperature increase generally increases reaction rate—a rule of thumb is that a 10°C temperature rise doubles the rate. Because of a temperature increase, the average kinetic energy of reactant molecules (or ions) also increases[1]; when the molecules (or ions) collide, this added kinetic energy becomes a part of the collision system. The distribution of the additional energy among the atoms in the colliding molecules increases the probability of bonds being broken; subsequent recombination of the fragments can result in the formation of product, resulting in an overall increase in the reaction rate.

Concentration Effects

An increase in the concentration of a reactant increases the probability of collision between the reactant molecules (or ions). As a result, most reaction rates increase when the concentrations of the reactants increase, although there are systems in which a decrease or no change in reaction rate occurs. Experiment 30 focuses on a **quantitative** investigation of the effect of concentration changes of the reactants on reaction rate.

Quantitative: measurements that are as exact as the chemical apparatus will allow.

Nature of Reactants

Some substances are naturally more reactive than others, and, therefore undergo more rapid chemical changes. To manufacture an automobile that does not rust, alternative construction materials, such as stainless steel or

---

[1]The kinetic energy-temperature relationship is expressed by the equation
$$K.E. = (1/2) mv^2 = (3/2) kT$$
k is called the Boltzman constant and **T** is the temperature in kelvins.

fiberglass can be used. The reaction between water and sodium is very rapid and exothermic, whereas with copper, no reaction occurs. Hydrogen and fluorine react explosively at room temperature; hydrogen and iodine react very slowly, even at elevated temperatures. Therefore, the nature of the reactants chosen for a particular process can affect a reaction rate.

Particle Size of Reactants

The smaller particle size of the reactant, the greater the reaction rate. A large piece of wood in the fireplace burns much more slowly than the kindling. Coal mine and grain elevator explosions result from the very finely divided (large surface area) dust particles being exposed to the oxygen of the air—a spark ignites the very rapid reaction.

Presence of a Catalyst

A **catalyst** increases the reaction rate without undergoing any *net* change in its properties for the reaction. Catalysts generally reroute the pathway (or mechanism) of the reaction in such a way that the reaction requires less energy and therefore proceeds more rapidly. The many biochemical reactions that occur in the body are catalyzed by specific proteins called enzymes.

PROCEDURE

Each portion of the experiment requires that you time the reaction. The time lapse should be recorded in seconds; therefore, before you begin this experiment, make sure you have available a clock or watch that can be read to the nearest second.

A. Temperature Effects

Disproportionation: a reaction in which a single substance undergoes oxidation and reduction.

The **disproportionation** of $Na_2S_2O_3$ in an acidic solution produces insoluble sulfur as a product. The time required for the cloudiness from the formation of sulfur is a measure of the reaction rate.

$$2\,HCl(aq) + Na_2S_2O_3(aq) \rightarrow S(s) + SO_2(g) + 2\,NaCl(aq) + H_2O(l)$$

Measure each volume with a pipet. Work in pairs to best complete this part of the experiment.

1. Pipet 5.0 mL of 0.1 *M* $Na_2S_2O_3$ into one set of three (clean) 150 mm test tubes. Into a second set of three 150 mm test tubes, pipet 5.0 mL of 0.1 *M* HCl. Do not intermix the pipets. Place a $Na_2S_2O_3$/HCl pair of test tubes in a salt/ice-water bath until thermal equilibrium is established (about 5–6 minutes). You may now want to proceed to Part A.2.

   O.K., get ready to time for the appearance of the sulfur cloudiness with a watch that reads to the nearest second. As one student pours the solutions together, the other notes the time . . . transfer one solution to the other and agitate the mixture vigorously for several seconds. Return the reaction mixture to the ice bath and record the time for the sulfur to appear. Record the temperature (±0.1°C) of the mixture.

2. Place a second pair of $Na_2S_2O_3$/HCl test tubes in a warm water bath adjusted to a temperature of ≤60°C. Repeat the mixing of the two solutions as in Part A.1 and record the time for the sulfur cloudiness to appear. You can predict the relative time even before the mixing. Record the actual temperature of the water bath.

3. Combine and time the remaining pair of $Na_2S_2O_3$/HCl solutions at room temperature.

4. Plot the time (seconds) *vs*. temperature (°C) for the appearance of sulfur on linear graph paper. Have your instructor approve your graph.

1. Pipet 2.0 mL of 1 M HCl, 3 M HCl, and 6 M HCl (**Caution:** HCl *causes a skin irritation*) into wells A1–A3 of a 24 well plate (see Figure 29.1). Polish a 30 mm strip of Mg metal, tear into three equal lengths, and determine the mass (±0.001 g) of each.

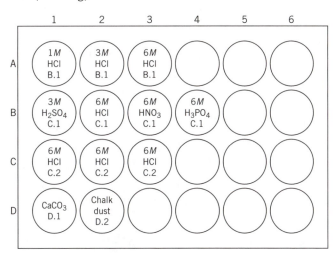

Figure 29.1
Arrangement of wells for observations in Parts B–D

2. Add a Mg strip (be sure you have recorded its mass) to the 1 M HCl and start timing (in seconds). When all traces of Mg ribbon disappear, stop timing. Record the elapsed time on the Data Sheet. Repeat the experiment with the 3 M HCl and 6 M HCl solutions.

3. Plot time (seconds) *vs.* (mol HCl/mol Mg) on linear graph paper. Have the instructor approve your graph.

C. Nature of Reactants

1. Place about 2 mL of 3 M H$_2$SO$_4$, 6 M HCl, 6 M HNO$_3$, and 6 M H$_3$PO$_4$ (**Caution:** *carefully handle acids*) into wells B1–B4 of the 24 well plate. Place a strip of polished Mg ribbon into each well and note the relative reaction rates. Record your observations.

2. Place about 1 mL of 6 M HCl into wells C1–C3. Add a polished 10 mm strip of Zn to the first, one of Fe to the second, and another of Cu to the third. Note the relative reaction rates and record your observations.

D. Particle Size of Reactants

1. Place a small piece of chalk, CaCO$_3$, in well D1 and add several drops of 6 M HCl. Place some chalk dust in an adjacent well D2, and repeat the addition. Account for your observations.

Dispose of the test chemicals from Parts A through D in the "Waste Acids" container.

The thermal decomposition of potassium chlorate generates small quantities of oxygen gas. Several tests of the chemical reactivity of oxygen are included in this experiment.

E. Presence of a Catalyst

Caution: This experiment can be dangerous if not performed properly. Before you begin, read the procedure carefully and adhere to the following guidelines:

- *Wear safety glasses*
- *Use a clean 200 mm Pyrex test tube—there should be no evidence of a black residue. Do not dry the inside of the test tube with a paper towel.*
- *Your instructor must approve your apparatus before you begin.*

1. Set up the apparatus in Figure 29.2. Fill a 50 mL graduated cylinder with water in the pneumatic trough and invert it over the gas collecting port. No air bubbles should be present in the cylinder. Into a clean, dry 200 mm Pyrex test tube, place about 0.5 g of $KClO_3$. Gently heat the $KClO_3$ and record the time (in seconds) required to collect 20 mL of O_2 gas.

$$2 KClO_3(s) \rightarrow 2 KCl(s) + 3 O_2(g)$$

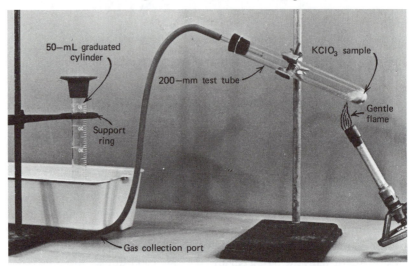

50–mL graduated cylinder

$KClO_3$ sample

200–mm test tube

Gentle flame

Support ring

Gas collection port

Figure 29.2
Apparatus for the collection of $O_2(g)$

2. Before removing the heat from the test tube, first disconnect the gas delivery tube from the test tube. (**Caution**: it is very important that the delivery tube be disconnected while the tube is hot; if it isn't, cool water will be drawn from the trough into the test tube and may cause it to break.) Then, allow the test tube and its contents to cool.

3. Repeat Parts E.1 and E.2, but in addition, add a pinch of MnO_2 to the 0.5 g $KClO_3$. Record the time required to collect 20 mL of O_2 gas.

4. Replace the inverted graduated cylinder with a 200 mm test tube; continue heating to collect the O_2 gas. Use additional 200 mm test tubes until no further oxygen is evolved; then follow the cool-down procedure described in Part E.2.

5. Ignite a wood splint, put out the flame so that only a glow remains. Place the glowing splint in one of the flasks.

Dispose of the chemicals from Part E in the "Waste Oxidants" container.

NOTES AND OBSERVATIONS

Reaction Rates

Date _____ Name _____ Lab Sec. _____ Desk No. _____

1. If the rate of a chemical reaction doubles for every 10°C temperature increase, by what factor will a chemical reaction increase if the temperature is increased over a 30°C range?

2. Explain why coal dust burns more rapidly than larger pieces of coal.

3. What volume (in mL) of oxygen gas (at STP) can be produced from the thermal decomposition of 0.50 g $KClO_3$? See Part E.

4. When hydrogen peroxide, H_2O_2, as an antiseptic, is placed on an open wound, bubbles form because of the reaction

$$2\,H_2O_2(aq) \rightarrow 2\,H_2O(l) + O_2(g)$$

However, H_2O_2 only slowly decomposes in the bottle. Why does the wound accelerate its decomposition?

5. Wood burns more rapidly in a fireplace that has a good draft of air. Explain.

6. List factors that can affect the rate of ammonia production by the Haber process.

$$N_2(g) + 3H_2(g) \rightarrow 2NH_3(g) + 46.1 \text{ kJ}$$

◇ Reaction Rates

Date _____ Name _____ Lab Sec. _____ Desk No. _____

A. Temperature Effects

Na$_2$S$_2$O$_3$-HCl Test Pair	Time Elapsed (seconds)	Temperature (°C)
1.		
2.		
3.		

Instructor's approval of graph _____

Based upon your data, what can you conclude about the effect of temperature on the rate of this reaction?

From the graph of your data, and assuming the same set of solutions, estimate the time for the

appearance of sulfur when the reaction occurs at 40°C _____; at 95°C _____.

B. Concentration Effects

Molar conc. of HCl	mol HCl	mass Mg (±0.01 g)	mol Mg	$\frac{\text{mol HCl}}{\text{mol Mg}}$	time (seconds)
1.0					
3.0					
6.0					

Instructor's approval of graph _____

How does a change in the molar concentration of HCl affect the time for a known amount of Mg to react?

From the graph of your data, estimate the time it would take for 0.2 g Mg to react in 3 mL of 0.40 M HCl.

C. Nature of Reactants

1. Record the relative reaction rates of the four acids with Mg in order of *increasing* activity.

   _____, _____, _____, _____

   What can you conclude about the relative chemical reactivity of the four acids?

2. List the three metals in order of the increasing reaction rate with 6 M HCl.

   _____, _____, _____

   What can you conclude about the chemical reactivity of the three metals?

D. Particle Size of Reactants

1. Explain how the physical state of calcium carbonate affects the rate of the chemical reaction.

E. Presence of a Catalyst

1. Instructor's approval of apparatus _____

2. Time to collect 20 mL of O_2 without a catalyst _____

   Time to collect 20 mL of O_2 with a catalyst _____

3. Describe the affect that a catalyst has on the rate of evolution of O_2 gas.

4. Describe the reaction of the glowing splint with oxygen. How can the observation be explained in terms of factors affecting reaction rates?

Experiment 30

◇ Rate Law Determination

A recent mishap at a fertilizer plant released large quantities of nitrogen dioxide, NO_2, into the atmosphere. A brown haze has formed over the area. What eventually happens to the brown haze? Physically the haze dissipates because of air currents and dilution; chemically, nitrogen dioxide reacts with water vapor to form nitrous acid, with ammonia gas (that occurs naturally in the atmosphere) and water vapor to form ammonium nitrite, and with limestone from buildings and the soil to form calcium nitrite.

Initially the chemical reactions of NO_2 are rapid because of its high concentration and the immediate presence of the other reactants, but as it and the other reactants combine chemically and as it dilutes in the atmosphere, the reactions slow; the quantity of NO_2 in the atmosphere steadily decreases to low and nearly undetectable levels. At these low levels we say that its concentration has reached an equilibrium with the concentration of NO_2 that existed prior to the mishap. The levels of NO_2 as a function of time can be plotted as shown in the figure.

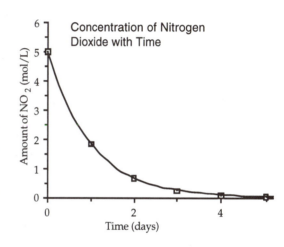

The progress of a chemical reaction can be measured quantitatively by any one of a number of methods described in Experiment 29. Nitrogen dioxide can even be measured spectrophotometrically since NO_2 is a red-brown gas.

- To determine the rate law for a chemical reaction
- To use a graphical analysis of experimental data

OBJECTIVES

PRINCIPLES

The rate of disappearance of a reactant or the rate of appearance of a product are indicative of the rate of a chemical reaction. The technique used to monitor and measure a reaction rate is dependent upon a unique property of one of the substances in the reaction. For example, a reactant may lose its color, a product may produce an odor, the reaction may generate heat or become more acidic, a product may be a gas or precipitate, etc., all of which can be monitored as a function of time. For our purposes, we will monitor the time required for a color to appear, the color appears after a quantitative amount of reactant has been consumed in the reaction.

The rate of a chemical reaction is often affected by the amount of reactant initially placed in the reaction system. Quantitatively, the rate is proportional to the molar concentration of each reactant raised to some power, called the **order** of the reactant. For the reaction $A_2 + 2B_2 \rightarrow 2AB_2$, this relationship between rate and concentration can be expressed as:

Order: the exponent (determined experimentally) of the molar concentration of a substance needed to express a proportionality of the molar concentration to the observed reaction rate.

$$\text{rate } \alpha \, [A_2]^p \; \text{and} \; \text{rate } \alpha \, [B_2]^q \; \text{or} \; \text{rate } \alpha \, [A_2]^p[B_2]^q$$

α: means "proportional to".

The superscripts, p and q, are the orders for reactants A_2 and B_2 respectively. When the proportionality sign, α, is replaced with a proportionality constant, k, the **rate law** for the reaction is

$$\text{rate} = k[A_2]^p[B_2]^q$$

The value of **k**, called the **specific rate constant,** is determined experimentally and varies with temperature and the presence of a catalyst, but is independent of reactant concentrations.

The orders, **p** and **q**, are also determined experimentally. The effect that a change in a reactant concentration has on a reaction rate is expressed by the order of that reactant. For example, suppose that in doubling the molar concentration of A_2 (while holding $[B_2]$ constant) the reaction rate is observed to increase by a factor of four: to maintain the proportionality between rate and $[A_2]$ (rate α $[A_2]^p$), **p** must equal 2. Experimentally the molar concentration of B_2 can remain virtually unchanged if a large excess of B_2 (compared to the amount of A_2) is initially placed in the reaction system. As a consequence, the $\Delta[A_2]$ is therefore much larger than the $\Delta[B_2]$ and therefore is more pronounced in affecting reaction rate.

In this experiment, the values of **k** and **p** are determined for the reaction of the iodate anion, IO_3^-, with the sulfite anion, SO_3^{2-}, in an acidic solution. The order of the IO_3^- in the reaction will be designated as **p**.

The reaction between the IO_3^- and SO_3^{2-} ions occurs in a series of steps.

(Step 1) $\quad IO_3^-(aq) + 3\,SO_3^{2-}(aq) \rightarrow I^-(aq) + 3\,SO_4^{2-}(aq)$
(Step 2) $\quad 5\,I^-(aq) + 6\,H^+(aq) + IO_3^-(aq) \rightarrow 3\,H_2O(l) + 3\,I_2(aq)$
(Step 3) $\quad 3\,I_2(aq) + 3\,SO_3^{2-}(aq) + 3\,H_2O(l) \rightarrow$
$$6\,I^-(aq) + 3\,SO_4^{2-}(aq) + 6\,H^+(aq)$$

(Net equation) $\;\; 2\,IO_3^-(aq) + 6\,SO_3^{2-}(aq) \rightarrow 2\,I^-(aq) + 6\,SO_4^{2-}(aq)$

How do we detect/monitor the reaction rate? Lets note the function and progress of the SO_3^{2-} anion in the reaction: we see that SO_3^{2-} reacts with not only the IO_3^- anion in step 1, but also with I_2 in step 3. What happens when all of the SO_3^{2-} anion has been consumed in the reaction? Once the SO_3^{2-} anion completely reacts, the I_2 that is produced in step 2 is "free" in solution; we detect its presence with starch, which together form a deep-blue solution due to the $I_2 \bullet$ starch complex.

$$I_2(aq) + \text{starch } (aq) \rightarrow I_2 \bullet \text{starch } (aq, \text{ deep blue color in solution})$$

This color appearance is the signal that detects the consumption of the SO_3^{2-} anion and therefore becomes our method for monitoring the reaction rate. If the amount of IO_3^- is increased in the reaction system, step 1 proceeds more rapidly and, accordingly, the SO_3^{2-} is consumed more rapidly. Therefore, we have a method for observing the effect that a change in the molar concentration of IO_3^- has on the reaction rate.

The rate law for the reaction between the IO_3^- and SO_3^{2-} ions is

$$\text{rate} = k[IO_3^-]^p[SO_3^{2-}]^q$$

In the experiment, the initial molar concentration of SO_3^{2-} is constant for each trial while the molar concentration of IO_3^- changes. This modifies the rate law to

$$\text{rate} = k[IO_3^-]^p \cdot \text{constant} \quad \text{or} \quad \text{rate} = k'[IO_3^-]^p$$

The reaction rate can be expressed as the amount (and/or concentration) of SO_3^{2-} consumed per measured time interval.

$$\text{rate} = \frac{-\Delta[SO_3^{2-}]}{\Delta t}$$

Because the same amount (and concentration) of SO_3^{2-} is used in several trials, the appearance of the deep-blue $I_2 \cdot$ starch complex (appearing only after the consumption of the SO_3^{2-}) will depend only upon the molar concentration of IO_3^-. Therefore, since $-\Delta[SO_3^{2-}]$ is constant throughout the collection of the data, the rate can also be expressed in reciprocal time; the depletion of the SO_3^{2-} is gauged by the appearance of the deep-blue $I_2 \cdot$ starch complex.

$$\text{rate (sec}^{-1}) = \frac{\text{appearance of } I_2 \cdot \text{starch}}{\Delta t}$$

Determination of the Reaction Order, p, and Rate Constant, k'

The order of $[IO_3^-]$ in the reaction rate is determined by changing its concentration in several trials while maintaining a constant $[SO_3^{2-}]$ and $[H^+]$. The data generated in the experiment is graphed to determine **p** and **k'**. The plot of the logarithmic form of the rate law, **rate = k'[IO_3^-]^p**, yields an equation with a straight line.

$$\text{log (rate)} = \text{log } k' + p \text{ log}[IO_3^-] \text{ or}$$
$$y = b + m x$$

Therefore, a plot of log (rate) vs. log $[IO_3^-]$ results in a straight line with a slope of **p**, the order of the reaction with respect to $[IO_3^-]$. The order, **q**, of $[SO_3^{2-}]$ in the reaction, could be similarly determined.

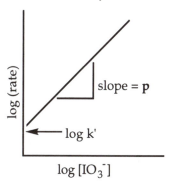

The reaction rate constant, **k'**, is calculated by substituting the rate, **p**, and $[IO_3^-]$ for a given trial into the rate law. It can also be determined from the y-intercept of the graphed data, where **b = log k'**.

Activation Energy

Reaction rates are temperature dependent. Higher temperatures increase the kinetic energy of the (reactant) molecules, such that when two reacting molecules collide, they do so with a much greater force (more energy is dispersed within the collision system) causing bonds to rupture, atoms to rearrange, and new bonds (products) to form more rapidly. The sum of these kinetic energies required for a reaction to occur is called the **activation energy** for the reaction.

The relationship between the reaction rate constant, **k'**, at a measured temperature, **T(K)**, and the activation energy, E_a, is expressed in the Arrhenius equation.

$$k' = Ae^{-E_a/RT}$$

A is a collision parameter for the reaction and **R** is the gas constant (=8.314 J/mol K). When a reaction is performed at two temperatures, $T_{(1)}$ and $T_{(2)}$, the ratio of the reaction rate constants can be determined from the equation

$$\frac{k_1'}{k_2'} = \frac{Ae^{-E_a/RT_{(1)}}}{Ae^{-E_a/RT_{(2)}}}$$

We can simplify the ratio of these two equations by using natural logarithms.

$$\ln \frac{k_1'}{k_2'} = -\frac{E_a}{R}\left[\frac{1}{T_{(1)}} - \frac{1}{T_{(2)}}\right]$$

Therefore, the activation energy for a reaction can be calculated from the determination of rate constants at two measured temperatures.

In this experiment you will determine the activation energy, E_a, for the $[IO_3^-]/[SO_3^{2-}]$ system by determining the k' at room temperature and again at 50°C.

PROCEDURE

Read the entire experimental procedure before beginning. You should work with a partner—one of you will need to determine the time lapse for the reaction and the other will need to mix the solutions.

A. Test Reactions

The test reactions are prepared according to Table 30.1.

Table 30.1. Composition of Test Reactions

Test Reaction	Solution A 0.02 M HIO_3	Starch	Deionized H_2O	Solution B 0.005 M H_2SO_3
1	5.0 mL	3 drops	0 mL	5.0 mL
2	4.0 mL	3 drops	1 mL	5.0 mL
3	3.0 mL	3 drops	2 mL	5.0 mL
4	2.0 mL	3 drops	3 mL	5.0 mL
5	1.0 mL	3 drops	4 mL	5.0 mL
6	choice	3 drops	choice	5.0 mL

1. **Test Reaction 1.** Prepare Solution A for Test Reaction 1 in a *clean* 150 mm test tube and Solution B in a second *clean* 150 mm test tube. The volumes of water and the HIO_3 and H_2SO_3 solutions should be measured with pipets.[1] Agitate each solution. Place a white sheet of paper behind Solution A so that the appearance of the deep-blue I_2•starch complex is more evident.

2. a. The reaction begins when the H_2SO_3 from Solution B is added to the HIO_3 from Solution A. Start timing, in **seconds**, the reaction when the two solutions are combined—be as exact as possible in starting and stopping when timing the duration of the reaction.[2]

   b. *O.K., get ready*; quickly pour Solution B into Solution A (B → A, Figure 30.1), **start** timing, and agitate for a good mix. The blue color appears suddenly; watch and **stop** timing at the *first* appearance of the deep-blue color (Figure 30.2). Record the time lapse to the nearest second. Record the temperature of the solution at $T_{(1)}$ on the Data Sheet; the data for the reaction times at $T_{(2)}$, the reaction time at a higher temperature, is collected in Part C.

---

[1] Do not intermix pipets from one trial to the next. Cleanliness is important, especially in this experiment.

[2] The time (seconds) elapsed from the initial mixing of the solutions (A and B) until the deep-blue color *first* appears is to be recorded.

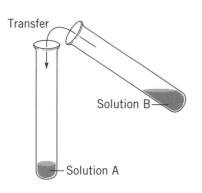

Transfer

Solution B

Solution A

Figure 30.1

Carefully and quickly transfer
Solution B to Solution A

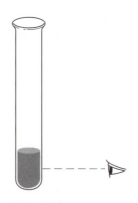

Figure 30.2

Closely observe the reaction
mixture to detect the *first*
appearance of color change

3. *Thoroughly* clean the 150 mm test tube used to observe the reaction.
    Prepare Solutions A and B for Test Reaction 2. Repeat the reaction
    procedure as described in Part A.2b.

4. Continue with the data collection for the other test reactions in Table
    30.1. If your instructor approves, repeat the tests or prepare and collect
    data for other volumes of HIO_3 in Solution A.

Dispose of the test solutions in the "Waste Salts" container.

1. Determine the reciprocal of the time (seconds) elapsed for the depletion
    of the SO_3^{2-} and appearance of the deep-blue $I_2 \bullet$ starch complex and
    record it as the reaction rate.[3] Calculate the logarithm of this rate and
    record on the Data Sheet.

**B. Data Analysis and
Calculations**

2. Calculate and record the initial molar concentration of the IO_3^- ion,
    $[IO_3^-]_i$, and the logarithm of $[IO_3^-]_i$ for each test reaction. Remember the
    total volume of each test reaction is 10 mL.

3. Plot on linear graph paper log (rate) *vs.* log $[IO_3^-]_i$. Draw a straight line
    that passes as close as possible to the 5 data points. Calculate the slope
    of the line; its value should approximate a whole number, but record its
    actual value on the Report Sheet. This is the order, **p**, of IO_3^- in the
    reaction. Ask your instructor to approve your graph.

4. Use the values of the rates of the test reactions (from Part B.1), $[IO_3^-]_i$
    (from Part B.2), **p** (from Part B.3), and the rate law, **rate = k'[IO_3^-]^p**, to
    calculate **k'** for the five test reactions. Calculate the average value of **k'**
    and express it with its proper units.

5. From the plot, extrapolate the straight line until it intersects the y-axis at
    x = 0 (or log $[IO_3^-]$ = 0). This y-value equals log k'; calculate **k'** from the
    plotted data.

---

[3]The reaction rate can also be expressed as $\dfrac{-\Delta[SO_3^{2-}]}{\Delta t}$. For each trial $\dfrac{0.025 \text{ mmol } SO_3^{2-}}{10 \text{ mL}}$
= 2.5×10^{-3} mol SO_3^{2-}/L are consumed. Therefore, 2.5×10^{-3} mol SO_3^{2-}/L multiplied
by reciprocal time is also an expression of the reaction rate. Consult with your
instructor to determine which expression of rate is to be recorded.

C. Activation Energy

1. Repeat Parts A and B but increase the temperature of solutions A and B to no more than 50°C. How are you going to do this? Prepare a water bath using a 600 mL beaker half-filled with water and heat to about 50°C. Place Solutions A and B in the hot water bath for 3–4 minutes until thermal equilibrium has been reached. At that point, quickly pour solution B into A, and record the time lapse as before. Record the temperature (±0.2°C) of the water bath[4] and use this time lapse for your calculations.

2. Perform the same data analysis as before. Using the values of k' (at $T_{(1)}$ and $T_{(2)}$), calculate the activation energy for the reaction. Remember to express temperature in kelvins and R = 8.314 J/mol K.

3. Repeat the calculation of the activation energy using the k' values (at $T_{(1)}$ and $T_{(2)}$) determined from the two graphs at x = 0.

NOTES, OBSERVATIONS, AND CALCULATIONS

---

[4]The temperature does not have to be exactly at 50°C, but the *actual* temperature should be recorded to ±0.2°C.

Rate Law Determination

Date _____Name _____ Lab Sec. _____Desk No. _____

1. a. If 3.5 hours is the time required to travel 165 miles, what is the *rate* of travel?

    b. If 4 minutes and 10 seconds (250 s) elapse in completing four laps of a 400 m track, what is the pace (running rate), in m/s, of the track star?

    c. A total of 27.3 seconds elapses before a 0.00344 mol/L of a reactant has been consumed. What is the reaction rate?

2. A reaction between the gaseous substances C and D proceeds at a measurable rate. When 1.0×10^{-3} mol G/L forms a sudden color change occurs. The following set of data was collected for the reaction:

$$3\,C(g) + 2\,D(g) \rightarrow 2\,F(g) + G(g)$$

$[C]_i$	$[D]_i$	Time for Color Change (*seconds*)	Rate = $\dfrac{\Delta[G]}{\Delta t}$
1.0×10^{-2}	1.0	30	
1.0×10^{-2}	3.0	10	
2.0×10^{-2}	3.0	1.3	
2.0×10^{-2}	1.0	3.7	
3.0×10^{-2}	3.0	0.37	

    a. What is the (whole number) order of C in the reaction? _____; of D in the reaction? _____

    b. What is the rate law?

    c. What is the specific rate constant, k?

d. At any given instant during the reaction,

the rate of appearance of F is _____ times the rate of disappearance of C.

the rate of disappearance of D is _____ times the rate of appearance of G.

the rate of appearance of G is _____ times the rate of appearance of F.

3. A set of rate data is plotted for the reaction of A at several concentrations while the B_2 concentration remains constant for $2 A + B_2 \rightarrow A_2B + B$
    a. From the graph, determine the order of A in the reaction.

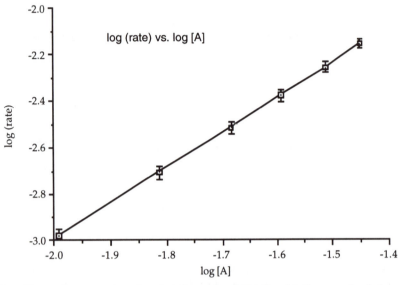

b. From the equation, log (rate) = p log [A] + log k', the graphed data, and the order of A in the reaction, calculate k', the specific rate constant for the reaction.

4. A 1:1 mixture of 0.040 M KIO_3 and 0.040 M H_2SO_4 is used to prepare the 0.020 M HIO_3 solution for today's experiment.
    a. Describe the procedure for preparing 150 mL of a 0.040 M KIO_3 solution, starting with solid KIO_3.

b. How do you prepare 150 mL of a 0.040 M H_2SO_4 solution, starting with 6.0 M H_2SO_4?

Rate Law Determination

Date _____ Name _____ Lab Sec. _____ Desk No. _____

A. Test Reactions

Test Reactions		1	2	3	4	5	6
1. Time elapsed (*sec*)	at $T_{(1)}$						
	at $T_{(2)}$						
2. Reaction temperature	$T_{(1)}$						
	$T_{(2)}$						

B. Data Analysis and Calculations

1. rate (sec^{-1}) or ($mol\ SO_3^{2-}/L{\cdot}sec$)	at $T_{(1)}$						
	at $T_{(2)}$						
2. log (rate)	at $T_{(1)}$						
	at $T_{(2)}$						
3. $[IO_3^-]_i$ (*mol/L*)	at $T_{(1)}$						
	at $T_{(2)}$						
4. log $[IO_3^-]_i$	at $T_{(1)}$						
	at $T_{(2)}$						

5. Plot log (rate) *vs*. log $[IO_3^-]_i$ Instructor's approval of graph   at $T_{(1)}$ _____   at $T_{(2)}$ _____

6. Value of **p** from graph

7. Value of **k'** for each trial	at $T_{(1)}$						
	at $T_{(2)}$						

8. Average value of **k'** (calculated)   at $T_{(1)}$ [ ]   at $T_{(2)}$ [ ]

9. Value of **k'** from graph   at $T_{(1)}$ [ ]   at $T_{(2)}$ [ ]

Calculations

C. Activation Energy

E_a from calculated k' values* _____

E_a from k' values from graph _____

* Show calculation of E_a here.

QUESTIONS

1. How would each of the following changes in the experimental procedure affect the reaction rate?
   a. a slight increase in the concentration of starch? Assume no volume change. Explain.

   b. a slight increase in the concentration of the HIO_3 in solution A. Explain.

   c. a slight increase in the amount of deionized water. Explain.

2. Two test reactions are the *minimum* needed to obtain the values of **p** and **k'** in this experiment. Explain the advantage of using additional test reactions, as were performed in this experiment.

3. What was happening in today's reaction between the times Solutions A and B were mixed and the appearance of the deep-blue $I_2 \bullet$ starch complex?

Preface

◇ Cation Identification

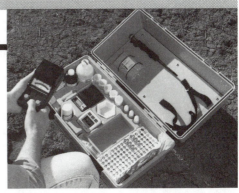

Quick and simple procedures for the identification of a substance are often necessary in forensic science. For example, rapid and easy tests have been developed for the detection of confiscated drugs and for the detection of drugs in a urinalysis. Home pregnancy test kits and tests for diabetes are currently available at the local pharmacy. A quick and easy identification of a cation (see photo) in a salt mixture or a drinking water sample is necessary in water laboratories when metal contamination is suspect and immediate action for public safety is necessary. Tests kits for determining the levels of chlorine, the alkalinity, and the pH in home swimming pools are available at pool supply stores. All of these quick and ready tests must be reproducible, reliable, and must use stable testing chemicals. These criteria are upheld, only if a very carefully outlined and researched procedure is followed.

The characteristic physical and chemical properties of a cation allow us, in the next series of experiments, to separate it from a mixture of cations and to characteristically identify its presence. For example, the Ag^+ ion is identified as being present by its precipitation as the chloride, i.e., $AgCl(s)$. While other cations precipitate as the chloride, AgCl is the only one that is soluble in an **ammoniacal solution**.[1] Similarly, from the solubility rules that you have learned earlier in the course, Ba^{2+} can be separated from a number of other cations with the addition of SO_4^{2-}; $BaSO_4$ forms a white precipitate, while other cations (except Pb^{2+}, Sr^{2+}, and Hg_2^{2+}) generally are soluble as sulfates.

PRINCIPLES

Ammoniacal solution: a solution that contains ammonia as the principal solute.

Therefore, with enough knowledge of the chemistry of the various cations, a unique separation and identification procedure can be developed. Some procedures are one step, quick tests, while others are more exhaustive. *The procedure, however, must systematically eliminate all other ions that may interfere with the test.* Cations can be classified according to groups that have similar chemical properties; cations within each group can then be further separated and identified. A procedure that follows this pattern of analysis is called **Qualitative Analysis**, one that we will follow in the next several experiments.

The separation and identification of cations use many chemical principles, many of which we will cite as we proceed. These principles will include precipitation, ionic equilibrium, acids and base properties, pH, oxidation and reduction, and complex ion formation. To help you to understand these test procedures, each experiment presents some pertinent chemical equations, and you are asked to write equations for other reactions that occur in the separation and identification of the cations. This usually appears in the Lab Preview.

The general procedure for the detection of a specific cation in a mixture of cations is to selectively separate and characteristically identify each cation.

PROCEDURES

---

[1] This was one test procedure that prospectors for silver used in the early prospecting days.

The common selective separation procedure in qualitative analysis is precipitation, although gas evolution is another procedure that is used. Characteristic identification of the cation is also commonly the formation of a precipitate, although gas evolution and color identification procedures are used.

Because precipitation procedures are so common to qualitative analysis, it is important to develop good techniques for handling precipitates. The techniques for compacting, breaking up, and washing a precipitate to minimize contamination are essential for a characteristic identification of a specific cation. Because adjustments in reaction conditions, such as pH, molar concentrations, etc., are often subtle for an efficient separation it is vital that the *chemistry* of the each step is understood as the test reagents are being added and the separation techniques are being applied.

Most of the tests for the cations will be performed in 3 mL (75 mm) test tubes and reagents will be added with medicine droppers ($\approx$20 drops/mL) or Beral pipets. Do not mix the medicine droppers (or Beral pipets) with the various reagents you will be using.

When mixing solutions in a test tube, agitate by tapping the side of the test tube (see Figure 13.1).

Break up a precipitate with a stirring rod or cork the test tube and invert, but *never* use your thumb (see Figure CI.1)!

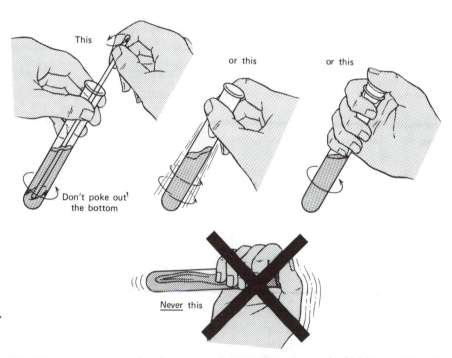

Figure CI.1
Technique for mixing solutions in a test tube
Occluded: absorbed or entrapped.

Washing a precipitate is often necessary to remove **occluded** impurities in your analysis. Add the water or wash liquid to the precipitate, mix thoroughly with a stirring rod, centrifuge, and decant. Generally two washings is satisfactory. Failure to properly wash precipitates often leads to errors in the analysis (and arguments with your laboratory instructor!).

Since you will be frequently using the centrifuge in the next several experiments, *be sure* to read Technique 14d very closely in the Techniques section of this manual.

A flow diagram is used to systematically organize the qualitative analysis test procedures for the various cations. A flow diagram is analogous to the *inverted* flow of a river: for example, *if*, at the mouth of the river, a variety of pollutants are in the water—from where did the pollutants originate? The environmental chemist systematically back tracks the flow of the water through the smaller rivers, creeks, and streams, analyzing the water at each stage of the investigation, to determine the point source of the pollutant. A qualitative analysis of cations follows a similar pattern: the sample has several cations—the sample is divided, analyzed, divided again, and analyzed again until confirmation of the presence or absence of the cation is determined.

Flow Diagrams

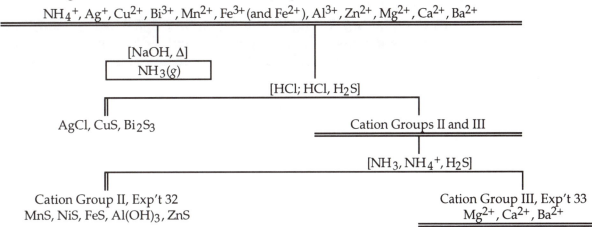

Flow diagram icons appear in the margins of the Procedure in each of the next three experiments as part of the complete flow diagram, the purpose of which is to maintain a sense of continuity in the many steps and techniques that need to be followed during the analyses.

A flow diagram for qualitative analysis utilizes several standard notations:

- Brackets, [ ], indicate the use of a test reagent, written in molecular form.

- A single horizontal line, _____, indicates that separation is made with a centrifuge.

- Two short vertical lines, ‖ , indicate that a precipitate has formed; these lines are drawn to the *left* of the single horizontal line.

- One short vertical line, | , indicates a supernatant and is drawn to the *right* of the single horizontal line.

- A double horizontal line, _____ , indicates soluble cations present.

- A square box, ☐ placed around the compound or test reagent confirms the presence of the cation.

Flow Diagram for Cation Identification (Experiments 31, 32, and 33)

NH_4^+, Ag^+, Cu^{2+}, Bi^{3+}, Mn^{2+}, Fe^{3+} (and Fe^{2+}), Al^{3+}, Zn^{2+}, Mg^{2+}, Ca^{2+}, Ba^{2+}

[NaOH, Δ]

$NH_3(g)$

[HCl; HCl, H_2S]

AgCl, CuS, Bi_2S_3

Cation Groups II and III

[NH_3, NH_4^+, H_2S]

Cation Group II, Exp't 32
MnS, NiS, FeS, Al(OH)$_3$, ZnS

Cation Group III, Exp't 33
Mg^{2+}, Ca^{2+}, Ba^{2+}

An unknown consisting of any number of the 12 cations analyzed in the next three experiments may be assigned at the conclusion of Experiment 33. Ask your instructor about that assignment. The preceding flow diagram is for these cations. A flow diagram for each group is presented in the corresponding experiment.

The following suggestions are offered while performing these "qual" experiments:

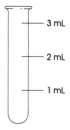

Figure CI.2
Transfer 1, 2, and 3 mL of water into a 75 mm test tube. and mark. Use these markings to estimate solution additions in qual tests.

- *Always* read the procedure in detail and with an understanding of the principles before lab. Why is this reagent added at this time? Is extra equipment necessary? Is a hot water bath needed? What precautions are to be taken?

- Closely follow simultaneously the Principles of each test, the flow diagram, the Procedure, and the Data Sheet, during the analysis.

- Mark with a file or water insoluble marker, 1, 2, and 3 mL intervals on a 75 mm test tube in order to quickly estimate volumes (Figure CI.2).

- Keep a wash bottle filled with deionized water at all times.

- Keep a number of *clean* dropper pipets (or Beral pipets) and 75 mm test tubes available; always rinse several times with deionized water immediately after their use.

- Maintain a "file" of confirmatory tests in the test tubes from the analysis of the *known* solution for each Group of ions, while you are conducting the analysis for the unknown sample; in that way, color and precipitate comparisons can be done quickly.

Caution: *In the next several experiments you will be handling a large number of chemicals (acids, bases, oxidizing and reducing agents, and, perhaps, even some toxic chemicals), some of which are more concentrated than others. Always be certain that your safety goggles are properly worn.*

*Carefully, handle all chemicals. Do not intentionally inhale the vapors of any chemical unless you are specifically told to do so; **avoid** skin contact with any chemicals—wash the skin immediately in the laboratory sink, eye wash fountain, or safety shower; **clean up** any spilled chemical. If you are uncertain of the proper cleanup procedure, flood with water, and consult your laboratory instructor; **be aware** of the techniques and procedures of neighboring chemists—discuss potential hazards with them.*

Dispose of the waste chemicals in the appropriately labeled waste containers. Consult your laboratory instructor to ensure proper disposal.

Experiment 31

◇ Cation Identification, I. NH_4^+, Ag^+, Cu^{2+}, Bi^{3+}

The world's largest active open pit copper mine is located near Salt Lake City, Utah. The copper is chemically combined in various compounds but its percent abundance in the ore is only about 0.6%! To determine the presence of copper in an ore that is that low in natural abundance requires some basic chemical skills to first detect its presence, a qualitative analysis, and then to determine its amount, a quantitative analysis. The copper ore is processed through a series of metallurgical processes until 98% pure copper is obtained prior to a final electrolytic refining process.

Similarly, most silver found in the Earth's crust is not found in the elemental state, but rather combined chemically with various components of its ore. Prospectors performed qualitative tests for the presence of silver by a procedure that is analogous to the one you will use in this experiment.

OBJECTIVE

* To separate and identify the presence of one or more cations from a mixture of the cations, NH_4^+, Ag^+, Cu^{2+}, and Bi^{3+}

PRINCIPLES

This experiment is a first in a series in which a set of reagents and a number of separation techniques are used to identify the presence of a specific cation among a group of cations that have similar chemical properties. We approach this study from that of an experimental chemist—we will perform some tests, write down observations, and then write a balanced equation that agrees with the data.[1]

The identification of a particular cation present in a mixture of cations is most often accomplished through a scheme of separation and precipitation steps. The optimal conditions for the separation and identification of a cation are very selective; the position of equilibrium (LeChatelier's Principle, see Experiment 19) is controlled by pH adjustment, occasionally with a buffer, by the presence of a limited concentration range of the precipitating anion, and/or by the presence of oxidizing or reducing conditions. For example, the Ag^+ is "insoluble" at low concentrations of chloride ion (Figure 31.1), but at higher concentrations of chloride ion the soluble complex anion, $[AgCl_2]^-$ forms.

Figure 31.1
Silver ion quickly forms a AgCl precipitate with the addition of chloride ion.

The sulfide salts of the cations in this group (exclusive of the ammonium ion) are insoluble in solutions that are at least 0.3 M H^+ (pH = 0.5). If the pH is greater than 0.5, sulfide precipitation of cations in the second group (Experiment 32) may also precipitate. Therefore, pH control is extremely important for both controlling the amount of sulfide ion present in solution

---

[1]For a more complete description and procedure for the qualitative analysis of cations, see Beran, *Laboratory Manual for Principles of General Chemistry, 5th Ed.*, John Wiley & Sons, Inc., 1994.

in situ: Latin for "in position" or in the natural or original position; in chemistry it implies the generation of a substance in solution for immediate reaction.

and in separating these two groups of cations—interference with the cation identification between these two groups can be problematic.

The sulfide ion for the precipitation of the cations is produced *in situ* from $H_2S(aq)$. The thermal degradation of thioacetamide, CH_3CSNH_2, in an acidic or basic solution generates small quantities of hydrogen sulfide, $H_2S(aq)$.

$$CH_3CSNH_2(aq) + 2\,H_2O(l) \xrightarrow{\Delta} CH_3CO_2^-(aq) + NH_4^+(aq) + H_2S(aq)$$

The slow generation of H_2S in the solution minimizes the amount of this foul-smelling, highly toxic gas in the laboratory. When H_2S is produced slowly, more compact sulfide precipitates form, making them easier to separate by centrifugation.

$H_2S(aq)$, as a weak, diprotic acid, ionizes slightly to produce the sulfide ion that is necessary for cation precipitation.

$$H_2S(aq) + 2\,H_2O(l) \rightleftarrows 2\,H_3O^+(aq) + S^{2-}(aq)$$

Applying LeChatelier's Principle to the dynamic equilibrium, we can infer that a high H_3O^+ concentration (low pH) shifts this equilibrium to the left, leaving a low sulfide concentration; conversely, a low H_3O^+ concentration (high pH) shifts the equilibrium to the right, increasing the sulfide concentration. Therefore, if cations precipitate as the sulfide salt at a low pH, then only small amounts of sulfide ion are needed for precipitation; this means that these sulfide salts have a very low solubility and also small K_s values. Sulfide salts having low solubilities are separated and identified in this experiment.

If a higher sulfide concentration is needed for the precipitation of the cation, then the pH of the solution must be adjusted upward; these sulfide salts are more soluble and have larger K_s values. The more soluble sulfide salts are separated and identified in Experiment 32.

The chemical principles and laboratory techniques that are used to separate and identify the four cations in this experiment require you to be a careful, conscientious chemist. Carefully read the Preface to Cation Identification, the Procedure, review your laboratory techniques, and complete the Lab Preview before beginning the analysis—it will save you time and minimize frustration.

The flow diagram outlines the procedure for the separation and the identification of the NH_4^+, Ag^+, Cu^{2+}, and Bi^{3+} cations. Review the Preface to Qualitative Analysis for an understanding of the symbolism in the flow diagram.

Ammonium Ion

Δ: the delta symbol is an accepted symbol for heat, placed over the reaction arrow in an equation.

The NH_4^+ ion is a weak acid and is stable in an acidic solution, but when its solution is made basic, NH_3 gas is evolved. With gentle heat, NH_3 gas is driven from the solution and can be detected by its odor or by a litmus test.

$$NH_4^+(aq) + OH^-(aq) \xrightarrow{\Delta} NH_3(g) + H_2O(l)$$

Silver Ion

The silver ion precipitates with low concentrations of chloride ion. In an ammoniacal solution, $AgCl$ dissolves to form the diamminesilver ion, $[Ag(NH_3)_2]^+$.

$$AgCl(s) + 2\,NH_3(aq) \rightarrow [Ag(NH_3)_2]^+(aq)$$

After removal of the silver ion from a sample containing a mixture of cations, the solution is made acidic with hydrochloric acid. Thioacetamide is added, generating the sulfide ion that precipitates the copper and bismuth cations and separates them from the more soluble sulfide cations of Experiment 32.

Addition of hot nitric acid to the CuS and Bi_2S_3 precipitates causes the sulfide ion of the salt to oxidize and release the Cu^{2+} and Bi^{3+} into solution.

$$3\,CuS(s) \ + \ 8\,H^+(aq) \ + \ 2\,NO_3^-(aq) \ \rightarrow$$
$$3\,Cu^{2+}(aq) \ + \ 3\,S(s) \ + \ 2\,NO(g) \ + \ 4\,H_2O(l)$$
$$Bi_2S_3(s) \ + \ 8\,H^+(aq) \ + \ 2\,NO_3^-(aq) \ \rightarrow$$
$$2\,Bi^{3+}(aq) \ + \ 3\,S(s) \ + \ 2\,NO(g) \ + \ 4\,H_2O(l)$$

Flow Diagram for Cations of Experiment 31

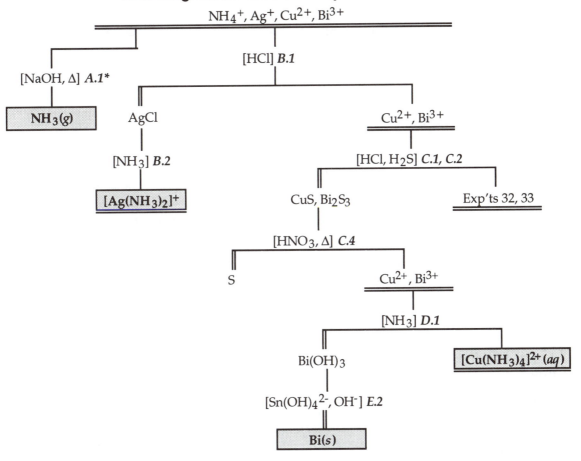

NH_4^+, Ag^+, Cu^{2+}, Bi^{3+}

[HCl] *B.1*

[NaOH, Δ] *A.1**

NH₃(g)   AgCl Cu^{2+}, Bi^{3+}

[NH₃] *B.2*   [HCl, H₂S] *C.1, C.2*

$[Ag(NH_3)_2]^+$ CuS, Bi_2S_3 Exp'ts 32, 33

[HNO₃, Δ] *C.4*

S Cu^{2+}, Bi^{3+}

[NH₃] *D.1*

Bi(OH)₃ $[Cu(NH_3)_4]^{2+}$ (aq)

$[Sn(OH)_4^{2-}, OH^-]$ *E.2*

Bi(s)

* Refers to Part A.1 of the Procedure

Aqueous NH_3 precipitates Bi^{3+} as a white hydroxide, but reacts with Cu^{2+} to form a deep blue $[Cu(NH_3)_4]^{2+}$ complex ion, a confirmation of the presence of Cu^{2+}.

Copper(II) Ion

$$Cu^{2+}(aq) \ + \ 4\,NH_3(aq) \ \rightarrow \ [Cu(NH_3)_4]^{2+}(aq)$$
(blue)

When a freshly prepared sodium stannite, $Na_2Sn(OH)_4$, solution is added to the bismuth hydroxide precipitate, the Bi^{3+} ion is reduced to black bismuth metal, confirming the presence of bismuth in the sample.

Bismuth(III) Ion

$$2\,Bi(OH)_3(s) \ + \ 3\,[Sn(OH)_4]^{2-}(aq) \ \rightarrow \ 2\,Bi(s) \ + \ 3\,[Sn(OH)_6]^{2-}(aq)$$
(black)

PROCEDURE

To become familiar with the identification of these cations, take a sample that contains the four cations and analyze it according to the procedure. At each numbered superscript (i.e., [#1]), STOP, and record on the Data Sheet. After the presence of each cation has been confirmed, SAVE the test tube so that its appearance can be compared to that for your unknown sample.

Caution: *Several concentrated and 6 molar acids and bases are used in the analysis of these cations. Handle each of these solutions with care. Read the Lab Safety section in Experiment 1 for instructions in handling acids and bases.*

> **Dispose of all washings and discarded solutions into a personal waste beaker. After completing the experiment discard the contents of the waste beaker in the "Waste Salts" container.**

A. Ammonium Ion Test

1. Place about 5 drops of 6 M NaOH (**Caution**!!) in a 75 mm test tube. Add 10–20 drops of the test solution containing the mixture of cations. Moisten a piece of red litmus paper with deionized water and place it over the mouth of the test tube. (Note: be careful *not* to let NaOH or the test solution to contact the litmus paper.) Warming with a *gentle* flame may be necessary.[#1]

2. Using the proper technique, check the odor of the gas evolved.

B. Silver Ion Test

1. Place 1 mL of the test solution containing the mixture of cations in a 75 mm test tube, add 2 drops of 6 M HCl (**Caution**: *handle acid with care*) (see Figure 31.2), and centrifuge.[#2] Ask the instructor about the safe operation of the centrifuge. Test the supernatant with drops of 6 M HCl for the additional formation of precipitate. Again centrifuge, if necessary. Decant the supernatant liquid and save for Part C.

2. To the precipitate, add 5 drops of 6 M NH$_3$ (**Caution:** *avoid inhalation and skin contact*).[#3]

> If it is determined that cations from Experiments 32 and 33 are *not* present in your sample or in your "test" solution, proceed to Part D.

C. Sulfide Precipitation of the Cations

Supernatant: the liquid/solution that remains above the precipitate after centrifuging.

1. To the supernatant from Part B.1, add several more drops of 6 M HCl. This should adjust the pH of the solution to about 0.5.

2. Add 10 to 15 drops of 1 M CH$_3$CSNH$_2$, heat in a hot water bath (≈95°C, Figure 31.3) for several minutes, cool, and centrifuge. Test for complete precipitation by repeating the thioacetamide addition to the supernatant. The precipitate contains the CuS and Bi$_2$S$_3$ salts;[#4] the **supernatant** contains cations that do not precipitate under these conditions (Experiments 32 and 33).[2]

3. Wash the precipitate twice with 2 M NH$_4$NO$_3$, stir, centrifuge, and discard each washing.

4. Add 15 drops of 6 M HNO$_3$ to the precipitate and heat to boiling until the precipitate dissolves (a direct flame may be necessary. **Caution**: *read Technique 6a for heating test tubes with a direct flame*). Cool and centrifuge. Transfer the supernatant[#5] to a 75 mm test tube; discard any free sulfur.

---

[2]If your sample is to be analyzed for additional cations in Experiments 32 and 33, save this supernatant for that analysis; otherwise discard it.

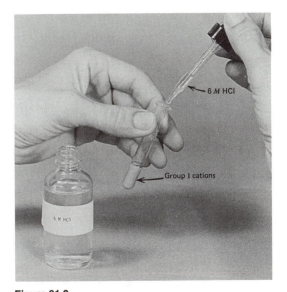

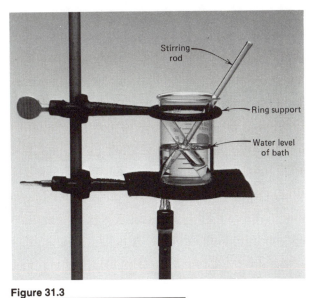

Figure 31.2
Precipitation of Ag⁺(*aq*)

Figure 31.3
The setup of a hot water bath

1. Add drops of conc NH₃ (**Caution**: *avoid inhalation or skin contact!*) to the supernatant of Part C.4 (or of Part B.1, if Part C is omitted). A deep blue solution confirms the presence of Cu^{2+} in the sample.[#6] Centrifuge, decant the solution, and save the precipitate[#7] for Part E.

D. Copper(II) Ion Test

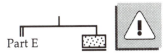

1. Prepare a fresh $Na_2Sn(OH)_4$ solution by placing 2 drops of 1 *M* $SnCl_2$ in a 75 mm test tube, followed by drops of 6 *M* NaOH until the $Sn(OH)_2$ precipitate just dissolves. Shake or stir the solution.

E. Bismuth(III) ion Test

2. Add several drops of this freshly prepared $Na_2Sn(OH)_4$ solution to the precipitate from Part D. The immediate formation of a black precipitate[#8] confirms the presence of bismuth in the sample.

1. Obtain an unknown "test" solution from your instructor. Record the number of the unknown on the Data Sheet. Determine which cation(s) is (are) present.

F. Unknown Test

Dispose of the test solutions in the "Waste Salts" container.

NOTES AND OBSERVATIONS

◇ Cation Identification, I. NH_4^+, Ag^+, Cu^{2+}, Bi^{3+}

Date _____Name _____ Lab Sec. _____Desk No. _____

1. Read the Preface to Cation Identification to answer the following:

    a. The approximate volume of a standard 75 mm test tube is _____.

    b. Small volumes of reagents are usually added using _____.

    c. A _____ is commonly used to break up a precipitate.

    d. A _____ is an instrument used to separate and compact a precipitate in a test tube.

    e. The clear solution above a precipitate is called the _____.

    f. The number of drops of water equivalent to 1 mL is about _____.

    g. A solution should be centrifuged for (how long?) _____.

    h. On a flow diagram, ‖ means _____.

    i. On a flow diagram, a single horizontal line means _____.

    j. On a flow diagram, ☐ means _____.

2. Describe the procedure for balancing a centrifuge.

3. a. Find the K_s values of AgCl and CuS in your text book.

    $K_s(AgCl) =$

    $K_s(CuS) =$

    b. Calculate the molar solubility of each salt. Which salt is the least soluble?

4. What reagent separates Cu^{2+} from Bi^{3+} in solution?

5. a. Explain why HNO_3 is added to the precipitate in Part C.4.

   b. What is the origin of the elemental sulfur in Part C.4?

6. What is the color of

   a. $[Cu(NH_3)_4]^{2+}$ _____

   b. bismuth metal _____

7. Write balanced equations for the following reactions cited in this experiment. Read the Principles and Procedure sections to assist you in writing the equations. Also use your textbook as resource information.

   a. The formation of ammonia gas from ammonium ion.

   b. The reactions of CuS and Bi_2S_3 with hot nitric acid.

   c. The reaction of Bi^{3+} with aqueous ammonia.

◇ Cation Identification, I. NH_4^+, Ag^+, Cu^{2+}, Bi^{3+}

Date _____ Name _____ Lab Sec. _____ Desk No. _____

Procedure Number and Ion	Test Reagent or Technique		Observation (Color or General Appearance)	Chemicals Responsible for Observation	Check (√) if Observed for Test Solution
#1 NH_4^+		gas			
#2 Ag^+		ppt			
#3		spnt			
#4		ppt			
#5		spnt			
#6 Cu^{2+}		spnt			
#7		ppt			
#8 Bi^{3+}		ppt			

Unknown number: _____

Cations present in unknown: _____

Instructor's approval: _____

QUESTIONS

1. Name and give the color for the following.

   a. AgCl _____ _____

   b. $[Ag(NH_3)_2]^+$ _____ _____

   c. CuS _____ _____

   d. $NaSn(OH)_4$ _____ _____

2. What happens in Part B.1 if 6 M HNO_3 is substituted for 6 M HCl?

3. What happens in Part B.2 if 6 M NaOH is substituted for 6 M NH_3?

4. What happens in Part D.1 if NaOH is substituted for NH_3?

5. What happens if conc HCl is added to the precipitate in Part D.1?

6. Identify a single reagent that separates

    a. Ag^+ from Cu^{2+} _____

    b. Cu^{2+} from Bi^{3+} _____

Experiment 32

◇ Cation Identification, II. Mn^{2+}, Ni^{2+}, Fe^{3+} (and Fe^{2+}), Al^{3+}, Zn^{2+}

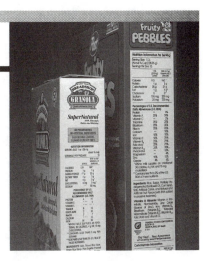

Trace (very small) amounts of metal ions are found throughout our bodies, the most abundant being iron, zinc, copper, and manganese. The daily requirement of metals designated as trace metals (or trace elements) is less than 100 mg/day. The trace metals are chemically bound to large molecules, such as proteins, which form the enzymes that function as catalysts for specific biochemical reactions, hormones, and vitamins. Iron, for example, is a part of the hemoglobin molecule that carries oxygen throughout the body and returns to the lungs with the carbon dioxide waste. Zinc is important to growth, copper deficiencies leads to anemia, and manganese is important to bone development. Excesses of aluminum have been linked to Alzheimer's disease. Over-the-counter mineral supplements are readily available containing various amounts of these elements.

The presence of trace amounts of metal ions in the body can be determined through procedures that are only slightly more complicated than what appear in this experiment.

OBJECTIVE

- To separate and identify the presence one or more cations from a mixture of the cations, Mn^{2+}, Ni^{2+}, Fe^{3+} (and Fe^{2+}), Al^{3+}, and Zn^{2+}

PRINCIPLES

Systematic separation: a series of sequential steps in a chemical analysis that are necessary for the successful separation of a substance.

For a thorough **systematic separation** and identification of cations in a mixture, the cations studied in this experiment are separated as insoluble salts from a solution containing sulfide ion at a pH near neutrality. The cations of this group do *not* precipitate as the chloride salts or as the sulfide salts in a solution with an approximate pH of 0.5 (Experiment 31). Therefore, a separation of the cations in this experiment from those cations in the previous experiment is rather easily accomplished through a pH adjustment.

Preparation of the Cations for Analysis

The Mn^{2+}, Ni^{2+}, Fe^{3+} (and Fe^{2+}), and Zn^{2+} cations precipitate as sulfide salts but Al^{3+} precipitates as the hydrated hydroxide in a basic solution containing sulfide ion. The sulfide ion is generated in the same manner as was done in Experiment 31; that is, from the decomposition of thioacetamide, CH_3CSNH_2, producing $H_2S(aq)$. The H_2S ionizes to produce low concentrations of sulfide ion in solution.

$$H_2S(aq) + 2 H_2O(l) \rightleftarrows 2 H_3O^+(aq) + S^{2-}(aq)$$

The addition of base, such as NH_3, causes this equilibrium to shift *right*, increasing the sulfide concentration to a level where the product of the molar concentrations of the cation and the sulfide ion exceeds the salt's K_s value and the salt precipitates. In this basic solution, the Al^{3+} precipitates as $Al(OH)_3$.

The $H_2S(aq)$ also serves as a **reducing agent**, reducing any Fe^{3+} ion to Fe^{2+} ion in solution; elemental sulfur forms as its oxidized product.

Reducing agent: a substance that decreases the oxidation number of another element or causes another substance to accept electrons in a redox reaction.

$$2\,Fe^{3+}(aq) \;+\; H_2S(aq) \;+\; 2\,H_2O(l) \;\rightarrow\; 2\,Fe^{2+}(aq) \;+\; S(s) \;+\; 2\,H_3O^+(aq)$$

Therefore, when a sample containing all the test cations in this experiment is treated with thioacetamide in a solution made basic with NH_3, a mixed precipitate of MnS, NiS, FeS, ZnS, and $Al(OH)_3$ forms.

The procedure for the separation and identification of these five cations is summarized in the following flow diagram. Follow it as you read through the Principles and Procedure of the experiment.

Flow Diagram for Cations of Experiment 32

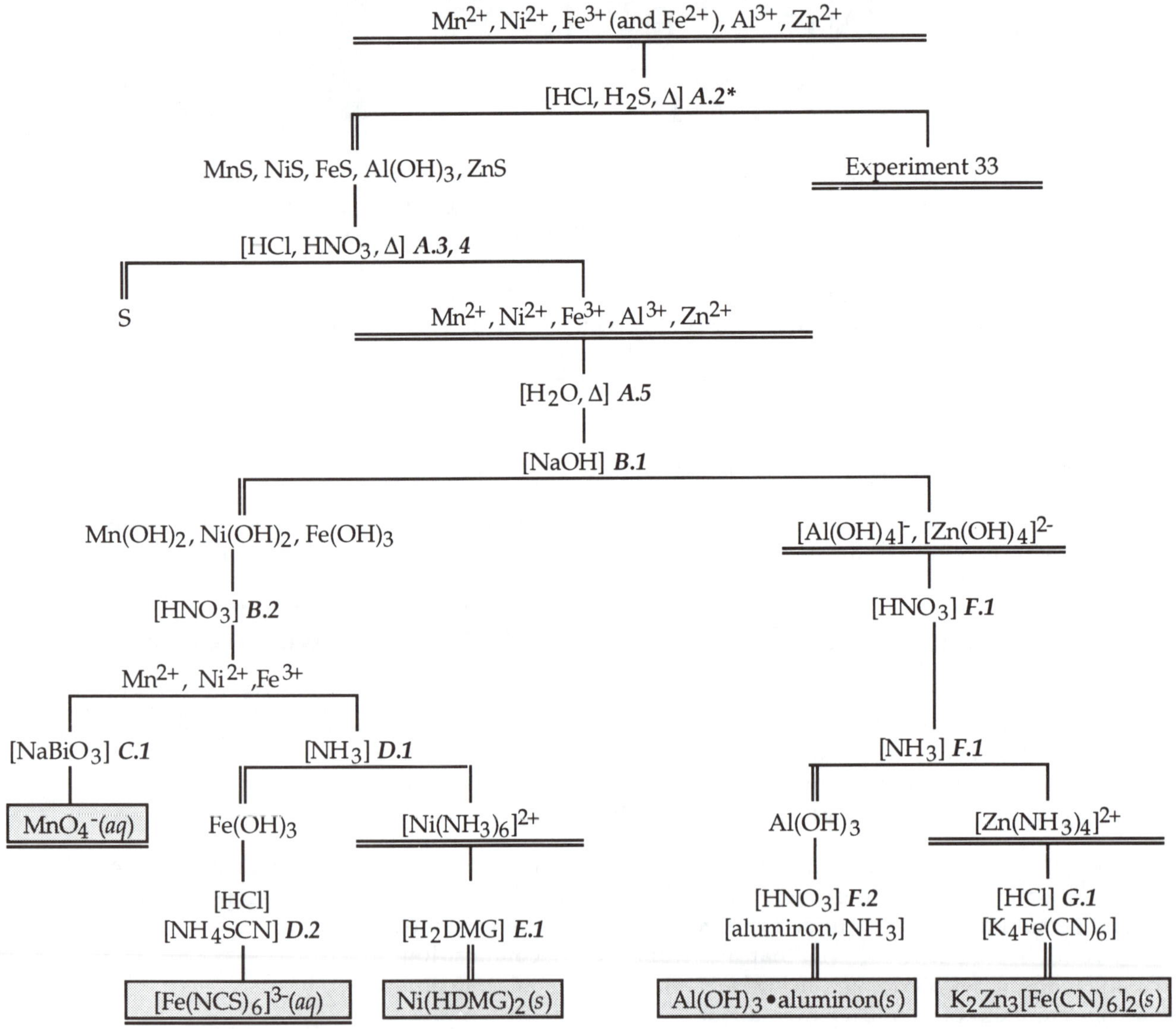

Mn^{2+}, Ni^{2+}, Fe^{3+} (and Fe^{2+}), Al^{3+}, Zn^{2+}

[HCl, H_2S, Δ] **A.2***

MnS, NiS, FeS, $Al(OH)_3$, ZnS   Experiment 33

[HCl, HNO_3, Δ] **A.3, 4**

S

Mn^{2+}, Ni^{2+}, Fe^{3+}, Al^{3+}, Zn^{2+}

[H_2O, Δ] **A.5**

[NaOH] **B.1**

$Mn(OH)_2$, $Ni(OH)_2$, $Fe(OH)_3$ [$Al(OH)_4]^-$, [$Zn(OH)_4]^{2-}$

[HNO_3] **B.2**   [HNO_3] **F.1**

Mn^{2+}, Ni^{2+}, Fe^{3+}

[$NaBiO_3$] **C.1**   [NH_3] **D.1**   [NH_3] **F.1**

| $MnO_4^-(aq)$ | $Fe(OH)_3$ | [$Ni(NH_3)_6]^{2+}$ | $Al(OH)_3$ | [$Zn(NH_3)_4]^{2+}$ |

[HCl]   [HNO_3] **F.2**   [HCl] **G.1**
[NH_4SCN] **D.2**   [H_2DMG] **E.1**   [aluminon, NH_3]   [$K_4Fe(CN)_6$]

| [$Fe(NCS)_6]^{3-}(aq)$ | $Ni(HDMG)_2(s)$ | $Al(OH)_3 \cdot$ aluminon(s) | $K_2Zn_3[Fe(CN)_6]_2(s)$ |

*Refers to Part A.2 in the Procedure

The MnS, FeS, ZnS, and $Al(OH)_3$ precipitates can be dissolved with a strong acid; however hot, conc HNO_3 is necessary to dissolve the NiS. The HNO_3 oxidizes the sulfide ion of the NiS, forming free, elemental sulfur and NO as products.

$$MnS(s) + 2H^+(aq) \rightarrow Mn^{2+}(aq) + H_2S(aq)$$
$$FeS(s) + 2H^+(aq) \rightarrow Fe^{2+}(aq) + H_2S(aq)$$
$$ZnS(s) + 2H^+(aq) \rightarrow Zn^{2+}(aq) + H_2S(aq)$$
$$Al(OH)_3(s) + 3H^+(aq) \rightarrow Al^{3+}(aq) + 3H_2O(l)$$
$$3NiS(s) + 8H^+(aq) + 2NO_3^-(aq) \rightarrow$$
$$3Ni^{2+}(aq) + 2NO(g) + 3S(s) + 4H_2O(l)$$

In addition, the HNO_3 oxidizes Fe^{2+} to Fe^{3+} in the solution. Therefore, HNO_3 dissolves the precipitates producing the soluble cations, sulfur, and $NO(g)$ as products.

Addition of strong base to an aqueous solution of the five cations precipitates Mn^{2+}, Ni^{2+}, and Fe^{3+} as gelatinous hydroxides; the Zn^{2+} and Al^{3+} hydroxides initially precipitate, but, being **amphoteric** hydroxides, dissolve in the strong base, forming the aluminate, $[Al(OH)_4]^-$, and zincate, $[Zn(OH)_4]^{2-}$, ions.

The gelatinous hydroxides of Mn^{2+}, Ni^{2+}, and Fe^{3+} dissolve in nitric acid.

A portion of the solution containing cations of the gelatinous hydroxides (now dissolved with nitric acid) is treated with sodium bismuthate, $NaBiO_3$, a strong oxidizing agent that oxidizes Mn^{2+} to the characteristic purple permanganate ion, MnO_4^-, confirming the presence of Mn^{2+} in the sample.

$$14H^+(aq) + 2Mn^{2+}(aq) + 5BiO_3^-(aq) \rightarrow$$
$$2MnO_4^-(aq) + 5Bi^{3+}(aq) + 7H_2O(l)$$
(purple)

A second portion of the same solution is treated with an excess of NH_3; this precipitates brown $Fe(OH)_3$ but forms the blue hexaammine complex of Ni^{2+}, $[Ni(NH_3)_6]^{2+}$. Acid dissolves the $Fe(OH)_3$ precipitate, whereupon the solution is treated with thiocyanate ion, SCN^-, forming a blood-red complex ion with Fe^{3+}, $[Fe(NCS)_6]^{3-}$, a confirmation of the presence of Fe^{3+} in the sample.

$$Fe^{3+}(aq) + 6SCN^-(aq) \rightarrow [Fe(NCS)_6]^{3-}(aq)$$
(blood-red)

The confirmation of Ni^{2+} is the appearance of the bright pink-red precipitate formed when dimethylglyoxime, H_2DMG,[1] is added to a solution of the hexaamminenickel(II) complex ion.

$$[Ni(NH_3)_6]^{2+}(aq) + 2H_2DMG(aq) \rightarrow$$
$$Ni(HDMG)_2(s) + 2NH_4^+(aq) + 4NH_3(aq)$$
(pink-red)

An excess of acid added to a solution containing the aluminate and zincate ions restores the formation of the Al^{3+} and Zn^{2+} ions. With the re-addition of NH_3, the Al^{3+} reprecipitates as the gelatinous hydroxide, but the Zn^{2+} forms the soluble tetraammine complex ion, $[Zn(NH_3)_4]^{2+}$.

The $Al(OH)_3$ precipitate is dissolved with HNO_3; aluminon reagent[2] and ammonia are added and $Al(OH)_3$ reprecipitates. Because $Al(OH)_3$ is a

Separation of Mn^{2+}, Ni^{2+}, and Fe^{3+} from Zn^{2+} and Al^{3+}

Amphoteric: capable of functioning as an acid or a base

Manganese(II) Ion

Iron(III) Ion

Nickel(II) Ion

Structure of $Ni(HDMG)_2(s)$

Aluminum Ion

---

[1]Dimethylglyoxime, abbreviated as H_2DMG for convenience in this experiment, is an organic complexing agent, specific for the precipitation of the nickel ion.

[2]The aluminon reagent is the ammonium salt of aurin tricarboxylic acid, a red dye.

bulky, gelatinous precipitate, the aluminon reagent, a red dye, adsorbs onto its surface giving the precipitate a pink/red appearance. This confirms the presence of Al^{3+} in the sample.

$$Al^{3+}(aq) + 3\,NH_3(aq) + 3\,H_2O(l) + aluminon(aq) \rightarrow$$
$$Al(OH)_3 \bullet aluminon(s) + 3\,NH_4^+(aq)$$
$$(pink/red)$$

Zinc Ion

When potassium hexacyanoferrate(II), $K_4Fe(CN)_6$, is added to an acidified solution of the $[Zn(NH_3)_4]^{2+}$ ion, a light green (blue-green) precipitate of $K_2Zn_3[Fe(CN)_6]_2$ forms, confirming the presence of Zn^{2+} in the sample.

$$3\,[Zn(NH_3)_4]^{2+}(aq) + 2\,K_4Fe(CN)_6(aq) + 12\,H^+(aq) \rightarrow$$
$$K_2Zn_3[Fe(CN)_6]_2(s) + 12\,NH_4^+(aq) + 6\,K^+(aq)$$
$$(light\ green)$$

PROCEDURE

To become familiar with the identification of these cations, take a sample that contains the five cations and analyze it according to the procedure. At each numbered superscript (i.e., [#1]), STOP, and record on the Data Sheet. After the presence of each cation has been confirmed, SAVE the test tube so that its appearance can be compared to that for your unknown sample.

Caution: *Several concentrated and 6 molar acids and bases are used in the analysis of these cations. Handle each of these solutions with care. Read the Lab Safety section in Experiment 1 for instructions in handling acids and bases.*

Dispose of all washings and discarded solutions into a personal waste beaker. After completing the experiment discard the contents of the waste beaker into the "Waste Salts" container.

A. Preparation of the Cations for Analysis

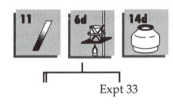

Expt 33

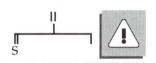

1. If your sample contains only the cations of this experiment, you will *not* need to precipitate the cations in a sulfide solution at the relatively high pH; instead you may proceed directly to Part B.

2. To 2 mL of a test solution that may contain cations for analysis in Experiment 33, add 1 mL of 2 M NH$_4$Cl. Add 6 M NH$_3$ until the solution is basic to litmus and then add 5 additional drops. Saturate the solution with H$_2$S by adding 6 drops of 1 M CH$_3$CSNH$_2$. Heat the solution in a hot water bath ($\approx$ 95°C, see Figure 31.2) for several minutes, cool, and centrifuge. Save the precipitate[#1] for Part A.3 and the supernatant for analysis in Experiment 33.

3. Wash twice the precipitate[3] with 2 mL of water and discard the washings. Add 5 drops of 6 M HCl and 5 drops of 6 M HNO$_3$ (**Caution**: *be careful in handling acids*) and again heat in the hot water bath until the precipitates dissolve. Cool and centrifuge; save the supernatant[#2] and discard any free, elemental sulfur.

4. Transfer the supernatant to an evaporating dish and heat gently until a moist residue remains (*not* to dryness). Add 1–2 mL of conc HNO$_3$ (**Caution**: *conc HNO$_3$ is a strong oxidizing agent—do not allow contact with the skin or clothing!*) and reheat with a "cool" flame (Figure 32.2) until a moist residue again forms.

---

[3]See the Preface to Cation Identification (page 343) for the proper technique in washing precipitates.

5. Dissolve the moist residue in 1–2 mL of deionized water and transfer the solution to a 75 mm test tube.

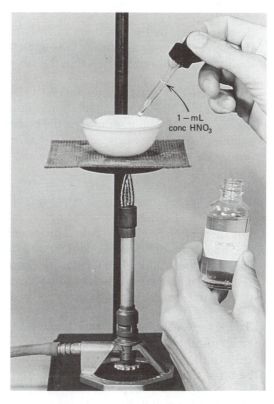

Figure 32.2
Addition of conc HNO₃ (Caution!!) to moist residue

Figure 32.2

Addition of conc HNO_3 (Caution!!) to moist residue

1. To the solution obtained from Part A.5 *or* from a test solution containing only the cations of Experiment 32 (in a 75 mm test tube), add 15 drops of 6 *M* NaOH (**Caution**: *avoid skin contact*). Centrifuge and save the precipitate.[#3] Decant the supernatant[#4] into a 75 mm test tube for Part F. Wash the precipitate with 5 drops of 6 *M* NaOH, centrifuge, and combine the wash with the supernatant.

2. Dissolve the precipitate[#5] with 1–2 mL of conc HNO_3 (**Caution**: *be careful!!*). A heating of the solution in the hot water bath may be necessary.

B. Separation of Ni^{2+}, Fe^{3+}, and Mn^{2+} from Zn^{2+} and Al^{3+}

Part F

1. Decant about 0.5 mL of the solution from Part B.2 into a 75 mm test tube. Add a pinch of solid $NaBiO_3$ to the solution, agitate, and centrifuge.[#6] The deep-purple MnO_4^- confirms the presence of manganese in the sample.[4]

C. Manganese(II) Ion Test

Part D

1. To the remaining solution from Part B.2, add 10 drops of 2 *M* NH_4Cl and then drops of conc NH_3 (**Caution**: *do not inhale—use a fume hood if available*) until the solution is basic to litmus and then a 2–3 drops excess to ensure the formation of $[Ni(NH_3)_6]^{2+}$. Centrifuge, save the precipitate,[#7] and transfer the supernatant[#8] to a 75 mm test tube for testing in Part E. Because $Fe(OH)_3$ is a bulky, gelatinous precipitate, additional centrifuge time may be necessary.

D. Iron(III) Ion Test

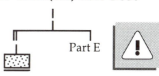

Part E

---

[4]If the deep-purple color forms and then fades, Cl⁻ is present and is being oxidized by the MnO_4^-; add more $NaBiO_3$.

2. Dissolve the precipitate with 6 M HCl and add 2 drops of 0.1 M NH$_4$SCN.[#9] The blood-red color confirms the presence of iron in the sample.

E. Nickel(II) Ion Test

1. To the supernatant solution from Part D.1, add 3 drops of dimethylglyoxime solution.[#10] The appearance of a pink-red precipitate confirms the presence of nickel in the sample.

F. Aluminum Ion Test

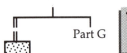

Part G

1. Acidify the supernatant from Part B.1 to litmus with 6 M HNO$_3$. Add drops of 6 M NH$_3$ until the solution is now basic to litmus; then add 5 more drops. Heat the solution in the boiling water bath for several minutes to digest the precipitate.[#11] Centrifuge and decant the supernatant[#12] into a 75 mm test tube and save for the Zn^{2+} analysis in Part G.

2. Wash the precipitate twice with 2–3 mL of hot water and discard each washing. Centrifugation may be necessary. Dissolve the precipitate with drops of 6 M HNO$_3$. Add 2 drops of the aluminon reagent, stir, and add drops of 6 M NH$_3$ until the solution is again basic to litmus and the gelatinous Al(OH)$_3$ precipitate reforms.[#13] Centrifuge the solution; if the precipitate is now pink or red and the solution is colorless, Al^{3+} is present in the sample.

G. Zinc Ion Test

1. To the supernatant from Part F.1, add 6 M HCl until the solution is acid to litmus; then add 3 drops of 0.2 M K$_4$Fe(CN)$_6$ and stir. A *very light green (blue-green) precipitate*[#14] confirms the presence of Zn^{2+} in the sample. Centrifugation may be necessary.

H. Unknown

1. Ask your laboratory instructor for a test solution containing one or more of the cations from this experiment and determine those present. Record the number of the test solution on the Data Sheet.

Dispose of the test solutions in the "Waste Salts" container.

NOTES AND OBSERVATIONS

◇ Cation Identification, II.
Mn^{2+}, Ni^{2+}, Fe^{3+} (and Fe^{2+}), Al^{3+}, Zn^{2+}

Date _____ Name _____ Lab Sec. _____ Desk No. _____

1. Read the Preface to Cation Identification and then describe the procedure for washing a precipitate.

2. Give the formula of a reagent that precipitates

   a. Ag^+ but not Ni^{2+} _____

   b. Mn^{2+} but not Zn^{2+} _____

   c. Bi^{3+} but not Al^{3+} _____

3. Give the formula of a reagent that dissolves

   a. $NiCl_2$ but not NiS _____

   b. FeS but not NiS _____

   c. $Al(OH)_3$ but not $Mn(OH)_2$ _____

4. $Al(OH)_3$ and $Zn(OH)_2$ are amphoteric. What does the word amphoteric mean?

5. What is the color of

   a. $[Ni(NH_3)_6]^{2+}$ _____

   b. $[Fe(NCS)_6]^{3-}$ _____

   c. $Ni(HDMG)_2$ _____

   d. MnO_4^- _____

6. Write balanced equations for the following reactions cited in this experiment. Read the Principles and Procedure to assist you in writing the equations. Also read your textbook as a resource for information.

a. The precipitation of the five cations in a sulfide solution with a pH near neutrality.

b. The dissolving of FeS, ZnS, MnS, and $Al(OH)_3$ with a strong acid.

c. The reaction of the five cations with an excess of strong base.

d. The reactions of Fe^{3+} and Ni^{2+} with NH_3.

e. The reactions of Al^{3+} and Zn^{2+} with NH_3.

◇ Cation Identification, II.
Mn^{2+}, Ni^{2+}, Fe^{3+} (and Fe^{2+}), Al^{3+}, Zn^{2+}

Date _____ Name _____ Lab Sec. _____ Desk No. _____

Procedure Number and Ion	Test Reagent or Technique		Observation (Color or General Appearance)	Chemicals Responsible for Observation	Check (√) if Observed for Test Solution
#1		ppt			
#2		spnt			
#3		ppt			
#4		spnt			
#5		spnt			
#6 Mn^{2+}		spnt			
#7 $Fe^{3+(2+)}$		ppt			
#8		spnt			
#9		spnt			
#10 Ni^{2+}		ppt			
#11 Al^{3+}		ppt			
#12		spnt			
#13		ppt			
#14 Zn^{2+}		ppt			

Unknown number: _____

Cations present in unknown: _____

Instructor's approval: _____

QUESTIONS

1.   What is the reducing agent that reduces Fe^{3+} to Fe^{2+} in Part A.2?

2. Nitric acid functions as an oxidizing agent and as an acid in Parts A.3 and A.4.

   a. Identify the two substances that HNO_3 oxidizes.

   b. How does it function only as an acid?

3. What ions would precipitate from solution in Part B.1 if NH_3 is substituted for NaOH? Explain.

4. What would happen in Part D.1 if NaOH is substituted for NH_3?

5. What changes in chemical reactions would be observed in Part F.1 if NaOH is substituted for NH_3?

6. Identify a reagent(s) that will

   a. precipitate Cu^{2+} but not Ni^{2+} _____

   b. dissolve $Al(OH)_3$ but not $Fe(OH)_3$ _____

   c. precipitate Fe^{3+} but not Ni^{2+} _____

   d. dissolve ZnS but not NiS _____

Experiment 33

◇ Cation
Identification, III.
Mg²⁺, Ca²⁺, Ba²⁺

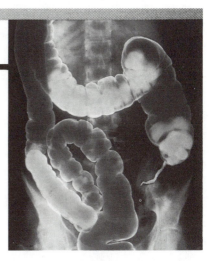

Calcium is the most abundant metallic element in the human body, present for the most part in bones and teeth generally as the phosphate salt known as an apatite. Calcium carbonate is the most common compound of calcium, being the principal component of limestone, stalactites, stalagmites, marble, and a few antacids. Calcium and magnesium ions are the major components of "hard" water. Magnesium is also a major component of bones and teeth, but it also assists in the transfer of electrical pulses between the cells. Magnesium is the trace metal in chlorophyll that is responsible for photosynthesis in plants. Barium sulfate is used as a drilling mud in oil exploration and for assisting in the x-ray analysis of the gastrointestinal tract (see photo).

OBJECTIVE

• To separate and identify the presence of Mg^{2+}, Ca^{2+}, and Ba^{2+} ions in a composite sample

PRINCIPLES

The chloride and sulfide salts (Experiments 31 and 32) of Mg^{2+}, Ca^{2+}, and Ba^{2+} ions are soluble. Many cations emit characteristic colors after being excited in a Bunsen flame. A "flame test" (Figure 33.1) is of assistance in our identification process of several metallic cations. When more than one cation that exhibits a positive flame test is present in the mixture, conflicting flame data renders flame tests inconclusive.

To separate and identify the alkaline-earth metal ions, Mg^{2+}, Ca^{2+}, and Ba^{2+}, the anions $NH_4PO_4^{2-}$, $C_2O_4^{2-}$, and SO_4^{2-} are used. The equations describing the separation and identification of each cation are to be written in the Lab Preview. The $NH_4PO_4^{2-}$ ion forms from the HPO_4^{2-} ion in an ammoniacal solution. The tests are reasonably straight-forward, but some care in pH adjustments is necessary.

A flow diagram for the analysis is to also be completed for the Lab Preview. Be sure to complete this prior to entering the laboratory.

Figure 33.1
A flame test is often used to determine the presence of a cation

PROCEDURE

To become familiar with the identification of these cations, take a sample that contains the three cations and analyze it according to the procedure. At each numbered superscript (i.e., #¹), STOP, and record on the Data Sheet. After the presence of each cation has been confirmed, SAVE the test tube so that its appearance can be compared to that for your unknown sample.

Caution: *Several concentrated and 6 molar acids and bases are used in the analysis of these cations. Handle each of these solutions with care. Read the Lab Safety section in Experiment 1 for instructions in handling acids and bases.*

Dispose of all washings and discarded solutions into a personal waste beaker. After completing the experiment discard the contents of the waste beaker into the "Waste Salts" container.

A. Flame Tests

1. Concentrate 2 mL of a sample solution of the three cations or the supernatant solution from Experiment 32, Part A.2 to a moist residue in an evaporating dish.

2. Clean a coiled platinum wire, sealed in an insulated handle, by heating it with the nonluminous flame of a Bunsen burner until the flame is colorless (Figure 33.2). Dip the platinum wire into the moist residue and return it to the flame. Note the color.[#1] Note that if your sample contains several cations, a mixture of colors in the flame will result and therefore may prove inconclusive.

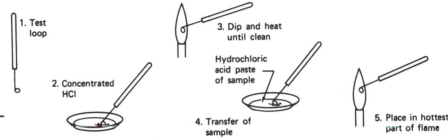

1. Test loop

2. Concentrated HCl

3. Dip and heat until clean

Hydrochloric acid paste of sample

4. Transfer of sample

5. Place in hottest part of flame

Figure 33.2
Technique for performing a flame test

B. Barium Ion Test

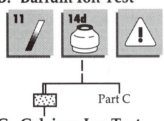

Part C

1. Dissolve the moist residue from Part A.1 with 1 mL of water. Acidify the solution to litmus with 6 M HCl (**Caution!**) and transfer to a 75 mm test tube. Add drops of 1 M K_2SO_4 until any precipitation appears complete.[#2] Centrifuge and save the supernatant for Part C.

2. Dissolve any precipitate with conc HCl (**Caution:** *avoid skin contact; avoid inhalation; clean up any spills.*) and perform a flame test.[#3]

C. Calcium Ion Test

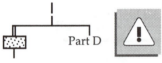

Part D

1. Adjust the supernatant from Part B.1 to be slightly basic to litmus with 6 M NH$_3$ (**Caution!**). Add 2–3 drops of 1 M $K_2C_2O_4$.[#4] Centrifuge and save the supernatant for Part D.

2. Dissolve any precipitate with 6 M HCl and perform a flame test.[#5]

D. Magnesium Ion Test

1. Add 1–2 drops of 6 M NH$_3$ to the supernatant from Part C.1. Add 2–3 drops of 1 M Na$_2$HPO$_4$, heat in a hot water (≈90°C) bath, and allow to stand. Any precipitate[#6] may be slow in forming; be patient.

2. Dissolve any precipitate with 6 M HCl and perform a flame test.[#7]

E. Unknown

1. Obtain a test solution for the cations in this experiment. Analyze it according to this procedure and report your findings to your laboratory instructor. Record the number of the test solution on the Data Sheet.

Dispose of the test solutions in the "Waste Salts" container.

NOTES AND OBSERVATIONS

◇ Cation Identification, III. Mg^{2+}, Ca^{2+}, Ba^{2+}

Date _____ Name _____ Lab Sec. _____ Desk No. _____

1. Record the K_s values (if they are considered insoluble salts) of the following salts. Use your textbook or other reference book, such as the CRC, *Handbook of Chemistry and Physics*.

K_s	Mg^{2+}	Ca^{2+}	Ba^{2+}
SO_4^{2-}			
$C_2O_4^{2-}$			
$NH_4PO_4^{2-}$			

2. Write balanced equations for the precipitation reactions of

   a. Mg^{2+} and $NH_4PO_4^{2-}$

   b. Ca^{2+} and $C_2O_4^{2-}$

   c. Ba^{2+} and SO_4^{2-}

3. Explain the difficulty that may arise if a flame test were used to identify a single cation in a mixture containing several cations.

4. Construct a flow diagram for the three cations in this experiment. Read the Principles and Procedure to help you in its construction.

◇ Cation Identification, III. Mg^{2+}, Ca^{2+}, Ba^{2+}

Date _____ Name _____ Lab Sec. _____ Desk No. _____

Procedure Number and Ion	Test Reagent or Technique		Observation (Color or General Appearance)	Chemicals Responsible for Observation	Check (√) if Observed for Test Solution
#1	flame				
#2 Ba^{2+}		ppt			
#3	flame				
#4 Ca^{2+}		ppt			
#5	flame				
#6 Mg^{2+}		ppt			
#7	flame				

Unknown number _____

Cations present in unknown: _____

Instructor's approval: _____

QUESTIONS

1. What is the color of the flame test for

   a. Ba^{2+} _____

   b. Ca^{2+} _____

   c. Mg^{2+} _____

2. What is the color of each salt?

   a. $MgNH_4PO_4$ _____

   b. CaC_2O_4 _____

   c. $BaSO_4$ _____

3. Write the formula of a reagent that separates

    a. Cu^{2+} from Ba^{2+} _____

    b. Ba^{2+} from Ca^{2+} _____

    c. Ba^{2+} from Mg^{2+} _____

    d. Mg^{2+} from Ca^{2+} _____

4. What will be observed if H_2SO_4 is used instead of HCl in Part B.2? Refer to the solubility rules in your textbook.

5. What happens if NaOH is used instead of NH_3 in Part D.1? (Use the solubility rules in your textbook to support your statement.)

Experiment 34

◇ Coordination Compounds

The Ragu Food Company of Rochester, NY produces a number of "ready-to-use" sauces for the quick preparation of chicken dinners. The sauces must have a relatively long shelf life; from the time the sauce is prepared until it is used may span several months without refrigeration. Various components of the sauce tend to promote the degradation (spoilage) of the food products; among those are the various trace metal ions that enter the food during the various stages of the preparation of the sauce, from the pans, ovens, kettles, etc. To guard against spoilage, Ragu adds a trace of calcium disodium ethylenediaminetetraacetate, $CaNa_2EDTA$. The EDTA is an anion that binds many of the trace metal ions, such as zinc, aluminum, and iron, into a coordination compound; the metal is bound so that it cannot act as a catalyst in the air oxidation of various food components leading to spoilage. Consequently, EDTA is also used in the processing of meats, fish products, dairy products, and vegetables.

The EDTA anion has the structure shown at right. It has six *sites* (designated with asterisks) which can donate an electron pair to a metal ion. Consequently, it has the effect of enveloping the metal ion so that it can offer no other chance of reaction. EDTA is also used as an antidote for metal poisoning, to bind hardening ions from water, and, in analytical chemistry, can be used for the analysis of nearly every metal.

OBJECTIVES

- To prepare several complex ions and compare their stabilities
- To synthesize an inorganic coordination compound
- To determine the purity and stability of the prepared compound

PRINCIPLES

A coordination compound consists of at least one complex ion; a complex ion consists of a metal ion, usually a transition metal ion, bonded to one or more Lewis bases. These bases, called **ligands**, generally form 2, 4, or 6 bonds to the metal ion; the number of bonds between the metal ion and the ligands is the **coordination number** of the metal ion. Most ligands donate only a single electron-pair to the metal ion, but others may donate as many as 2, 3, or, more rarely, 4 or 6.

In the complex ion, $[Ag(NH_3)_2]^+$, the two ammonia molecules are the ligands, and the lone electron pair on each nitrogen bonds to the Ag(I) ion. The structure is linear.

$$[H_3N : Ag : NH_3]^+$$

The iron(III) ion in the complex ion, $[Fe(CN)_6]^{3-}$, has a coordination number of 6; the complex ion forms an octahedral structure, with the electron-pair donor site on the C atom of the $:C \equiv N:^-$ ligand. The $[Fe(CN)_6]^{3-}$ complex ion (Figure 34.1) carries a 3^- charge and each CN^- ligand has a 1^- charge; the Fe in the complex ion is therefore Fe(III). Since each $:C \equiv N:^-$ is a monodentate ligand, the coordination number of the iron(III) ion is six.

Figure 34.1

The structure of the $[Fe(CN)_6]^{3-}$ complex ion

In this experiment, you will prepare a number of Cu(II), Ni(II), and Co(II) complex ions and compare their stability using a number of different ligands. You will then synthesize the coordination compound [Co(NH₃)₄CO₃]NO₃ (Figure 34.2). Four :NH₃ molecules and one CO₃²⁻ ion form bonds to Co(III). Two oxygen atoms on the carbonate ion serve as Lewis bases to the Co(III) ion; note again that the coordination number of the cobalt(III) ion is six.

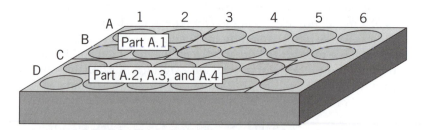

The complex ion has a 1⁺ charge and the NO_3^- serves as the neutralizing anion to the complex cation.

The coordination compound, $[Co(NH_3)_4CO_3]NO_3$, dissolves in water to produce two ions, $[Co(NH_3)_4CO_3]^+$ and NO_3^-, like that of $NaNO_3$.

Figure 34.2

The structure of the $[Co(NH_3)_4CO_3]^+$ complex ion

$$[Co(NH_3)_4CO_3]NO_3(aq) \xrightarrow{H_2O} [Co(NH_3)_4CO_3]^+(aq) + NO_3^-(aq)$$
$$NaNO_3(aq) \xrightarrow{H_2O} Na^+(aq) + NO_3^-(aq)$$

Addition of Ca^{2+} to the $[Co(NH_3)_4CO_3]NO_3$ solution yields *no* CaCO₃ precipitate because the CO_3^{2-} ion remains bound to the cobalt(III) ion in the complex ion. How do you suppose the presence of NH₃ ligands affects the pH of the solution? We'll find out.

PROCEDURE

A. A Look at Some Complex Ions

1. **Chloro Complexes.** Set up wells A1–A3 and B1–B3 in a 24 well plate (Figure 34.3). Place 10 drops of 0.1 *M* CuSO₄ in wells A1 and B1. Repeat with 0.1 *M* Ni(NO₃)₂ in wells A2 and B2, and 0.1 *M* CoCl₂ wells A3 and B3. Add 1 mL (≈20 drops) of conc HCl (**Caution:** *handle carefully, do not allow conc HCl to contact skin or clothing. Flush immediately with water.*) to wells B1–B3. Swirl the solutions. Compare the color of the two 1 x 3 arrays, the original and the one with the added HCl. Record your observations on the Data Sheet.

Slowly add drops of deionized water (totally ≈1 mL) to wells B1–B3. Compare the solution colors to those of the original solution. Does the original color return? What does this tell you about the stability of the chloro complex ions?

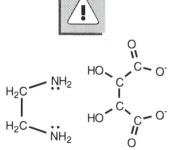

Figure 34.3

Arrangement of wells for observation of chemical change

2. **Copper and Cobalt Complex Ions.** Place 10 drops of 0.1 *M* CuSO₄ in wells C1–C5. In wells D1–D5, place 10 drops of 0.1 *M* CoCl₂.

3. Place 5 drops of conc NH₃ (**Caution:** *avoid breathing its vapors*) in wells C1 and D1, 5 drops of ethylenediamine—abbreviated, en (**Caution:** *avoid breathing its vapors*)—in wells C2 and D2, 5 drops of 0.2 *M* ammonium tartrate, (NH₄)₂C₄H₄O₆ in wells C3 and D3, and 5 drops of 0.1 *M* KSCN in wells C4 and D4. If a precipitate forms in any one of the solutions, add an excess of the ligand-containing solution. Compare the solutions with those in wells C5 and D5.

ethylenediamine   tartrate ion

4. Add 3 drops of 1 *M* NaOH to wells C1–C4 and to wells D1–D4. Observe the changes closely as the NaOH is added. Try to account for your observations.

Dispose of the test solutions in the "Waste Salts" container. Rinse the 24 well plate with deionized water and discard in the "Waste Salts" container.

B. Preparation of [Co(NH$_3$)$_4$CO$_3$]NO$_3$

1. Dissolve about 2 g (±0.01 g) of (NH$_4$)$_2$CO$_3$ in 6 mL of deionized, boiled H$_2$O. Cautiously add 6 mL of conc NH$_3$. (**Caution !**)[1]

2. Dissolve 1.6 g (±0.01 g) of Co(NO$_3$)$_2$•6H$_2$O in 3 mL of H$_2$O contained in a 50 mL beaker. While stirring, add the (NH$_4$)$_2$CO$_3$ from Part B.1. Cool the mixture to 10°C in an ice bath and slowly add 1 mL (20 drops) of 30% H$_2$O$_2$ (**Caution**: 30% H$_2$O$_2$ *causes severe skin burns in the form of white blotches; wash the affected skin immediately with water*) to this solution.

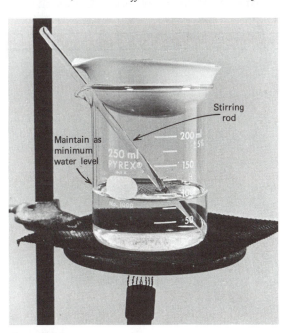

Figure 34.4
The steam bath is used for reducing the volume of the reaction mixture

Figure 34.5
The apparatus for testing the stability of the complex ion

3. Transfer the solution to an evaporating dish; using a steam bath (Figure 34.4) reduce the total volume by about one-half. During the evaporation, periodically add a total of 0.5 g (±0.01 g) of (NH$_4$)$_2$CO$_3$ in small amounts. Clean a vacuum filter flask and then vacuum filter the *hot* solution. Cool the **filtrate** in an ice bath.

Filtrate: the solution that passes through the funnel.

4. Red crystals should form as the filtrate cools. Vacuum filter the crystals, wash the crystals *at least* three times with 3–5 mL of methanol. Air-dry the crystals. Determine the mass of product and calculate the percent yield.

---

[1]If a fume hood is available, use it while transferring the conc NH$_3$.

C. Testing the [Co(NH₃)₄CO₃]⁺ Ion

Convex: the outward curvature of a piece of glassware.

1. **Litmus test and CaCl₂ test.** Dissolve a small pinch of the coordination compound in H_2O on a watchglass. Test the solution with red and blue litmus paper. Record your observation. Add several drops of 0.1 M CaCl₂ solution. Record your observation.

2. **Na₂CO₃ test.** Most carbonate salts are insoluble, including those of Co^{2+} and Co^{3+}. To an equal-size sample dissolved in H_2O in a 75 mm test tube, add several drops of 0.1 M Na₂CO₃ solution. Centrifuge and record your observations. What can you conclude about the stability of the $[Co(NH_3)_3CO_3]^+$ complex?

3. **Stability test.** Place a pinch of the compound in an evaporating dish and add 5 mL of water. Add 20 drops (1 mL) of 1 M HCl. Adhere a piece of red litmus paper to the **convex** side of a watchglass (Figure 34.5). Now add 2 mL of 0.05 M NaOH to the solution and cover with the watchglass, convex side down. Gently heat.[2] What do you observe? Comment on the stability of the $[Co(NH_3)_4CO_3]^+$ ion.

> **Dispose of the test solutions in the "Waste Salts" container.**

NOTES, OBSERVATIONS, AND CALCULATIONS

---

[2]See Experiment 31: this is test for the presence of NH_4^+, converted to $NH_3(g)$.

Coordination Compounds

Date _____ Name _____ Lab Sec. _____ Desk No. _____

1. a. What is a complex ion?

    b. What is a ligand?

    c. What is meant by the statement, "the coordination number of the cobalt(II) ion is usually six" in a complex ion?

2. Write the formula of the complex ion formed between

    a. the ligand, Cl⁻, and copper(II) ion, assuming a coordination number of four.

    b. the ligand, H_2O, and nickel(II) ion, assuming a coordination number of six.

3. Identify the ligand(s) in the following coordination compounds.

   a. $[Pt(NH_3)_2Cl_2]$ _____

   b. $[Cu(NH_3)_4]SO_4$ _____

   c. $[Cr(NH_3)_4(H_2O)_2]Cl_3$ _____

4. What is the coordination number of the metal ion in each complex ion in Question 3?

   a. _____; b. _____; c. _____

5. Tartaric acid is an acid found in grapes. Write the structural formula of tartaric acid.

6. The CO_3^{2-} that is a ligand in the $[Co(NH_3)_4CO_3]^+$ complex ion is called a bidentate. Refer to your text and define bidentate.

7. A 1.6 g sample of $Co(NO_3)_2 \cdot 6H_2O$ is used to prepare the complex, $[Co(NH_3)_4CO_3]NO_3$. Assuming that all "other" reactants used in the synthesis of the complex are in excess, calculate the theoretical yield of the complex.

Coordination Compounds

Date _____ Name _____ Lab Sec. _____ Desk No. _____

A. A Look at Some Complex Ions

Solution	Color/H_2O	Color/HCl	Formula of Complex Ion	Effect of H_2O
0.1 M $CuSO_4$				
0.1 M $Ni(NO_3)_2$				
0.1 M $CoCl_2$				

Cu^{2+} and ligand:	NH_3	en	tar^{2-}	SCN$^-$	H_2O
color					
formula					
effect of OH$^-$					

Co^{2+} and ligand:	NH_3	en	tar^{2-}	SCN$^-$	H_2O
color					
formula					
effect of OH$^-$					

B. Preparation of $[Co(NH_3)_4CO_3]NO_3$

1. Mass of $Co(NO_3)_2 \cdot 6H_2O$ (g) _____

2. Mass of $[Co(NH_3)_4CO_3]NO_3$ (g) _____

3. Theoretical yield of $[Co(NH_3)_4CO_3]NO_3$ based on $Co(NO_3)_2 \cdot 6H_2O$ (g)* _____

4. Percent Yield (%) _____

C. Testing the $[Co(NH_3)_4CO_3]^+$ ion

Test	Observation	Conclusion
Litmus test		
$CaCl_2$ test		
Na_2CO_3 test		
Stability		

QUESTIONS

1. a. Compare the relative stability of the chloro complex ions of Cu^{2+}, Ni^{2+}, and Co^{2+}.

   b. Compare the relative stability of the NH_3 and ethylenediamine complex ions of Co^{2+}.

2. a. What is the purpose of the 30% H_2O_2 in Part B.2 of the experimental procedure?

   b. Write an equation for the reaction. Ask your instructor for assistance.

3. In the stability test (Part C.3), why was it first necessary to add 1 M HCl?

4. For the same concentration as $[Co(NH_3)_4CO_3]NO_3$, which of the following salts show the same conductivity when dissolved in water? Explain your answer.

   a. $Co(NO_3)_3$

   b. KCl

   c. Na_2CO_3

   d. $NaNO_3$

Experiment 35

Organic Compounds

Enzymes, leaves on trees, sharks, humans, pesticides, carpeting, plastics and synthetic fibers . . . it seems as though most matter with which we are familiar is constructed of compounds we call **organic**. The shoes we wear, the clothes we buy, the furniture in our homes, the sex pheromones of insects, the smell of an orange blossom (see photo), and the gourmet hamburger are made of organic compounds. The breadth, depth, and versatility of the element carbon, an element of all organic compounds, extend far beyond that of any other element in the Periodic Table. Organic chemistry alone is a specialized field of study that also includes biochemistry and many subdisciplines of biology.

But what is organic gardening? Virtually all of the compounds in plants and vegetables are organic, but it is the absence of synthetic organic pesticides in the growth of the garden vegetables that label the gardening practice "organic".

OBJECTIVE

- To study the chemical behavior of hydrocarbons, alcohols, organic acids and bases, and esters

PRINCIPLES

Carbon has the unusual property of bonding to itself. Although other atoms of the elements do this, carbon does so much more extensively. Because of this unique property, over 5 million **organic compounds** have been reported in the literature. As a result, a complete knowledge of the chemical characteristics of each organic compound would indeed be impossible to acquire. The complexity is lessened somewhat by the characteristic chemical behavior of a large segment of compounds. Organic compounds group into natural classes because members of each class possess similar physical and chemical properties. We will study the chemical properties of five of these classes: the hydrocarbons, alcohols, the organic acids and bases, and the esters.

Hydrocarbons

Hydrocarbons are compounds that contain only the elements carbon and hydrogen. According to their chemical properties, the hydrocarbons are subdivided into three subgroups: the saturated hydrocarbons, the unsaturated hydrocarbons, and the aromatic hydrocarbons.

Inert: substances that react chemically only under extreme or specified conditions.
Combustion: any reaction where oxygen and another substance are the only reactants.

The **saturated hydrocarbons** are commonly referred to as alkanes. All C–C and C–H bonds are single bonds in the alkanes and are relatively **inert** to chemical attack. By far, their most significant reaction is **combustion** (e.g., propane C_3H_8), forming CO_2 and H_2O as products.

$$C_3H_8 + 5\,O_2 \rightarrow 3\,CO_2 + 4\,H_2O$$

The alkanes slowly react with chlorine and bromine whereby a halogen atom substitutes for a hydrogen atom.

$$C_3H_8 + Br_2 \rightarrow C_3H_7Br + HBr$$

Figure 35.1

A propane torch is a handy, portable source of intense heat

H₃C–C(H)=CH₂

propene

benzene

The **unsaturated hydrocarbons** have two subgroups: the alkenes and the alkynes. The alkenes have *at least* one C=C bond and the alkynes have at least one C≡C bond in the compound. These double and triple bonds, called unsaturated bonds, are chemically quite reactive. For example, bromine quickly reacts across the C=C bond in propene, C_3H_6, to form 1,2-dibromopropane.

$$CH_3–CH=CH_2 + Br_2 \rightarrow CH_3–CHBr–CHBr$$

The unsaturated hydrocarbons also react with oxidizing agents, such as $KMnO_4$, to produce alcohols (Baeyer's test). The $KMnO_4$, a purple reagent, is reduced to MnO_2, a brown precipitate, in the reaction. The brown precipitate may appear as a red-brown solution.

$$3\,CH_2=CH_2(l) + 2\,MnO_4^-(aq) + 4\,H_2O(l) \rightarrow$$
(colorless) **(purple)**
$$3\,CHOH–CHOH(aq) + 2\,MnO_2(s) + 2\,OH^-(aq)$$
(colorless) **(brown)**

Aromatic hydrocarbons are characterized by the presence of the six-membered carbon ring called benzene, C_6H_6. Although somewhat resistant to chemical attack, aromatic compounds are usually more reactive than the alkanes, but far less reactive than the alkenes or alkynes. For example, bromine reacts very slowly with the alkanes but readily displaces hydrogen from the benzene ring with a catalyst.

Many organic compounds are substituted hydrocarbons where an atom or group of atoms displace a hydrogen or carbon atom. The atom or group of atoms is commonly referred to as a **functional group**, which imparts characteristic properties to the compound.

Alcohols

Primary alcohol: the –OH group is attached to a carbon atom that is bonded only to one carbon atom or hydrogen atom.

Secondary alcohol: the –OH group is attached to a carbon atom that is bonded to two other carbon atoms.

Tertiary alcohol: the –OH group is attached to a carbon atom that is bonded to three other carbon atoms.

An **alcohol** is a hydrocarbon in which an –OH group (called a **hydroxyl** group) replaces a hydrogen atom in a hydrocarbon. An alcohol is also like water in that an alkyl group[1] replaces a hydrogen atom in the water molecule. Therefore, alcohols have physical properties that are intermediate between those of hydrocarbons and water.

Depending upon which carbon atom the –OH group is attached in the hydrocarbon, alcohols are classified as 1°, 2°, or 3°. Examples include

- $CH_3–CH_2–OH$, a **1° alcohol**, called ethanol (grain alcohol)
- $CH_3–CHOH–CH_3$, a **2° alcohol**, called isopropanol (rubbing alcohol)
- $CH_3–COH(CH_3)–CH_3$, a **3° alcohol**, called tertiary butanol

---

[1]An alkyl group is simply a hydrocarbon from which a hydrogen atom has been removed; for example, the ethyl group, $C_2H_5–$, results when a hydrogen atom is removed from ethane, C_2H_6.

Each alcohol reacts differently with a mild oxidizing agent, such as $K_2Cr_2O_7$. Only 1° and 2° alcohols react to form aldehydes and ketones respectively, as the primary product. A color change from a brilliant orange, due to $Cr_2O_7^{2-}$, to green, due to Cr^{3+}, and/or a change in odor signifies the chemical reaction.

The iodoform test distinguishes between the 1°, 2°, and 3° alcohols. The test is positive when iodine in the presence of a base, such as NaOH, oxidizes the alcohol, producing the acid (with one less carbon atom) and iodoform, CHI_3. A very characteristic odor and a light yellow precipitate confirms the reaction.

Organic **acids** are hydrocarbons in which a –COOH group (called a **carboxyl** group) substitutes for a –CH_3 group of a hydrocarbon. The acids readily react with $NaHCO_3$ releasing CO_2 gas.

$$R–COOH(aq) + Na^+HCO_3^-(aq) \rightarrow R–COO^-Na^+(aq) + H_2O(l) + CO_2(g)$$

Organic acids can easily be prepared by the oxidation of an alcohol with a strong oxidizing agent, such as $KMnO_4$.

Organic **bases** are hydrocarbons in which an –NH_2 group (called an **amine** group) substitutes for a hydrogen atom. The lone electron pair on the nitrogen atom serves as a Lewis base. As a Lewis base, an amine can be a **ligand** in a complex ion. A pH test of an aqueous solution of many organic compounds shows that most are neutral; however, the acids have a low pH and the bases have a high pH. Organic bases tend to have a "fishy" odor as well.

An organic **ester** is the product of the reaction between an organic acid and an alcohol. Esters tend to be volatile and have a pleasant odor. The natural scents of many flowers and flavors of many fruits are due to one or more esters. The presence of the R–COO–R′ bond structure is present in all esters. Table 35.1 list some common esters along with their flavor or aroma.

Important natural esters are fats (e.g., butter, lard, and tallow) and oils (e.g., linseed, cottonseed, peanut, and olive) used in the synthesis of oleomargarine, peanut butter, and vegetable shortening.

Table 35.1 Chemical Formulas for Natural Flavors and Aromas

Formula	Flavor/Aroma
$C_3H_7–COO–C_2H_5$	pineapple
$C_2H_5–COO–C_5H_{11}$	apricot
$CH_3–COO–C_8H_{15}$	orange
$H_2N–C_6H_4–COO\ CH_3$	grape
$CH_3–COO–C_5H_{11}$	banana
$H–COO–C_2H_5$	rum

Caution: *Many organic compounds are volatile and flammable. Therefore, do not have any open flames away from the hood or in designated "flame" areas.*

Obtain a test sample from your instructor. Perform the following tests on the test sample at the same time you are testing a sample known to have the functional group being studied. Use this procedure to identify the class of organic compounds in your unknown.

Acids and Bases

R–: represents a bonded H atom or any organic group.

Ligand: a Lewis base that bonds to a metal ion for the formation of a complex ion.

Esters

PROCEDURE

The sequence for the completion of Parts A through E can be in any desired order.

A. Chemical Reactivity of Hydrocarbons

A number of suggested organic compounds, listed on the Data Sheet, are to be tested for unsaturation. Unsaturated compounds show a positive test with both Br_2 and $KMnO_4$. On occasion however, a test is positive even though the compound has no double or triple bonds. Refer to your text for the structural formulas and note the exceptions on the Data Sheet.

1. **Br_2 test**. In a 150 mm test tube, add 4–5 drops of liquid or dissolve 0.1 g of solid organic compound in 2 mL of CH_2Cl_2 (**Caution**: *avoid breathing CH_2Cl_2 vapors—use a fume hood if available*)[2]. Add drops of 5% Br_2/CH_2Cl_2 (**Caution**: *avoid contact with the Br_2/CH_2Cl_2. Br_2 can cause skin burns.*) and agitate after each drop.

   An immediate disappearance of the bromine color indicates the presence of an alkene. If the color of bromine slowly fades, add drops (5 drops maximum) and set the reaction mixture aside for the remainder of the laboratory period and make an occasional observation. The slow color fade results from a substitution reaction rather than an addition reaction. Observe and record.

2. **$KMnO_4$ test**. In a 150 mm test tube, add 4–5 drops of liquid or dissolve 0.1 g of solid organic compound in 2 mL of H_2O or acetone. Add drops of 2% $KMnO_4$ (**Caution**: *$KMnO_4$ is a strong oxidizing agent; avoid skin contact*) and agitate. If a brown precipitate forms within 3 minutes, an alkene is likely present.

B. Chemical Reactivity of Alcohols

Four alcohols and your unknown are listed on the Data Sheet. Use these while completing the following procedure.

1. **Iodoform test.** a. Dissolve 10 drops of each alcohol in 5 mL of H_2O in separate 150 mm test tubes. Add 1 mL of 10% I_2/KI. While shaking, add drops of 10% NaOH (**Caution**: *avoid skin contact, a concentrated base solution removes layers of skin*) until the brown color of I_2 is destroyed, but leaving a slightly yellow solution. If the solution becomes colorless, add a drop (or two) of I_2/KI until the yellow color persists. Continue to add I_2/KI and NaOH as long as the brown color of I_2 is consumed, and the yellow solution persists.

   b. Look for the formation of iodoform, CHI_3, a yellow crystalline precipitate.

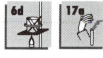

   c. If the iodoform does not form, then warm gently in a warm water bath, but do not exceed 60°C. Allow the test tube and contents to cool. Half-fill the test tube with deionized water and allow the mixture to stand for 10 minutes.

   d. Note the color and characteristic odor of iodoform.[3] Summarize the comparative results for the alcohols on the Data Sheet.

---

[2] 1-Octanol can be substituted for methylene chloride, CH_2Cl_2.

[3] Whenever the amount of alcohol is small, iodoform may not separate; however, the characteristic odor establishes its formation.

2. **Alcohol Oxidation.** In a 150 mm test tube, place 2 mL of 0.1 $MK_2Cr_2O_7$ and *slowly* add 1 mL of conc H_2SO_4 (**Caution**: *conc H_2SO_4 is a severe skin irritant and causes clothes to disappear; wash the affected area with large amounts of water*). Swirl to dissolve the $K_2Cr_2O_7$ and cool with tap water. Slowly add 2 mL of the test alcohol. Note any color change and odor. Compare its odor with the test alcohol.

C. Chemical Reactivity of Acids and Bases

1. Place 4–5 drops of each organic acid test solution listed on the Data Sheet in a 75 mm test tube. Test the pH of each sample with litmus paper.

   Repeat the litmus test with aniline. Test the odor of the aniline.

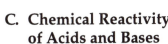

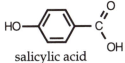
aniline

2. Place 4–5 drops of each acid listed on the Data Sheet in a 75 mm test tube. Add 1 mL of 10% $NaHCO_3$. A distinct "fizzing" sound is detectable even if the visual evolution of CO_2 is questionable. Record.

3. Place 2 mL of 0.1 M $KMnO_4$ in a 150 mm test tube. Slowly add 1 mL of ethanol. Watch for a color change. Compare its odor with that of acetic acid. What is your conclusion?

4. Add several drops of 1 M $CuSO_4$ to aniline. Compare the color of the $CuSO_4$ solution with the Cu^{2+}/aniline mixture.

D. Preparation of an Ester

1. **Banana Oil.** Place 2 mL of glacial CH_3COOH (**Caution**: *do not inhale*) and 3 mL of n-$C_5H_{11}OH$ (called pentanol) in a 150 mm test tube. Slowly and cautiously add 1 mL of conc H_2SO_4 (**Caution**: *conc H_2SO_4 causes severe skin burns*). Heat the mixture gently in a hot water bath using a hot plate or very carefully over a low flame in the hood. If the odor changes, the product is an ester. Have the instructor assist you in writing the balanced equation for the reaction.

or

2. **Oil of Wintergreen.** In a 75 mm test tube, place a pinch of salicylic acid, HO–C_6H_4–$COOH$. Add 1 drop of 3 M H_2SO_4 and 3 drops of water. After about 30 s, add 3–4 drops of methanol, CH_3OH. Place a loose cotton plug into the mouth of the test tube; place the test tube in a hot water bath (about 60°C) for 20–30 minutes. Note the odor. Have the instructor assist you in writing the balanced equation for the reaction.

salicylic acid

3. Have your instructor approve your successful synthesis of an ester.

Dispose of all organic test solutions in the "Waste Organics" container.

Functional Groups in Organic Compounds

Class of Compounds	Structure

alkane

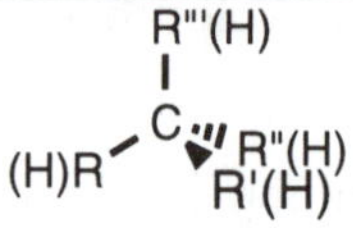

$$\begin{array}{c} R'''(H) \\ | \\ (H)R-C-R''(H) \\ | \\ R'(H) \end{array}$$

alkene

$$\begin{array}{c} (H)'R \qquad R'''(H) \\ \diagdown \qquad \diagup \\ C = C \\ \diagup \qquad \diagdown \\ (H)R \qquad R''(H) \end{array}$$

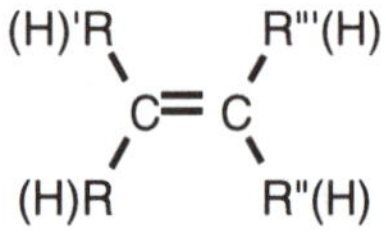

alkyne

$$(H)R - C \equiv C - R'(H)$$

aromatic

alcohol

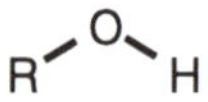

$$R - O - H$$

acid

$$\begin{array}{c} O \\ \| \\ (H)R - C - O - H \end{array}$$

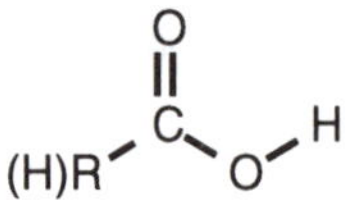

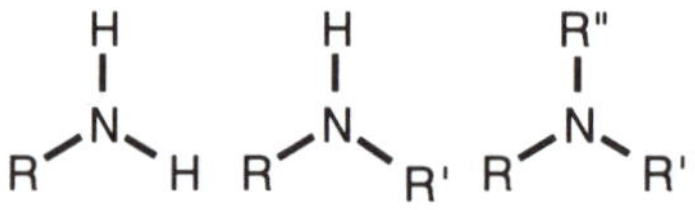

base

$$\begin{array}{ccc} H & H & R'' \\ | & | & | \\ R-N-H & R-N-R' & R-N-R' \\ 1°\ amine & 2°\ amine & 3°\ amine \end{array}$$

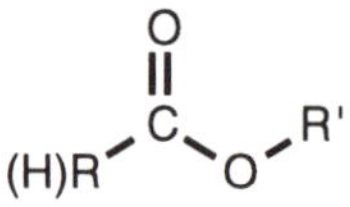

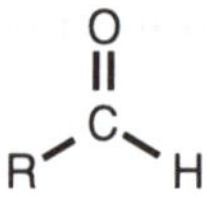

ester

$$\begin{array}{c} O \\ \| \\ (H)R - C - O - R' \end{array}$$

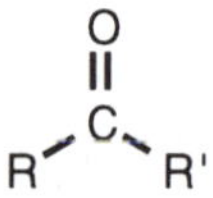

aldehyde

$$\begin{array}{c} O \\ \| \\ R - C - H \end{array}$$

ketone

$$\begin{array}{c} O \\ \| \\ R - C - R' \end{array}$$

amide

$$\begin{array}{c} O \\ \| \\ (H)R - C - N - R''(H) \\ | \\ R'(H) \end{array}$$

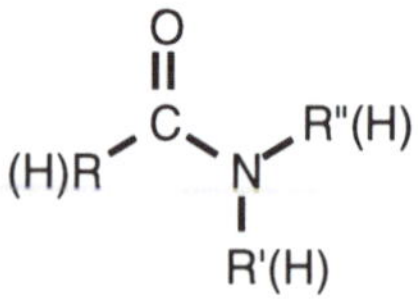

The feature R(H)— means that the group can be either H or a hydrocarbon (including aromatic) group. A prime, double prime, or triple prime means that the R groups may be alike or different.

◇ Organic Compounds

Date _____ Name _____ Lab Sec. _____ Desk No. _____

1. Distinguish between

    a. a saturated and an unsaturated hydrocarbon.

    b. an alkene and an alkyne.

    c. an alcohol and a hydrocarbon.

2. a. Write the chemical equation for the combustion of heptane, C_7H_{16}.

    b. Write the chemical equation for the reaction of Br_2 with $CH_3–CH=CH–CH_2Br$.

3. Describe a general procedure for preparing organic acids in the laboratory.

4. What is the principal product in the oxidation of ethanol with $KMnO_4$?

5. How does a chemist quickly distinguish an organic acid from an organic base?

◇ Organic Compounds

Date _____ Name _____ Lab Sec. _____ Desk No. _____

A. Chemical Reactivity of Hydrocarbons

Identify the tests as positive (+) or negative (-).

Compound	Structural Formula[1]	Br_2/CH_2Cl_2 (+ or -)	$KMnO_4$ (+ or -)
n-hexane or cyclohexane			
1-pentene or 2-pentene			
cyclohexene			
1-hexyne			
ethanol			
naphthalene			
p-xylene			
methane (see Figure 35.2)			
test solution	to be determined		

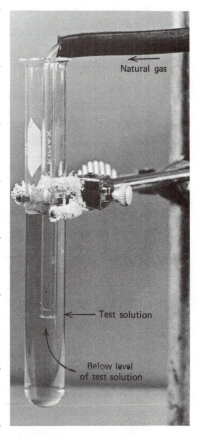

Figure 35.2
Apparatus for testing natural gas for unsaturation

Natural gas
Test solution
Below level of test solution

---

[1]Use your text to assist you in writing the structural formulas.

B. Chemical Reactivity of Alcohols

1. Iodoform test

Alcohol	Subclass 1°, 2°, 3°	Observation	Oxidation Product
methanol			
ethanol			
i-propanol			
t-butanol			
unknown			

Conclusion of observations on alcohols.

2. Oxidation of

Alcohol	Subclass 1°, 2°, 3°	Observation	Oxidation Product
methanol			
ethanol			
i-propanol			
t-butanol			
unknown			

C. Chemical Reactivity of Acids and Bases

1.

Name of Acid/Base	Results of Litmus Test	Check (√) for Test Solution
acetic acid		
propionic acid		
aniline		

2.

Name of Acid	Results of NaHCO$_3$ Test	Check (√) for Test Solution
acetic acid		
propionic acid		
benzoic acid		
phenol		

3. Oxidation of Ethanol

Observation _____

Conclusion _____

4. $CuSO_4$ with aniline

Observation _____

E. Preparation of an Ester

1. **Banana Oil**. Write an equation for the reaction.

2. **Oil of Wintergreen**. Write the formula for the compound.

3. Instructor's approval of ester synthesis: _____

QUESTIONS

1. How can you distinguish between the following compounds?

   a. CH_3NH_2 and CH_3OH

   b. C_2H_5OH and C_4H_9OH, a 3° alcohol

   c. C_3H_8 and C_3H_7OH

   d. $CH_3CH_2CH_3$ and $CH_3CH_2CH_2OH$

2. What reactants make the ester characteristic of the following flavors? (See Table 35.1)

   a. pineapple

   b. orange

   c. grape

3. Acetic acid is prepared by the oxidation of ethanol with permanganate ion in an acidic solution. Use the half-reaction method for writing a balanced (redox) equation for the reaction. The reduction product of the permanganate ion in an acidic solution is Mn(II).

Appendix A

◇ Glassworking

Chemists need a working knowledge of the simple manipulations involved in cutting, bending, and fire polishing glass. Custom-made glass rods and tubing are used for handling laboratory chemicals and for the construction of laboratory apparatus.

Hot glass and cold glass have the same appearance; therefore **never** place hot glass tubing on anything combustible and **never** touch glass that has recently been in a flame. Above all, do not hand hot glass to anyone, especially the laboratory instructor, unless you are trying to end an friendship (or not improve your grade).

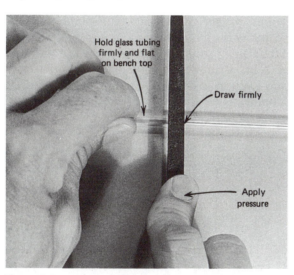

Figure A.1
Scratch the glass tubing with a triangular file

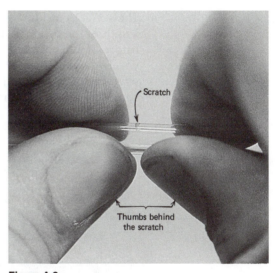

Figure A.2
Position the thumbs *behind* the scratch

The proper technique for cutting glass tubing or glass rod is to place a piece of the glass tubing or rod on the lab bench. Use a triangular file or glass scorer to make a deep scratch—draw the file *once* across the glass; this requires some pressure and should not be a sawing action (Figure A.1). Place a drop of water or glycerin on the scratch. Place the thumbs, about 1 cm apart, on each side and *behind* the scratch; hold the tubing at the waist with the scratch facing *outward* from the body (Figure A.2). Bend the tube backward as you pull it away from center (Figure A.3). With practice, the glass breaks evenly. If not, try again making the scratch a bit deeper.

Cutting Glass

Any cut piece of glass (tubing or rod) has a rough, sharp edge. Fire-polishing smoothes the edge to guard against cuts and the scratching of glassware. Hold and rotate the cut glass in the flame until the edge is

Fire-Polishing

fire-polished (Figure A.4). Do not overheat; if glass tubing is heated too much, the end closes.

Figure A.3
Pull and bend—*simultaneously*—to break the tubing

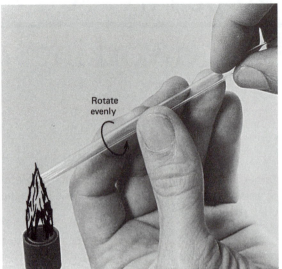

Figure A.4
Fire polish to *just* round the edges

Bending Glass

The secret to bending glass tubing properly is to heat it sufficiently over the length of the bend. Heat and rotate the glass tubing (cut and fire-polish as before) in the hottest part of the flame (using a wing tip on the burner) until you feel a slight sag (Figure A.5). Remove the softened glass, hold it for *several seconds* to give the temperatures inside and outside the tube time to equilibrate. Slowly and smoothly bend the glass upward to the desired angle (Figure A.6). A "good" bend has the same diameter throughout with no constriction at the bend; poor bends are shown in Figure A.7.

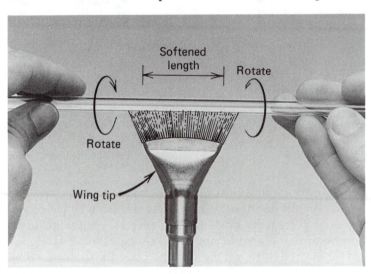

Figure A.5
Slow rotation in the hottest part of the flame

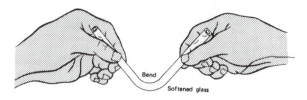

Figure A.6
Bend the softened glass tubing slowly upward

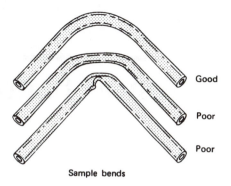

Sample bends

Figure A.7
Good and poor glass bends

Appendix B

◇ Conversion Factors

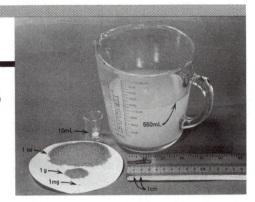

SI Prefixes

Prefix	Meaning (*Power of Ten*)	Abbreviation	Example using "grams"
femto-	10^{-15}	f	10^{-15} g = fg
pico-	10^{-12}	p	10^{-12} g = pg
nano-	10^{-9}	n	10^{-9} g = ng
micro-	10^{-6}	μ	10^{-6} g = μg
milli-	10^{-3}	m	10^{-3} g = mg
centi-	10^{-2}	c	10^{-2} g = cg
deci-	10^{-1}	d	10^{-1} g = dg
kilo-	10^{3}	k	10^{3} g = kg
mega-	10^{6}	M	10^{6} g = Mg
giga-	10^{9}	G	10^{9} g = Gg

SI and English Conversions

Physical Quantity	SI unit	Conversion Factors
Length	meter (*m*)	1 km = 0.6214 mi 1 m = 39.37 in 1 in = 0.02540 m = 2.540 cm
Volume	cubic meter (m^3) [liter (*L*)]	$1 \text{ L} = 10^{-3} \text{ m}^3 = 1 \text{ dm}^3 = 10^3 \text{ mL}$ 1 L = 1.057 qt 1 oz (fluid) = 29.57 mL
Mass	kilogram (*kg*) [gram (*g*)]	1 lb = 453.6 g 1 kg = 2.205 lb 1 Pa = 1 N/m^2
Pressure	pascal (*Pa*) [atmosphere (*atm*)]	1 atm = 101.325 kPa = 760 torr 1 atm = 14.7 lb/in^2 (psi)
Temperature	kelvins (*K*) [degrees Celsius (°C)]	K = 273 + °C
Energy	joule (*J*) [calorie (*cal*)]	1 cal = 4.184 J 1 Btu = 1054 J

Length

1 meter (*m*) = 39.37 in = distance light travels in $\frac{1}{299\,792\,548}$ of a second

1 inch (*in*) = 2.540 cm = 0.02540 m

1 kilometer (*km*) = 0.6214 (statute) mile

1 angstrom (*Å*) = 1 x 10^{-10} m = 0.1 nm

1 micron (*μm*) = 1 x 10^{-6} m

1 gram (g) = 0.03527 oz = 15.43 grains   **Mass**
1 kilogram (kg) = 2.205 lb = 35.27 oz
1 metric ton = 1×10^6 g = 1.102 short ton = 0.9843 long ton
1 pound (lb) = 453.6 g = 7000 grains
1 ounce (oz) = 28.35 g

1 liter (L) = 1.057 fl qt = 1×10^3 mL = 1×10^3 cm^3 = 61.02 in^3 **Volume**
1 fluid quart ($fl\ qt$) = 946.4 mL = 0.250 gal
1 fluid ounce ($fl\ oz$) = 29.57 mL
1 cubic foot (ft^3) = 28.32 L = 0.02832 m^3

1 atmosphere (atm) = 760.0 torr = 760.0 mm Hg = 29.92 in Hg   **Pressure**
        = 14.696 lb/in^2 = 1.013 bars = 101.325 kPa
1 kilopascal (kPa) = 1 N/m^2
1 torr = 1mm Hg = 133.3 N/m^2

1 joule (J) = 0.2389 cal = 9.48×10^{-4} Btu = 1×10^7 ergs   **Energy**
1 calorie (cal) = 4.184 J = 3.087 ft lb
1 British thermal unit (Btu) = 252.0 cal = 1054 J = 3.93×10^{-4} hp hr
        = 2.93×10^{-4} kw hr
1 liter atmosphere ($L\ atm$) = 24.2 cal = 101.3 J
1 electron volt (eV) = 1.602×10^{-19} J

velocity of light (c) = 2.998×10^8 m/s = 186,272 mi/s   **Constants and Other**
gas constant (R) = 0.08205 L atm/(mol K) = 8.314 J/(mol K)   **Conversion Data**
        = 1.986 cal/(mol K) = 62.36 L torr/(mol K)
Avogadro's number (N_0) = 6.023×10^{23} /mol
$°F = 1.8°C + 32$
$K = °C + 273.15$
Planck's constant (h) = 6.625×10^{-34} J s/photon

Appendix C

◈ Vapor Pressure of Water

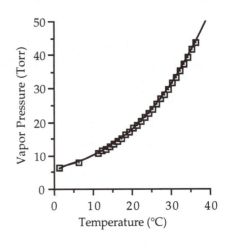

Temperature (℃)	Pressure (*torr*)
0	4.6
5	6.5
10	9.2
11	9.8
12	10.5
13	11.2
14	12.0
15	12.5
16	13.6
17	14.5
18	15.5
19	16.5
20	17.5
21	18.6
22	19.8
23	21.0
24	22.3
25	23.8
26	25.2
27	26.7
28	28.3
29	30.0
30	31.8
31	33.7
32	35.7
33	37.7
34	39.9
35	42.2
–	–
100	760.0

Appendix D

◇ Concentrations of Acids and Bases

Concentrated Reagent	Approximate Molar Concentration	Approximate Mass %	Specific Gravity	mL to dilute to 1 L for a 1.0 M Solution
acetic acid	17.4	99.7	1.05	57.5
hydrochloric acid	12.1	37.0	1.19	82.6
nitric acid	15.7	70.0	1.41	63.7
phosphoric acid	14.7	85.0	1.69	68.1
sulfuric acid	17.8	95.0	1.84	56.2
ammonia (aq) (ammonium hydroxide)	14.8	29% (NH_3)	0.90	67.6

Caution: *When diluting reagents, add the more concentrated reagent to the more dilute reagent (or solvent). Never add water to a concentrated acid!*

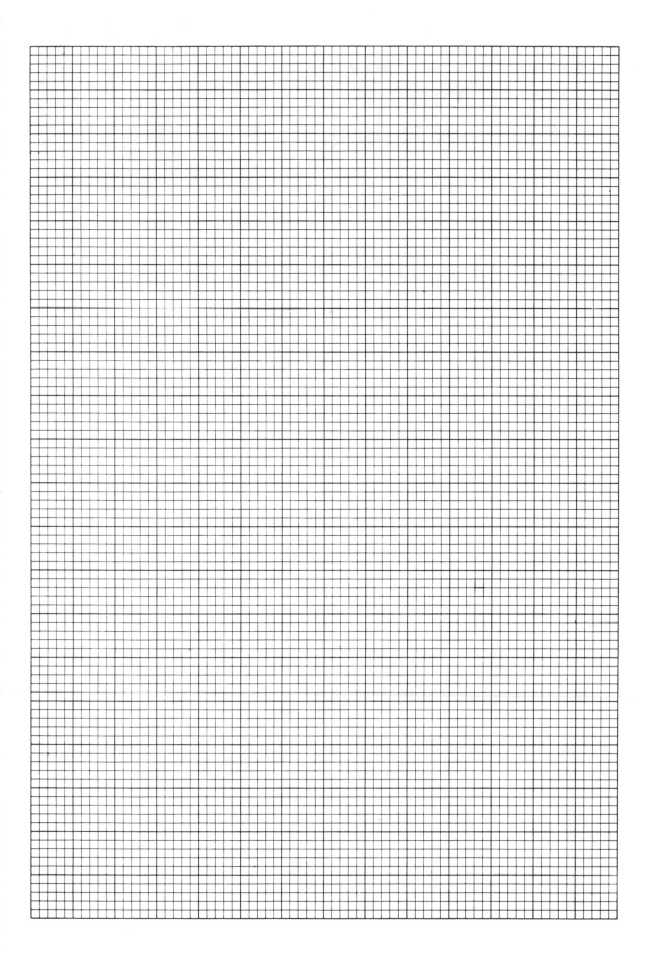

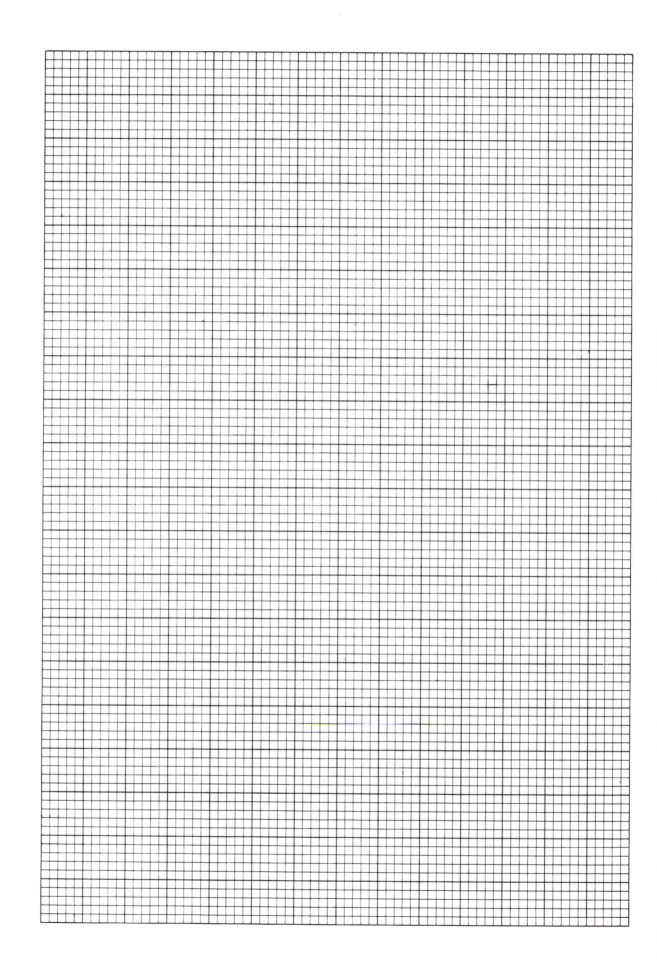

AAO-8312

AAO-8312